Social and Ecological Interactions in the Galapagos Islands

Series Editors

Stephen J. Walsh, Center for Galapagos Studies, University of North Carolina at Chapel Hill, Chapel Hill, NC, USA

Carlos F. Mena, UNC-USFQ Galapagos Science Center, Universidad San Francisco de Quito, Quito, Ecuador

More information about this series at http://www.springer.com/series/10427

Stephen J. Walsh • Diego Riveros-Iregui
Javier Arce-Nazario • Philip H. Page

Editors

Land Cover and Land Use Change on Islands

Social & Ecological Threats to Sustainability

Springer

Editors
Stephen J. Walsh
Center for Galapagos Studies
University of North Carolina at Chapel Hill
Chapel Hill, NC, USA

Diego Riveros-Iregui
Department of Geography
University of North Carolina at Chapel Hill
Chapel Hill, NC, USA

Javier Arce-Nazario
Department of Geography
University of North Carolina at Chapel Hill
Chapel Hill, NC, USA

Philip H. Page
Center for Galapagos Studies
University of North Carolina at Chapel Hill
Chapel Hill, NC, USA

ISSN 2195-1055   ISSN 2195-1063   (electronic)
Social and Ecological Interactions in the Galapagos Islands
ISBN 978-3-030-43975-0   ISBN 978-3-030-43973-6   (eBook)
https://doi.org/10.1007/978-3-030-43973-6

This Springer imprint is published by the registered company Springer Nature Switzerland AG
The registered company address is: Gewerbestrasse 11, 6330 Cham, Switzerland

Series Preface

In May 2011, the University of North Carolina (UNC) at Chapel Hill, USA, and the Universidad San Francisco de Quito (USFQ), Ecuador, jointly dedicated the Galapagos Science Center, an education, research, and community outreach facility on San Cristobal Island in the Galapagos Archipelago of Ecuador. The building dedication was the culmination of an ongoing partnership between UNC and USFQ that began several years earlier through a 2006 invitation to Carlos Mena and Steve Walsh to assist the Galapagos National Park and The Nature Conservancy in a remote sensing assessment of land cover/land use change throughout the archipelago. Leveraging-related work in the Ecuadorian Amazon, Carlos Mena (USFQ Professor of Life and Environmental Sciences) and Steve Walsh (UNC Lyle V. Jones Distinguished Professor of Geography), Co-Directors of the Galapagos Science Center, traveled throughout the islands using satellite imagery and spectral and geospatial equipment to validate preliminary analyses of the Galapagos with a focus on invasive plant species. Since that project, Mena and Walsh have continued to regularly engage the Galapagos Islands on topics important to science and society and to coordinate research, education, and outreach programs conducted at the Galapagos Science Center by faculty, staff, and students from both campuses as well as by collaborating scientists from institutions around the globe. Together the UNC-USFQ Galapagos Science Team seeks to understand the complex social, terrestrial, and marine subsystems in the Galapagos Islands and their linked and integrative effects to address fundamental questions on the Galapagos and on similarly challenged island settings around the globe.

Now with over 60 park-permitted projects operating at the Galapagos Science Center and a diversity of scientific topics being studied using a host of theories and practices, innovative and transformative work continues in compelling and vital ways. The state-of-the-art facilities at the Galapagos Science Center include 20,000 square-feet of space that supports four laboratories (i.e., Microbiology and Genetics, Terrestrial Ecology, Marine Ecology, and Geospatial Modeling and Analysis), operated by a permanent administrative and technical staff to support science, conservation, and sustainability in the Galapagos Islands. In addition, students enroll in classes taught by UNC and USFQ faculty as well as conduct research to complete their undergraduate honors theses, graduate theses, and doctoral dissertations. And several scientists at the Galapagos Science Center engage the community on topics that include water and pathogens, nutrition and public health, tourism and community development, marine ecology and oceanography, and invasive species and land cover/land use change.

From these beginnings and with the general intention of developing a Galapagos Book Series to document our findings, highlight special needs, and describe novel approaches to unravel the social-ecological challenges to the conservation and sustainability of the Galapagos Islands, the Galapagos Book Series with Springer Nature was launched through its inaugural book, *Science and Conservation in the Galapagos Islands: Frameworks & Perspectives (2013)*, edited by Steve Walsh and Carlos Mena. Since 2013, the Book Series has continued to expand, with books now covering several important topics, see below:

Denkinger, J., & Vinueza, L. (2014). *The Galapagos Marine Reserve: A Dynamic Social-Ecological System*. Social and Ecological Interactions in the Galapagos Islands (S. J. Walsh & C. F. Mena, Series Editors). Springer Nature.

Kvan, T., & Karakiewicz, J. (2019). *Urban Galapagos: Transition to Sustainability in Complex Adaptive Systems*. Social and Ecological Interactions in the Galapagos Islands (S. J. Walsh & C. F. Mena, Series Editors). Springer Nature.

Parker, P. G. (2018). *Disease ecology: Galapagos Birds and their Parasites*. Social and Ecological Interactions in the Galapagos Islands (S. J. Walsh & C. F. Mena, Series Editors). Springer Nature.

Quiroga, D., & Sevilla, A. (2017). *Darwin, Darwinism and Conservation in the Galapagos Islands*. Social and Ecological Interactions in the Galapagos Islands (S. J. Walsh & C. F. Mena, Series Editors). Springer Nature.

Torres, M., & Mena, C. F. (2018). *Understanding Invasive Species in the Galapagos Islands: From the Molecular to the Landscape*. Social and Ecological Interactions in the Galapagos Islands (S. J. Walsh & C. F. Mena, Series Editors). Springer Nature.

Trueba, G., & Montufar, C. (2013). *Evolution from the Galapagos: Two Centuries after Darwin*. Social and Ecological Interactions in the Galapagos Islands (S. J. Walsh & C. F. Mena, Series Editors). Springer Nature.

Tyler, M. E. (2018). *Sustainable Energy Mix in Fragile Environments: Frameworks and Perspectives*. Social and Ecological Interactions in the Galapagos Islands (S. J. Walsh & C. F. Mena, Series Editors). Springer Nature.

Walsh, S. J., & Mena, C. F. (2013). *Science and Conservation in the Galapagos Islands: Frameworks & Perspectives*. Social and Ecological Interactions in the Galapagos Islands (S. J. Walsh & C. F. Mena, Series Editors). Springer Nature.

Walsh, S. J., Riveros-Iregui, D., Acre-Nazario, J., & Page, P. H. (In Press). *Land Cover and Land Use Change on Islands: Social & Ecological Threats to Sustainability*. Social and Ecological Interactions in the Galapagos Islands (S. J. Walsh & C. F. Mena, Series Editors). Springer Nature.

Now with considerable pleasure we welcome, *Land Cover and Land Use Change on Islands: Social & Ecological Threats to Sustainability*, edited by Stephen J. Walsh, Diego Riveros-Iregui, Javier Arce-Nazario, and Philip H. Page.

In short, the general goals of the Galapagos Book Series are to examine topics that are not only important in the Galapagos Islands, but also vital to island ecosystems around the globe. Increasingly, viewing islands as a coupled human-natural system offers a more holistic perspective for framing the many challenges to island conservation and sustainability. The perspectives used to study islands need to acknowledge the important context of history, human population, migration of plants, animals, and people, economic development, social and ecological disturbances, resource limitations, such as freshwater, and the evolution and adaptation of species (including humans) on islands to changing circumstances and conditions, both endogenous and exogenous. This book offers new and compelling insights that further adds to the Galapagos Book Series in important and fundamental ways.

Chapel Hill, NC, USA   Stephen J. Walsh
Quito, Ecuador   Carlos F. Mena

Contents

Prologue: Geographies of Hope and Despair: Land Cover and Land Use on Islands

Godfrey Baldacchino

Introduction: The Age of Islands

Islands have long fascinated scholars, but perhaps never more so than in the current epoch of the early twenty first century, gripped as it is by the contradictory dynamics of scientific and technological progress on one hand and a viral pandemic and environmental catastrophe on the other (Bonnett, 2020). Artificial islands are built as enticing, exclusive sites of pricey real estate (Jackson & Della Dora, 2009); while other islands, and their communities, succumb to the slow yet steady threat of saltwater intrusion or sea level rise (Farbotko, 2010a). Enclave/island spaces are the new frontline spaces of development, and the emblematic sites of the Anthropocene (Pugh, 2018; Sidaway, 2007).

If islands did not exist, we would simply have to invent them. They entice outsiders: as synecdoches (whereby a part is made to represent the whole): as "prototypical ethnoscapes" (Baldacchino, 2007a, p. 9); and as handy, manageable and scaled-down reproductions of (larger and messier) continents (Kirch, 1997). The smaller islands get, the simpler and the greater the imputed convenience of this 'island-mainland' correlation. No wonder, therefore, that scientists—often outsiders—descend Gulliver-like upon (smaller) islands to identify, witness, observe and then depart, while inferring and deducing cause-effect relationships, which they acknowledge as writ large in larger (read mainland) contexts (Baldacchino, 2008, p. 42). It is as if islands have been ordained and disposed to act as "outposts of globalisation" (Ratter, 2018); and as advance indicators or extreme reproductions of what is present or future elsewhere (Baldacchino, 2007b). No discipline has been spared from this exercise; but zoologists (think Charles Darwin, Rosemary Grant),

G. Baldacchino (✉)
Department of Sociology, Faculty of Arts, University of Malta,
Msida Campus, MSD, Malta
e-mail: godfrey.baldacchino@um.edu.mt

© Springer Nature Switzerland AG 2020
S. J. Walsh et al. (eds.), *Land Cover and Land Use Change on Islands*, Social and Ecological Interactions in the Galapagos Islands,
https://doi.org/10.1007/978-3-030-43973-6_1

bio-geographers (Jared Diamond, Rosemarie Gillespie) and anthropologists (Bronislaw Malinowski, Margaret Mead) probably lead the pack with their procliv- ity and enthusiasm for such island fieldwork and *in situ* observation (Baldacchino, 2006).

Illusionary Beacons of Stability

The self-evident physicality of an island offers a beguiling expression of stability: a piece of land surrounded by water, crafted by God and/or Nature. And yet, this staid condition of islandness is illusory; the picture-perfect image is transient (Kelman, 2018). First, this is because of the natural cycles of geological and environmental change, which sculpted the island in the first place: from volcanic eruptions, coral growth, or the erosion of erstwhile connected peninsulae and promontories. Cycles of vegetation, and their accompanying fauna, are replaced in succession. The same forces can and do eventually lead to the wholesale disappearances of such islands, though not necessarily in our lifetime (Whittaker, Fernández-Palacios, Matthews, Borregaard, & Triantis, 2017).

A second cause is the impact of the human species on its natural environment, readily visible in island features. Land is reclaimed to extend surface area; sand, stone and gravel are shifted to design or better protect harbours and coastlines; bridges are built to connect islands, and to connect islands to mainlands; in which case, some might say that they are no longer islands (Royle, 2002). Swamps drained, mines quarried, hills levelled, forests felled, river courses dammed and altered … with modernity, history has transitioned into one continuous and open-ended strug- gle to force landscape and geography to succumb to human intent. (Some would add greed.) As with French writer Albert Camus when he visited the island of Manhattan, it would be easy to forget that this "desert of iron and cement" is actually an island (Camus, 1989, p. 51). In this mission of "culture as development," humans play a significant part in transposing or abetting the movement of species from one ecosys- tem (where they may have evolved naturally) to another (where they may find them- selves in different predicaments, ranging from being hugely disadvantaged to finding themselves in dominant positions and with fewer or no natural predators) (Quammen, 2012).

A human-mediated spread of invasive, non-native species drives biodiversity loss and habitat degradation all over the planet, but these consequences are nowhere as stark as on small islands, with their fragile ecosystems (with native and endemic species having evolved in splendid isolation from predators, diseases and competi- tors) as well as with strained and limited human resource skill and expertise pools. Already in the heydays of colonialism, islands were savagely transformed into plat- forms for monocrop economies (think tobacco, sugarcane, banana, pineapple), or sites for the planned transfer of invasive species, driven by the whim to reproduce, say, the idyllic English countryside (Grove, 1995; Royle, 2007). Such small islands may be hotbeds of biodiversity; but, barring extreme measures of access limitation or prevention—not easy to impose, as the recent history of the Galápagos archipelago reminds us—they are *not* likely to withstand or escape the impact of humanity over time. References to a 'balance' between conservation and development are often euphemisms disguising serious issues of ecological degradation (Mathis & Rose, 2016). Whatever traces of 'nature' can be found in such disturbed enisled spaces, the best we can hope for are "human gardens": seemingly natural, but actually constructed scapes (Picard, 2011). On most small islands, we need to acknowledge that we live in a "post wild world" (Marris, 2013).

Size and scale conspire to make such changes appear even more dramatic (Fordham & Brook, 2010; Hay, Forbes, & Mimura, 2013; Kerr, 2005; Kier et al., 2009; Kueffer et al., 2010; Pelling & Uitto, 2001; Spatz et al., 2014), a dynamic also described as "articulation by compression" (Brinklow, 2013). For the first time ever, cityscapes now represent the homes of the majority of humanity (Berry, 2015); but, on small islands, such urbanisation has led to exceptionally heavy population densities, and therefore a greater propensity to sprawl and physically connect island urban zones with contiguous islands or mainlands (Grydehøj, 2014, 2015). Many of the world's capital cities, built originally on islands to afford better protection from attack, have outgrown their protective defensive walls and possibly eliminated the aquatic border, now an irritating barrier to expansion, that separated them from nearby land (Baldacchino, 2014).

Islands and Density

Islands that are political units are also geographical enclaves that tend to have higher population densities than mainlands, also because offloading people across the sea remains a more problematic, and definitely more dangerous venture than distributing them across land borders onto a neighbouring land mass. Moreover, around half of humankind dwells on or near coastal regions, because continental interiors are disadvantaged locations for settlement. Amongst island states and territories, subnational island jurisdictions (SNIJs) tend to be even more attractive spaces for in-migration than sovereign island states, even though they tend to have a much smaller land area (Armstrong & Read, 2003; McElroy & Pearce, 2006).

At the risk of serving as a paean to positivism, the much higher mean population density for islands than for continents is supported by the statistical evidence. Excluding the large (but practically empty) land mass of Greenland—for all its land area of 2 million km^2, its resident population is around 55,000—the world's island units have a mean population density of 144 persons per km^2: this is *three times* the mean value of 48 persons per km^2 that works out for Eurasia, America, Africa and Australia combined; and excluding Australia would only make a marginal difference (see Table 1).

Islands occupy just 1.86% of the Earth's surface area; and this percentage drops down to just 1.47% if one again excludes Greenland. However, they are the

Table 1 Population density on islands and continents compared (2010 data)

Land Mass	Population (A)	Land Area (km²) (B)	Population Density (A/B)
1. Four continents	6,550,400,000	136,071,330	48
2. As (1) above, less Australia	6,530,000,000	128,453,330	51
3. All island states and territories	588,800,000	6,263,612	94
4. As (3) above, less Greenland	588,700,000	4,088,000	144

Source: Baldacchino (2011, p. 168)

Table 2 The 13 states and *territories* (in italics) with the highest population density (of more than 2,000 persons per square mile) for base year 2010 (*rounded figures*)

Rank	Jurisdiction	Of which, islands	Resident Population	Area(mi²)	Density(/ mi²)
	World (land area only)		7,100,000,000	57,510,000	123
1	*Macau (People's republic of China)*	Partly	546,200	11.3	48,450
2	Monaco		33,000	0.75	44,000
3	Singapore	Fully	5,077,000	274.2	18,510
4	*Hong Kong (People's republic of China)*	Partly	7,008,900	428	16,380
5	*Gibraltar (UK)*		31,000	2.6	13,260
6	Vatican City /Holy See		1000	0.17	5880
7	Malta	Fully	410,000	122	3360
8	*Bermuda (UK)*	Fully	65,000	20	3250
9	Bangladesh		164,425,000	55,598	2960
10	Bahrain	Fully	807,000	280	2880
11	Maldives	Fully	314,000	115	2730
12	*Guernsey (British Isles)*	Fully	65,700	30	2180
13	*Jersey (British Isles)*	Fully	91,500	45	2040

Source: Baldacchino (2011). Jurisdictions *in italics* above are self-governing units and not independent states

collective home to some 10% of the world's population: almost 600 million people (Baldacchino, 2006, p. 3).

Gross mean figures of population density—calculated as the mid-year resident population per unit of land area—can be misleading, since various regions in the world are inhospitable to human life and populations tend anyway to cluster and aggregate around coastal regions, riverbanks, ports and sources of fresh water. Still, several of the most densely populated territories in the world are city-states and small jurisdictions (see Table 2). Their residents share a relatively small land area, high levels of urbanisation, relatively high levels of economic prosperity but accompanied by relatively high levels of environmental degradation. Many tend to be peninsular or island units, preventing a natural spillover of population across contiguous

Table 3 Discrete (unbridged) islands with very high population densities (over 2000 persons per square mile)

Population Density (per square km)	Island Unit
820	Oreor (Palau)
830	Losap (Federated States of Micronesia)
840	Kili (Marshall Islands)
840	New Providence (Bahamas)
900	Moen (Federated States of Micronesia)
920	Java (Indonesia)
1000	Tarawa (Kiribati)
1000	Funafuti (Tuvalu)
1130	San Andrés (Colombia)
1260	Malta (main island of Maltese islands)
2460	Majuro (Marshall Islands)
5180	Malé (Maldives)

Source: Baldacchino (2011)

borders. The glaring exception is Bangladesh, the only country in the world with a large (and relatively poor) population and a high population density: at least 100 million people there are at risk from the effects of (even moderate) sea level rise (Islam & Van Amstel, 2018).

Some of these jurisdictions, like Jersey, are single island entities. Others boast a number of island units, in which case a mean national population density often conceals more extreme statistics at the sub-state level. This is most evident in the cases of Malé, capital island of the Maldives, and home to some two-thirds of that country's population. Others include New Providence (capital island of the Bahamas, and location of Nassau), Moen (capital island within Chuuk, one of the four Federated States of Micronesia), Majuro (capital island atoll of the Marshall Islands), South Tarawa (main atoll settlement within sprawling Kiribati), Malta (main island within the Maltese islands) and San Andrés (a sub-national island jurisdiction of Colombia). In each of these cases, population densities are much higher than their respective national mean figures. Many of the world's most densely populated islands are to be found amongst South Pacific archipelagic states (see Table 3). All of these, except Java, Indonesia, are small island units.

Empty Islands

If one is looking for extreme cases of population density, islands offer ample examples from *both ends* of the density continuum. Indeed: island jurisdictions do not just provide scenarios of very high population density, with places like Bermuda, Malta and Singapore topping the list. They also throw up examples of delineated land areas with very low or zero population density: islands - including the island continent of Antarctica - offer the *only* examples of completely de/unpopulated, geographically discrete and self-identifiable areas on the globe: every other type of landform—montane, steppe, desert, valley, forest, river delta, taiga, tundra …—is at some point physically connected to another. Not islands: "'uninhabited' is a word attached only to islands" (Birkett, 1997, p. 14). In their 'emptiness', such island locales are attractive, and in sometimes very contrasting ways. One enticement could be the exploitation of their (often unique) natural qualities and apparent 'underdevelopment' or 'pristine' state for the purpose of identifying, and then protecting, nature reserves, possibly harbouring rare, threatened and/or endemic species. After all, nature reserves are "habitat islands" in any case (Pickett & Thompson, 1978). Such island spaces are easier to protect from the curious or adventurous. Another, contrasting attraction could be the use of such islands, especially depopulated ones, as locales for offshoring undesirable "waste" (human or material) and dangerous experiments: an "enforcement archipelago" that includes detention centres for refugee claimants, high security prisons, quarantine stations, nuclear waste dump sites and high-risk scientific test facilities (Mountz, 2011).

Islands as Tourism Destinations

Pressure on land is greatly exaggerated on islands, also because many of them have transitioned organically into tourism destinations (Carlsen & Butler, 2011). Many islands come with unique cultural or natural specificities; and so these locales become attractive places to visit (Harrison & Hitchcock, 2005). The obligatory crossing over water (by air or by ship/boat) becomes part of the catharsis associated with the spiritually or mentally cleansing journey over water to an island 'paradise' (Patton, 2007). It is no wonder, therefore, that almost a sixth of UNESCO's World Heritage Sites—115, at the latest count—are found on islands, or are islands *in toto* (World Heritage Sites, 2019). And yet, the pressure of visitor numbers threatens the sustainability of the tourism industry, especially on small islands (Apostolopoulos & Gayle, 2002; Lim & Cooper, 2009). Tourism aggravates the crowding and pressure on basic resources (transport, water, energy, foreshores …) and introduces an additional and different set of land use and sea/landscape stakeholders into the bargain. Overwhelmed by their own galloping success in attracting visitors, and miffed by the failed promises of mega-projects gone horribly wrong (Lippert & McCarty, 2016), small islands scramble to manage tourism numbers as best they can: encouraging small scale eco-operations; closing tourist sites for 'maintenance' (Dickinson, 2019); and mounting hostile displays against tourists, while claiming the right to 'take back' their island (Dodds & Butler, 2019). In pursuing the mantra of eco-tourism, small islands may also invest in inefficient or ineffective renewable energy and sustainability initiatives so as to hold on to an illusory eco-island status, thereby ensnaring themselves in an eco-label (Grydehøj & Kelman, 2017).

There are many initiatives underway in the name of small island sustainability; but progress is slow and may shift scarce resources and policy attention from other, more pressing concerns (Baldacchino & Kelman, 2014). Working towards sustainable development can be elusive in small islands (as well as in small island and archipelagic states) because this is fraught with multi-scalar challenges. These include limited biodiversity, extensive in and/or out-migration, pressure of tourism visitations, external interventions and protocols, scarce human resources, weak management systems, inadequate data (and problems of interpretation), social divisions and tensions (often invisible to outsiders) and simultaneous quests for modernity and conservation (Connell, 2018). Moreover, small islands by definition thrive and survive by inputs (including in-migrating species) derived from beyond their shores: it comes as no surprise that Cuba, long subjected to a trade embargo, was feted by the World Wildlife Fund in 2006 as the only country on the planet anywhere close to sustainability (Guevara-Stone, 2008).

Prospects

Nearly a quarter of all sovereign states are islands, and islands have taken the lead in the development of innovative forms of governance (Felt, 2003; Stratford, 2006), environmental management, and in the development of alternative energy technologies (Hay, 2006, p. 20). Meanwhile, from Tuvalu to the Venice Lagoon, islands have become the nostalgic targets of a sadistic streak of 'dark' tourism, invaded by visitors attracted to such places while they remain accessible, and indirectly contributing to and hastening their demise with their carbon footprint (Farbotko, 2010b; Hindley & Font, 2017).

If any traces of optimism are to be found in the pages of this book, then it may be the sophisticated capture of data that steals the show. From the Hawaiian islands and the Galápagos, to Montserrat and Sulawesi, more powerful and yet more affordable technology has provided important datasets that capture the state of environmental degradation, ecosystem service disruption, loss of forest cover, increase of land dedicated to agriculture, and the penetration of non-native invasive species. One can also better overlay and integrate different classes of data to approximate the multifaceted and integrative nature of environmental change, and at various spatial scales. It is already possible to compare the state of today's islands with their condition in the distant, or not so distant, past: again, small islands can demonstrate radical landscape changes over relatively short periods of time. The expectation is that, armed with the science and the data, and the visual 'before and after' imagery that they permit, policy makers are better convinced and equipped to make the case and to implement measures that brake, or perhaps even revert, the consequences of rampant globalisation and consumerism. Islands may yet present themselves as geographies of hope, rather than of despair.

References

Apostolopoulos, Y., & Gayle, D. J. (Eds.). (2002). *Island tourism and sustainable development: Caribbean, pacific and mediterranean experiences.* New York: Praeger.

Armstrong, H. W., & Read, R. (2003). Small states, islands and small states that are also islands. *Studies in Regional Science, 33*(1), 237–260.

Baldacchino, G. (2006). Island, island studies, island studies journal. *Island Studies Journal,* 1(1), 3–18. Retrieved from https://www.islandstudies.ca/sites/vre2.upei.ca.islandstudies.ca/files/u2/ISJ-1-1-2006-Baldacchino-pp3-18.pdf

Baldacchino, G. (2007a). Introducing a world of islands. In G. Baldacchino (Ed.), *A world of islands: An island studies reader* (pp. 1–29). Charlottetown, Canada and Luqa, Malta: Institute of Island Studies, University of Prince Edward Island and Agenda Academic.

Baldacchino, G. (2007b). Islands as novelty sites. *Geographical Review, 97*(2), 165–174.

Baldacchino, G. (2008). Studying islands: On whose terms? Some epistemological and methodological challenges to the pursuit of island studies. *Island Studies Journal, 3*(1), 37–56. Retrieved from https://www.islandstudies.ca/sites/default/files/ISJ-3-1-2008-Baldacchino-FINAL.pdf

Baldacchino, G. (Ed.). (2011). *Extreme heritage management: The practices and policies of densely populated islands.* New York: Berghahn Books.

Baldacchino, G. (2014). Capital and port cities on small islands sallying forth beyond their walls: A Mediterranean exercise. *Journal of Mediterranean Studies, 23*(2), 137–151.

Baldacchino, G., & Kelman, I. (2014). Critiquing the pursuit of island sustainability: Blue and green, with hardly a colour in between. *Shima: The International Journal of Research into Island Cultures, 8*(2), 1–21.

Berry, B. J. (2015). *The human consequences of urbanisation.* New York: Macmillan International Higher Education.

Birkett, D. (1997). *Serpent in paradise.* London: Picador.

Bonnett, A. (2020). *The age of islands: In search of new and disappearing islands.* London: Atlantic Books.

Brinklow, L. (2013, June 22). The ABCs of island living. Opinion. *The Guardian,* Charlottetown, Canada, A15.

Camus, Albert (1989). *American journals.* (H. Levick, Trans.). London: Hamish Hamilton.

Connell, J. (2018). Islands: balancing development and sustainability?. *Environmental Conservation, 45*(2), 111–124.

Carlsen, J., & Butler, R. (Eds.). (2011). *Island tourism: Towards a sustainable perspective.* Wallingford: CABI.

Dickinson, G. (2019, November 12). The Faroe Islands to 'close for maintenance' in 2020. *The Telegraph (UK).* Retrieved from https://www.telegraph.co.uk/travel/destinations/europe/faroe-islands/articles/faroe-islands-overtourism-closed-for-maintenance/

Dodds, R., & Butler, R. (Eds.). (2019). *Overtourism: Issues, realities and solutions.* Amsterdam: De Gruyter Oldenbourg.

Farbotko, C. (2010a). Wishful sinking: Disappearing islands, climate refugees and cosmopolitan experimentation. *Asia Pacific Viewpoint, 51*(1), 47–60.

Farbotko, C. (2010b). 'The global warming clock is ticking so see these places while you can': Voyeuristic tourism and model environmental citizens on Tuvalu's disappearing islands. *Singapore Journal of Tropical Geography, 31*(2), 224–238.

Felt, L. F. (2003). *Small, isolated and successful: Lessons from small, isolated societies of the North Atlantic.* St John's NL: Report presented to the Royal Commission on 'Renewing and strengthening our place in Canada'. Retrieved from www.gov.nl.ca/publicat/royalcomm/research/Felt.pdf

Fordham, D. A., & Brook, B. W. (2010). Why tropical island endemics are acutely susceptible to global change. *Biodiversity and Conservation, 19*(2), 329–342.

Grove, R. (1995). *Green Imperialism: Colonial expansion, tropical island Edens and the origins of environmentalism*. Cambridge: Cambridge University Press.

Grydehøj, A. (2014). Guest editorial introduction: Understanding island cities. *Island Studies Journal, 9*(2), 183–190.

Grydehøj, A. (2015). Island city formation and urban island studies. *Area, 47*(4), 429–435.

Grydehøj, A., & Kelman, I. (2017). The eco-island trap: Climate change mitigation and conspicuous sustainability. *Area, 49*(1), 106–113.

Guevara-Stone, L. (2008). Viva la revolucion energetica: In two short years, energy-smart Cuba has bolted past every country on the planet. *Alternatives Journal, 34*(5–6), 22–25. Retrieved from https://www.renewableenergyworld.com/2009/04/09/la-revolucion-energetica-cubas-energy-revolution/#gref

Harrison, D., & Hitchcock, M. (Eds.). (2005). *The politics of world heritage: Negotiating tourism and conservation*. Clevedon: Channel View Publications.

Hay, J. E., Forbes, D. L., & Mimura, N. (2013). Understanding and managing global change in small islands. *Sustainability Science, 8*(3), 303–308.

Hay, P. (2006). A phenomenology of islands. *Island Studies Journal, 1*(1), 19–42.

Hindley, A., & Font, X. (2017). Ethics and influences in tourist perceptions of climate change. *Current Issues in Tourism, 20*(16), 1684–1700.

Islam, M. N., & van Amstel, A. (Eds.). (2018). *Bangladesh: Climate change impacts, mitigation and adaptation in developing countries*. Heidelberg, Germany: Springer International.

Jackson, M., & Della Dora, V. (2009). "Dreams so big only the sea can hold them": Man-made islands as anxious spaces, cultural icons, and travelling visions. *Environment and Planning A, 41*(9), 2086–2104.

Kelman, I. (2018). Islandness within climate change narratives of small island developing states (SIDS). *Island Studies Journal, 13*(1), 149–166.

Kerr, S. A. (2005). What is small island sustainable development about? *Ocean & Coastal Management, 48*(7–8), 503–524.

Kier, G., Kreft, H., Lee, T. M., Jetz, W., Ibisch, P. L., Nowicki, C., … Barthlott, W. (2009). A global assessment of endemism and species richness across island and mainland regions. *Proceedings of the National Academy of Sciences, 106*(23), 9322–9327.

Kirch, P. (1997). Epilogue: Islands as microcosms of global change. In P. V. Kirch & T. L. Hunt (Eds.), *Historical ecology in the Pacific Islands* (pp. 284–286). New Haven CT: Yale University Press.

Kueffer, C., Daehler, C. C., Torres-Santana, C. W., Lavergne, C., Meyer, J.-Y., Otto, R., & Silva, L. (2010). A global comparison of plant invasions on oceanic islands. *Perspectives in Plant Ecology, Evolution and Systematics, 12*(1), 145–161.

Lim, C. C., & Cooper, C. (2009). Beyond sustainability: Optimising island tourism development. *International Journal of Tourism Research, 11*(1), 89–103.

Lippert, J., & McCarty, D. (2016, January 4). The ghosts of Baha Mar: How a $3.5 Billion Paradise went bust. *Bloomberg Pursuits*. Retrieved from https://www.bloomberg.com/news/articles/2016-01-04/the-ghosts-of-baha-mar-how-a-3-5-billion-paradise-went-bust

Marris, E. (2013). *Rambunctious garden: Saving nature in a post-wild world*. New York: Bloomsbury.

Mathis, A., & Rose, J. (2016). Balancing tourism, conservation, and development: A political ecology of ecotourism on the Galapagos Islands. *Journal of Ecotourism, 15*(1), 64–77.

McElroy, J. L., & Pearce, K. B. (2006). The advantages of political affiliation: Dependent and independent small island profiles. *The Round Table: Commonwealth Journal of International Affairs, 95*(386), 529–540.

Mountz, A. (2011). The enforcement archipelago: Detention, haunting and asylum on islands. *Political Geography, 30*(3), 118–128.

Patton, K. C. (2007). *The sea can wash away all evils: Modern marine pollution and the ancient cathartic ocean*. Columbia, NY: Columbia University Press.

Pelling, M., & Uitto, J. I. (2001). Small island developing states: Natural disaster vulnerability and global change. *Global Environmental Change B: Environmental Hazards, 3*(1), 49–62.

Picard, D. (2011). *Tourism, magic and modernity: Cultivating the human garden.* New York: Berghahn Books.

Pickett, S. T., & Thompson, J. N. (1978). Patch dynamics and the design of nature reserves. *Biological Conservation, 13*(1), 27–37.

Pugh, J. (2018). Relationality and island studies in the Anthropocene. *Island Studies Journal, 13*(2), 93–110.

Quammen, D. (2012). *The song of the dodo: Island biogeography in an age of extinctions.* New York: Random House.

Ratter, B. M. (2018). *Geography of small islands: Outposts of globalisation.* New York: Springer.

Royle, S. A. (2002). *A geography of islands: Small island insularity.* Oxford: Routledge.

Royle, S. A. (2007). *The Company's island: St Helena, company colonies and the colonial endeavour.* London: IB Tauris.

Sidaway, J. D. (2007). Enclave space: A new metageography of development? *Area, 39*(3), 331–339.

Spatz, D. R., Newton, K. M., Heinz, R., Tershy, B., Holmes, N. D., Butchart, S. H. M., & Croll, D. A. (2014). The biogeography of globally threatened seabirds and island conservation opportunities. *Conservation Biology, 28*(5), 1282–1290.

Stratford, E. (2006). Technologies of agency and performance: *Tasmania Together* and the constitution of harmonious island identity. *Geoforum, 37*(2), 273–286.

Whittaker, R. J., Fernández-Palacios, J. M., Matthews, T. J., Borregaard, M. K., & Triantis, K. A. (2017). Island biogeography: Taking the long view of nature's laboratories. *Science, 357*(6354). Retrieved from https://science.sciencemag.org/content/357/6354/eaam8326.abstract

World Heritage Sites. (2019). *World Heritage List.* Search results for 'islands'. Retrieved from: https://whc.unesco.org/en/list/?search=islands&order=property

Economic and Related Aspects of Land Use on Islands: A Meta Perspective

Richard E. Bilsborrow

Preface

This is the first of two studies that seek to distill the recent extensive and growing socio-economic literature on island land use and factors related to it and its changes over time, focusing on methodological aspects and contributions. This chapter will briefly describe the full scope of the eight topics covered, including the four larger ones focused upon in this chapter plus the four topics to be covered in a separate but companion study (covering population, urbanization, tourism, and climate). We recognize, however, that they there are inherently *many interrelationships* across and within not only the four broad topics focused upon in this chapter but with the other four topics as well—all relevant to understanding the evolution of changes in land use on islands over time in the recent past and in the future and factors underlying these changes.

Although my personal interest and experience is on developing countries, I do attempt to include here references on islands which are not developing countries if they have broader methodological interest. Islands reviewed here include almost continent-sized islands such as Madagascar and large islands of Indonesia and the Philippines, though most will be small in size, from all of the non-polar oceans. I generally exclude the literature focused on the measurement of land use itself or the technology of remote sensing/satellite-based data and methodologies employed, which is reviewed separately in this volume (see Chapter 3 by Walsh, et al. and case studies). Exceptions are studies on LCLUC measured from remote sensing which have a significant substantive, social science focus. Hawaii is an integral part of the United States so is not covered here, and coverage of Puerto Rico and the Galapagos Islands is also limited as they are covered directly in other chapters of this book. Finally, none of the extensive, so-called grey literature nor government documents nor publications of international agencies is covered in this review.

R. E. Bilsborrow (✉)
Carolina Population Center, University of North Carolina, Chapel Hill, NC, USA
e-mail: richard_bilsborrow@unc.edu

© Springer Nature Switzerland AG 2020
S. J. Walsh et al. (eds.), *Land Cover and Land Use Change on Islands*, Social and Ecological Interactions in the Galapagos Islands, https://doi.org/10.1007/978-3-030-43973-6_2

Methodology for and Scope of Literature Review

The procedure for conducting the literature search and evaluations began with computer searches by topic (using multiple key words combined with land use) of various academic literature databases, using criteria for initial filtering to delete short/popular literature references and a wealth of specialized scientific literature from the natural and physical sciences. Finally, from what remained, more subjective criteria based on the topics focused upon here and my own multi-disciplinary interests albeit with inevitable biases were used to pare it down to an almost manageable body of relevant publications since 1990, and available in English (see Appendix for details). Abstracts were then obtained for as many of these as possible (over 90%)—around a thousand for the eight topics altogether, about 600 for those covered in this chapter), and then skimmed to identify those for which full texts were sought and usually obtained, which constituted the tier 1 references, with the rest constituting tier 2 (see table in Walsh et al., chapter 3 above). However, upon careful reading of the full articles or further reading of the abstracts at the time of writing for each topic, many abstracts stimulated me to re-categorize them as tier 1 (and therefore of inclusion in the text here as well as the references), while a number of the tier 1 items were downgraded to tier 2. Moreover, some items were re-classified as more appropriate for the second, companion study focusing on population, urbanization, tourism and climate (in preparation, but not part of this book). These time-consuming processes resulted in somewhat over 100 publications covered in this meta-analysis, and cited at the end of the chapter. The full bibliography for all eight topics, after further paring down at the second time of reading abstracts stage, eliminating duplicates, and cleaning up references to be of similar format, comprises about 700 items (vs. over 950 originally found and shown in the Walsh chapter above), and is available from the author or editors on request.

The time reference is generally from 1988 to 2018, when this review began. Doubtless important new publications have come out since then, which should be borne in mind by readers of this book. The overall goal was to identify analytical studies that include insightful descriptions of linkages between land use and socio-economic factors as well as those using interesting or novel methodologies to link the particular topic to land cover use and land use change (*LCLUC*).

For further details, see the Appendix to this chapter.

The map below (Figure 1) illustrates the location of the islands and the number of studies on each from tiers 1 and 2 combined for all eight topics from the initial classification of over 900 studies. It shows the concentration on various islands as expected (Puerto Rico, Hawaii, Madagascar, with 55, 49 and 18 studies on each, respectively), but more than expected on some islands, including more on Indonesia (especially Sumatra with 18) than on the Philippines, and a fair amount on Jamaica and St. Lucia in the Caribbean; on Hainan (China), the Galapagos Islands, the Solomon Islands, Fiji and Vanuatu, in the Pacific (16 down to 8 studies on each); Mauritius, nearby Reunion, and Zanzibar, in the Indian Ocean; the Canary Islands, Madeira and the Azores, in the Atlantic; and Lesvos (Greece), Malta and Mallorca (Spain), in the Mediterranean.

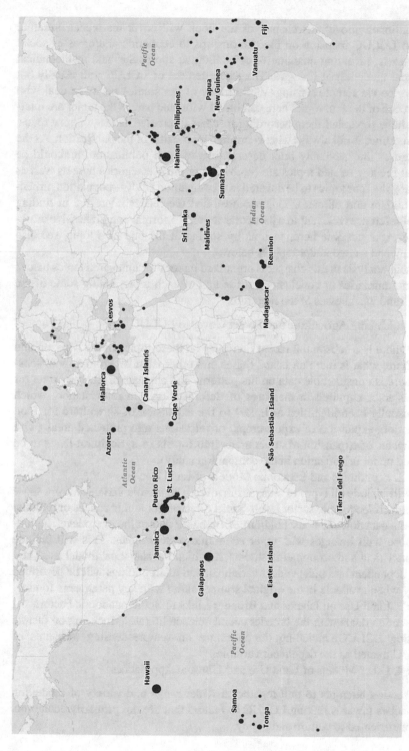

Fig. 1 Study area location

As mentioned above, this chapter of the book will cover the recent literature relating to LCLUC focusing on its relationships to economic and/or other social science factors, involving quantitative or qualitative approaches and mathematical or statistical modeling of linkages. The measurement of LCLUC will usually but not always be via remote sensing. While many of the studies reviewed deal with more than one of the four topics here combined with land use, as the topics are inter-related, I have separated them according to what I perceive as the *primary* focus, which sometimes is not always what is implied by the title of the publication. As the reader peruses the inevitably brief descriptions of each publication, it should be clear what the key related topics are (both among the 4 categories here as well as from among the four topics to be covered in a companion study—population, urban-ization, tourism and climate). This should assist the reader interested in finding additional references relevant to a particular topic. No formal comprehensive cross-referencing was possible here, such as by stating at the end of a topic, see also authors x, y and z cited under topic X below.

The meta-analysis in this chapter is organized under four numbered topics below, with short summaries or conclusions at the end of each topic, noting some of the highlights and weaknesses of the literature.

- Topic 1. Linking Agriculture and Forest Cover to LCLUC

- While there is a substantial recent literature on these topics in the Amazon and elsewhere, what is there on Island States and other islands? From remote sens-ing, there is considerable data on the patterns and changes in LU over time for most islands, captured in measures of deforestation and reforestation, which have usually been identified as linked to the expansion of agriculture for food production, expansion of export crops, or intrusions into protected areas (with the impacts of expansion of urban areas into forests or agricultural lands to be covered under urbanization in the companion study).
- Topic 2. Economics and Economic Models of Land Use
- This will include all types of analyses involving economic variables, from those at the macro/economic sector levels based often on economic census or national accounts data down to household/farm/firm based data at the micro level, includ-ing those with linkages with other economic sectors/topics. This will include both statistical and econometric studies, in which the statistical model used, rel-evant dependent and independent variables, and main findings will be described briefly when available in the original source, along with key parameters found.
- Topic 3. Land Use on Islands and Broader Links to Socio-economic Factors
- This section deals with the broader social-science literature focusing on factors affecting LCLUC, including the literature on remote sensing measures of LCLUC treated as the dependent variable.
- Topic 4. Other Models of Land Use and Unusual Approaches

- This section attempts to pull together a wider range and variety of modelling approaches towards linking LCLUC to factors that are not primarily economic, or more oriented to culture and policy.

Topic 1. Linking Agriculture and Forest Cover to LCLUC

Over the course of the past 10,000 years since the beginnings of agriculture with the domestication of wheat in Mesopotamia and corn in the central valley of Mexico, the expansion of agriculture has been central to human progress and population growth. It made possible converting from hunting and gathering to agriculture as the main source of sustenance, converting forests, shrub and grasslands to food crops, to the settling of populations in small stable communities and ultimately towns and cities. This in turn led to the creation and advance of civilizations over the recent millennia. With the industrial revolution, over the past two centuries or so, this has involved considerable rural-urban migration, with vast shifts in population from rural to urban areas, with the urban population now reaching 55% of the world total. Increasing urban and total populations require more food, from more agricultural products, as well as greater exploitation of forests for fuel, housing and food, both transforming land use across the planet. Much of the expansion of the land area planted in agricultural products has historically come from clearing forests (along with clearing grasslands, shrubs, etc.,) constituting the *extensification* of agriculture (which neo-Malthusians associate with Malthus). With the advancements of technology including the science of agriculture, agricultural production has also expanded via the *intensification* of agriculture (Bilsborrow & Carr, 2001; Bilsborrow & Geores, 1992, 1994; Boserup, 1965). This expansion has made possible the vast increase in human populations over the past two centuries, *pari passu* with the expansion of industry, services and technology, including vast advances in transportation, communications and desires for better lives, involving vast trade in goods and services across countries and oceans including the international migration of people and capital. Islands have been a part of all these processes as well. This section will focus on recent shifts in agricultural and forest areas in islands and their interrelationships.

A host of related topics and socio-economic and biophysical variables related to either or both agricultural changes and changes in forest cover are found in the island literature, but have been filtered out of the discussion in this chapter due to their being less centrally linked to land use *sui generis*. These include (unless explicitly linked to land use) water supply/hydrology/ground water/aquifer depletion; studies of a particular crop or tree species; invasive species and species extinctions; coral reefs and mangroves; mining; gender roles; fire; volcanos; soil erosion; river sedimentation; cultural and political studies, etc. There do exist investigations focusing on all of these topics and more that we encountered in the literature search, with many in the tier 2 references not generally included in the review here (see Appendix). On the other hand, we did seek to cover under this topic logging and fuelwood use in relation to land use; small farm and plantation agriculture; reforestation; protected areas (forests, not reefs); geographic and geophysical aspects of land; locations of fields, farm houses and forests relative to transportation networks, market towns and cities; population size, growth and density in relation to land use; rural labor opportunities and unemployment (under topic 2). Alas, we shall see that on many of these topics there is little recent analytical/quantitative research on islands found here.

As is well known, by now there has emerged a considerable recent literature on agriculture, land use and deforestation in developing countries, especially in the Americas and Asia, most examining impacts of agricultural expansion on forest cover. These studies were summarized in an earlier meta-analysis by Geist and Lambin (2002) and Lambin et al. (2001); have included a number on the Amazon region, including Brazil and Ecuador (e.g., Keller, Bustamente, Gash, & Silva Dias, 2009; Pan & Bilsborrow, 2005; Walker, 2004), Thailand (Walsh, Entwisle, Rindfuss, & Page, 2006). Several studies of economists have examined causes of tropical deforestation with economic and econometric models (e.g., see Brown & Pearce, 1994; Kaimowitz & Angelsen, 1998). But what evidence is there on Island States and other islands?

The literature review here is organized by geographic region, beginning with the Caribbean region, followed by Asia (mostly the island countries of Southeast Asia and islands next to China), then the many mostly small Pacific islands spread across the vast Pacific), islands along the two ocean coasts of Africa (led by Madagascar), and finally a category Other referring to relevant special studies on islands of developed countries. This is the geographic sequence followed for this topic as well as the other three topics in this chapter.

Caribbean

We begin with Jamaica, then cover other islands generally also experiencing deforestation, and conclude with the exceptional case of Puerto Rico, which has been undergoing an extraordinary long-term process of reforestation.

Two particularly intriguing, quantitative studies on Jamaica are by Tole (2001, 2002)and Newman, McLaren, and Wilson (2014), both combining and integrating data from satellite imagery with socio-economic-population data, covering different parts of Jamaica and at different times. Both also seek to ascertain the causes of its substantial deforestation. As Newman et al. note (p. 186), "despite the concentration of land cover change studies from the Latin America region, drivers of change in Caribbean islands have been largely understudied". Jamaica has had one of the highest rates of deforestation in the region over the past 30 years or so. Tole (2002) combines MSS satellite imagery from 1987 and 1992 with population census data to examine human factors contributing to loss of forest cover in 1987–92. She describes her approach as a meso-study, intermediate between analysis of household or plot level data and macro or province-regional level data, saying it has the advantage of greater detail than macro studies while providing more generalizable results than data from households in a small geographic area. During her study period, Jamaica was experiencing serious economic and social problems (inflation, high unemployment, growing debt, declining public spending), together with almost 1% annual population growth. MSS data with pixel sizes of 57m × 57m, or about 4 times those of Landsat TM, for its 51 constituencies, were used as units of analysis since they were available for the whole island, including urban centers. Data for characterizing the populations of the 51 areas were compiled for households from the 1992 population census and a 1998 World Bank-supported Jamaica Survey of Living Conditions. Euclidean distance to the nearest of the five principal urban areas from the constituency centroids were computed; population density was estimated from census data divided by land area; and other variables were computed for mean dependency ratios, house quality, education, fuel use, and poverty. OLS (Ordinary Least Squares) linear regression results are not very strong (R^2 from 0.2 to 0.4), with deforestation positively linked to "social welfare" (more deforestation where people were poor, more dependent on fuelwood, and had higher population density and age-dependency), but also linked to higher male education. Despite the weak results, the study was innovative for the region in integrating use of satellite and population census data, making use of what data were available. This could evidently be emulated in many other islands as most have some data of both types--satellite imagery and population census data.

Newman et al. (2014) deals with a far smaller area, of only 68,024 ha in Cockpit Country in the west-central highlands, where forest reserves constitute half of the area. The site is one of very high biodiversity in plants, frogs and birds. They investigate not only socio-economic drivers but also biophysical conditions; by controlling for the latter, they can isolate effects of the former. The area overall is not densely populated, has mostly small farms under 2 ha, with farming the main occupation, growing some subsistence crops but mostly yams for export. Biophysical features including forest cover were available from area photographs at 6m resolution for a series of years from 1942 to 2010. They sought out as many potentially useful explanatory variables as possible, based on the discussion in Geist and Lambin (2002), but were limited in the end to population density (census population, as in Tole, but divided by the district area, for much smaller areas than constituencies); Euclidean distances to towns and capital cities; relative wealth/poverty level (based on primary water source, sanitation, lighting source, fuel use and house tenure); plus education and employment/unemployment. Using logistic regression to measure whether forest cover rose or fell in each district in various 10 to 20- year periods, the authors concluded that *long-term deforestation* (over the whole 68 year period) was primarily determined by *biophysical and geographic* factors, specifically the terrain (slope), existing forest fragmentation at the beginning of the period (forest contiguity index), and Euclidean distance from nearest road to forest edge. These results are not surprising. At the same time, socio-economic factors were more important in the shorter-term decadal periods, notably employment status, population density, age-structure of community population, and main water source as an indicator of wealth: Thus lower densities of households were associated with *higher* land clearing, surprisingly. The authors noted that land clearing was also linked positively to the proportion of adult women, and therefore absent men, with the latter engaged in non-farm work, returning to clear forest for farming when they could not find jobs. The linkage between clearing and lower population density may be explained as follows: when men work in non-farm occupations, they earn higher incomes, some of which is taken back to the origin households and invested in acquiring more land and/or clearing existing land they have in the origin, mainly to expand cash crop agriculture. Overall, the statistical results are not particularly strong and sometimes surprising for economic and demographic variables, they are strong for biophysical-geographic factors, with similar results for factors determining *reforestation.*

Before getting to Puerto Rico, several studies note an unusual, relevant land characteristic of much of the Caribbean islands referred to as *Caribbean karst*, common throughout the Greater Antilles as well as the Bahamas, Antigua, Cayman Islands, Virgin Islands, Guadeloupe, Barbados, Trinidad & Tobago, and the Dutch Antilles. The karst is difficult to farm, and includes towers, sinkholes (*cenotes*, in Yucatan, Mexico), caves, etc. Karst lands are also prone to both drought and flooding, fragile and subject to environmental change, though with care can provide long run agricultural goods (Day, 2010). Their particular vulnerable characteristics should be taken into account in land use planning. Unfortunately, humans have instead damaged them greatly, destroying natural vegetation, contaminating water, and using them as quaries. Climate change, increasing temperatures and changing weather with overall decreasing precipitation is also disrupting the karst hydrological cycle, leading to increasing desertification. Population growth and economic growth are likely to further threaten the karst-based ecosystems, unless improvements are made in land management and planning to ensure long-term sustainability.

Chopin and colleagues (2014, 2015) have undertaken intriguing research on the French island of Guadeloupe. In the 2014 study, seeing farmers having a poor understanding of desirable crop rotation patterns, the authors first assess the evolution of crop acreages over time and space, to develop a dynamic typology of farmers to classify their multi-year use of land in seven steps, 3 based on farm typology, 3 on landscape changes, and one on factors influencing those changes. They apply the method to 3591 farms, identifying 8 farm types according to crop acreage size, and then use it to describe the process of diversification of 111 sugarcane growers into fruits and vegetables, noting their dependence on water availability. They say their 7-step method could be used to measure ecosystem services associated with changes in agriculture at the landscape level as well.

In the second paper, Chopin et al. develop a multi-scale bioeconomic model for the classification and assessment of cropping systems at the field (plot), farm, sub-regional/community and regional scales to provide cropping system mosaics for regional optimization of the sum of individual farmer's utilities under field, farm and regional biophysical and socio-economic constraints. Dubbed MOSAICA, they say it can be used in different regions with data on the location of the field and farm, on cropping system yields, on locational characteristics and on policy schemes. The model is applied in Guadeloupe (Chopin et al., 2015) to test the impacts of three alternative policy scenarios of agricultural subsidies, producing three cropping system mosaics in which the area under traditional bananas and sugarcane was transformed into breeding better varieties plus other higher value crops. The entire cropped area of the three small islands constituting Guadeloupe is only 32,948 ha, containing 7749 farms with a mean size of 4 ha, ranging from under 1 to over 100 ha. The model requires extensive data down to each plot size, which was available for over 70% of the farms from a geodatabase of fields from the government agency providing farm subsidies. Data are needed on input use and yields ideally for all plots to generate data for optimization. 36 cropping systems for Guadeloupe were identified from previous work, synthesized into the 8 farm types of the topology cited above. Risk aversion coefficients were postulated for each type. The geographic database includes field biophysical traits and farm structure data from the agricultural census, and the Delphi method is used to develop other parameters based on seeking and synthesizing expert opinions. The objective function is based on Markowitz-Freund to maximize regional utility over the whole population of farms, subject to certain biophysical limitations due to karst soil conditions, etc. Several interesting policy scenarios were then experimented with, including the effects of eliminating subsidies for sugarcane and bananas, as required by the World Trade Organization by 2017 (within 3 years). Results generally predicted sharp declines in those two crops along with increases in areas in pasture, fallow, fruit orchards and vegetables. Differential effects of scenarios on agricultural revenues, food self-sufficiency, crop diversification, employment, soil and water quality, etc., were investigated. Strengths lie in the model's capacity to provide rich results for forecasting/planning, but the data needs are huge, farmers' decisions have to be simplified, parameters are static, etc. The authors note also that the model would benefit from being linked to other models incorporating markets and economic sectors. Ones I would also specifically suggest, include tourism and the external sector (exports, imports, foreign aid, and remittances from migrants). But these suggestions result in far more additional complexity. While the model is not construed as a research model, MOSAICA could be used, as is characteristic of models, to identify needs for further research to obtain more reliable model parameters for itself, as well as other models (e.g., input-output, risk parameters).

Finally, the island which has been far and away the most studied in the Caribbean is Puerto Rico, which, following massive deforestation in the first half of the twentieth century due to human population growth and clearing for agriculture, has undergone a sustained, major *reforestation* during the second half. While reforestation was known to be going on before, we begin with the article of Rudel, Perez-Lugo, and Zichal (2000) as it provides an excellent descriptive analysis, beginning with the striking fact that the forest cover, which had shrunk to 9% of the island by 1950, rose to 37% by 1990. To put this in context, as elsewhere in the region, the island was largely covered in forest (majority in sub-tropical moist forests) at the time of European contact. The land was then converted primarily to export agriculture—coffee for Europe, sugar for North America, and tobacco for the US cigar market. By 1930, the country was largely cleared of forests with a dense rural population and considered a Malthusian problem, with a population of 2.2 million in 1950—denser than almost all European countries. 80% of the rural residents were landless, working as wage laborers on export crop plantations and farms. Between 1950 and 1990, the population grew by 1.3 million, while the country reforested. How did this happen?

The authors draw on data for individual land plots from a Forest Service inventory in 1990: the country was overlaid with a grid of 3km × 3km pixels from which a systematic sample of aerial photographs were taken of 1 ha plots at intersection lines of the grid. This was followed up with ground truthing (not explained beyond "visits"), eliminating urban land and primary forests, leaving 675 plots of land. The other, principal sources of baseline data on human activities were from the 1950 census of population and the 1959 agricultural census, which, like most, collected data on farm size, tenure, crops grown, off-farm employment and the most valuable crop in the community closest to the intersection point. Data were aggregated to characterize each *municipio* as primarily coffee, sugarcane or tobacco communities, or reforesting. Logistic regression was used to determine the factors responsible for re-forestation, of both one ha plots and *municipios*. Independent variables examined were elevation, mean farm size in the *municipio* in 1959, proportion of farms with off-farm income or mean off-farm income, percent change in population, poverty (% households with under $500 annual income in 1959 or 1978), mean expenditures on fertilizer, and an interaction term of coffee x precipitation. In the regression based on regional measures (n = 69 *municipios*), reforestation was found positively linked to elevation, the coffee-precipitation linkage (proportion of existing land in coffee and high precipitation), off-farm work, and poverty; and negatively to farm size, with the results similar in a parallel regression for small one-ha plots, except for an additional strong negative effect of population growth and no effects of poverty or off-farm work. Thus rural population decline was a major factor in reforestation (see also discussion of agricultural abandonment in Parés-Ramos et al. 2008) —consistent with the observed substantial rural-urban migration in the 40 year period, linked directly to rapid industrialization and higher wages. This latter finding is consistent with (and perhaps inspired?) the "forest transition" hypothesis associated with Rudel et al. (2002), in which reforestation is expected to occur as a country develops through industrialization which stimulates rural-urban migration over time. However, as noted by the authors, Puerto Rico also has an unusual relationship with the United States, facilitating migration to the mainland and the sending of substantial remittances back, which doubtless contributed to the process.

Among notable other studies that followed are those of Helmer (2004), Lugo (2002), Marín-Spiotta, Ostertag, and Silver (2007), and Yackulic et al. (2011). Helmer used data from air photos in 1977–78 and Landsat TM images (30m × 30m pixels) to examine the biophysical and geographic factors associated with "land development", or conversion from non-urban use (forest, shrub, agriculture) to urban use. Using logit models (pixel converted or not), she found the major factors to be the type of initial land use including type of forest cover if forested, elevation, and, especially, accessibility and proximity to an existing urban area. She predicted that ecological zones near the coast are most at risk of being converted unless new, strictly protected reserve areas are established to preserve them. No socio-economic data or variables were examined. In a second paper, Lugo (2002) examines the evolution of tree species (new, alien as well as native species) in Puerto Rico over time during the process of reforestation, finding new species achieving dominance early in the process, but providing some protection for regeneration of native species (as "refugia"), so that by 60–80 years, reforested stands have similar species richness as native stands. However, they have fewer endemic species and lower soil carbon and litter, and accumulate above-ground biomass as well as soil carbon more slowly. This is also the topic of study in Marin-Spiotta et al., who examine changes in plant species composition as abandoned pastureland in southeastern Puerto Rico was naturally regenerated by secondary forest species (5 types) over periods of 10, 20, 30, 60 and 80 years. The highest accumulation of above ground biomass (carbon) was during the first 20 years, and by 80 years secondary forests had greater biomass than primary forests due to replacement of woody species by palms. Overall, the new ecosystems had similar species richness, and higher potential for carbon sequestration than remnant primary forests, perhaps a small positive sign for climate change amelioration if generalizable to other secondary forests.

The final study here on Puerto Rico, of Yackulic et al. (2011), examines the effects of both biophysical and socio-economic factors on the forest transition, but again based on ecological (micro area—not household or farm-level) data, over the periods from 1977–1991 and 1991–2000. They found biophysical factors (slope, rainfall, surrounding land cover) most important in determining reforestation at the *municipio* level (n=78) in the first time period, with reforestation driven by abandonment of less productive, steeper farmlands, mostly in west-central parts of the island. But these factors had little explanatory power at the smaller scale *barrio* (n=875) level. The importance of socio-economic factors (only variables studied being mean income and population density) was stated by the authors to be higher in the second time period, but still neither variable reached statistical significance on either reforestation or deforestation. The process of rural-outmigration to suburban and urban areas—a central aspect of population dynamics, according to other studies above—was not captured by these two static variables. More important, the model might have investigated the effects of changes in population and incomes *over time* on the dynamic process of changes in forest cover, though perhaps data were sought but not found. Finally, areas with protected areas experienced significantly less deforestation, as is to be hoped.

The last study is on the Pacific coast of Costa Rica (Kull, Ibrahim, & Meredith, 2007), which has also successfully undergone reforestation, for totally different reasons (due to good policy, not forced by endogenous processes of development as in Puerto Rico) reflecting successful synergies between international conservation ideologies, tourism which stimulated real estate investment for infrastructure, and modest migration for livelihood diversification. Though these same forces involving tourism have been present in, for example, Madagascar—noted in the publication— such a positive evolution has not occurred there, as we see later in reviewing forest cover and its changes in the context of Africa. Thus, it is possible for globalization to stimulate positive changes in the form of reforestation linked to tourism, though this may simultaneously increase social marginalization and inequality, as noted in this paper.

One overall conclusion from the Caribbean studies is that the quantitative ones either only studied biophysical variable effects on land use change or, when they attempted to incorporate socio-economic factors, examined only very limited ones, e.g., population size or density, from census data—often from just a single census—to reflect socio-economic factors. This drastically limits the variables being investigated, and can also lead to ecological fallacies in drawing conclusions from associations at higher scales from limited variables when more detailed micro-level data might reveal relationships better (see Robinson, 1950). Thus, having much more detailed data on socio-economic factors, such as from household surveys, provides far more potentially relevant explanatory variables, although it usually requires greater budgetary resources to design, finance and analyze the survey. But often there are useful data available from an existing survey on the island, which should always be checked for carefully.

Asia

Asia, in this chapter refers to islands along the coast areas of continental Asia, from the Red Sea east in the Indian Ocean to South and Southeast Asia and East Asia, including the nearly continental size islands of Indonesia and the Philippines. The many smaller Pacific islands (Small Island Developing States, or SIDS) are reviewed in a separate sub-section below, which also will include, following customary practice, Papua New Guinea. Not surprisingly, it appears that the Philippines and Indonesia are the developing countries which have the most high-quality island studies in the Asia-Pacific region.

A convenient place to start is with the excellent theoretical discussion of Henley (2005), which includes a case study historical review of processes of agrarian change over time in 1850–1950 in three small areas of Sulawesi (Indonesia). Henley begins by noting that his time reference covers a period "before…industrialization of the region, the Green Revolution or modern medicine and birth control" (p. 153), and draws on Brookfield's critique (Blaikie & Brookfield, 1987; Brookfield, 1984, 2001) of Boserup's (1965) view of the intensification of agriculture induced by population pressure. Henley points out that Brookfield explained that even though periods of intensification of agriculture (e.g., additional labor for weeding, transplanting rice, or shorter fallow times) were associated with population increase, they more often were induced by price increases, as it is well known that farmers "pay close attention to market signals" (Brookfield, 2001, p. 215). Henley also questions the direction of causality, that once output rises (for whatever reason), that facilitates population increase, a reversion to Malthusian theory! In fact, Henley finds a strong correlation between soil fertility and population density (comparing Java with other islands of Indonesia)—also observed by Malthus (1798, 1960 ed.). Finally, Henley quotes Brookfield (2001) as saying that terracing is not only associated with wet rice agriculture on hilly slopes in densely populated Asia, but that it also evolved spontaneously in many other areas of the world as a logical method to retain water for crops and reduce soil erosion on sloping lands (without, however, considering whether linked to population growth). Finally, agroforestry, including commercial tree products for export (copra, rubber, fruits…), has evolved in recent decades in Southeast Asia, facilitating increases in population density, as has in some areas consolidation of private individual land tenure, which is conducive to investments in sustainable agriculture. But none of these arguments disproves that Boserupian intensification is often linked to increasing population density, moreover, examination of the three different farm systems in the three areas of Sulawesi differing in soil fertility showed that population density and ease of market access/ exports were linked to favorable soils and market opportunities as possible *causes* of population growth. Interestingly, the areas experienced progressive deforestation with *either* low or high population density, except in the Sangir volcanic soil islands where natural forests were replaced over time by shifts from subsistence rice and other food crops to coconut palm trees, then copra and nutmeg trees (all for export), via agro-forestry, maintaining a level of forest cover.

A second Indonesia study examined the socio-economic effects of large pulp-wood plantations in Sumatra and Kalimantan (Pirard & Mayer, 2009), focusing on impacts on local communities in South Sumatra. They find that it is possible for such plantations to improve rather than threaten local communities' livelihoods by providing wage labor opportunities. For this to work well, local communities need to retain about one-sixth of all the land in use in their traditional forms of local agriculture and agro-forestry for food subsistence, with no more than five sixths in pulpwood; pulpwood has to allow for its use of local labor to not conflict with seasonal labor needs of traditional agriculture; and pulpwood has to give priority to hiring local labor and respect customary land and resource rights.

In an interesting study on the Philippines, at both a national scale and for a small island in the middle of the archipelago, Sibuyan, Verburg and Veldkamp (2004) use both biophysical and socio-economic data to explore the causes of land use change and deforestation. Reviewing first the historical past, they note that the Philippines was 90% covered in forests when the Spaniards arrived in the middle of the sixteenth century, decreasing to 70% by 1890 and then rapidly to 23% by 1987. Starting in the 1970s, worries about the loss of forest cover led to efforts to reduce logging and protect watersheds on steep slopes and areas above 1000 m. Upland agriculture had increased by six-fold in 1960–87 (see also Cruz, 1999, on how population growth stimulated migration from the lowlands to the uplands in the Philippines). The model of LULUC used is CLUE-s (Conversion of Land Use and its Effects at Small scale: Veldkamp & Fresco, 1996; Verburg, Soepboer, Limpiada, Espaldon, & Sharifa, 2002), based on matching land use suitability and accessibility (supply) and aggregate land use demand to determine land use change, as affected by the policy context, incorporating path dependence (past land use inertia). It essentially assesses the likelihood of a change in land use of grid cells (using a logit model), taking into account plausible transition sequences (from a land use conversion matrix: e.g., allowing forest to agriculture to grass, but not urban to agriculture, etc.) and conversion elasticities. Biophysical factors studied include altitude, slope, aspect, distance to nearest road and city over 100,000, and whether hilly, mountainous, coastal or beach/tidal; the only socio-economic factors used are population density at the sub-district (barangay) level and land tenure from cadastral maps. They apply the model to the whole country for forecasting changes in 1990–2010 at a spatial resolution of 2.5km × 2.5km grids and to Sibuyan at a resolution of 100m × 100m, based on three scenarios: (1) continuation of previous year practices; (2) incorporation of government spatial policies based on suitability of areas for agriculture vs. ecologically fragile and better for incorporation in national protected areas; (3) "sustainable agriculture", decreasing shifting cultivation and replacing it by intensified permanent agriculture. They conclude that the coarse model is useful only for identifying "hot spots" of land clearing and not land use or its changes, since satellite imagery shows land cover (forest vs. agriculture) but often cannot distinguish well primary forest from secondary forest, nor secondary forest from tree crops, nor different types of agricultural crops, nor capture much fragmentation. The higher resolution model is therefore more useful, especially if accompanied by fieldwork to ground-truth forest and crop types. It is thus seen as useful for studying specific factors involved in land-use change and developing better policies, but they do not mention the limitations of their model in having only two socio-economic factors: much more is needed to better understand human impacts and improve policies.

Another useful study on the Philippines is by Shively (2001), examining relationships between agricultural intensification, local labor markets, and deforestation in a frontier area. Panel data collected for 1994–2000 show the effects of widespread expansion of irrigation in lowland agricultural lands on increasing the demand for labor, thereby reducing demographic pressures on forest clearing in *upland* areas which has dominated land clearing in recent decades. In this virtuous circle, poverty has also been reduced.

It is remarkable how much work has been done by Chinese scholars on Hainan (see also other topics in this chapter below), with one of the first relevant ones being by Xu (1990) on the processes of agricultural change after 1950 under socialism and how these changed starting in the 1980s. First, there was rapid growth of agriculture and forest clearing, to establish rubber estate farms with central government support (selling internally at three times the world price), as well as of sugarcane and other tropical tree crops, while traditional rice and sweet potato production fell. Rubber production was the most productive land use (though distorted by the internal price), with rice and rice with other crops and horticulture also existing on small private farms. Starting in the 1980s, rubber estate farms paid laborers by contract, while the villages with small privatized farm plots developed a more diversified market-oriented agriculture than before, including the minority Loe population switching from traditional slash and burn cultivation. A second study on Hainan (Dong, Xiao, Sheldon, Biradar, & Xie, 2012) mapped the forests and rubber plantations by integrating PALSAR and MODIS imagery. A third (Wang et al., 2013) examines the effectiveness of nature reserves for controlling deforestation. While it was found that areas inside the reserves experienced an increase in forest cover overall while those outside continued deforesting, the difference was noted as being exaggerated as increasing fragmentation occurred within the reserve and the reserves are generally located anyway in less accessible, higher locations (see also Wang et al., 2012).

Pacific Islands

Four islands in the region had studies that passed the screening processes for this topic (many more will be found under tourism, for example, in the companion study)—for Papua-New Guinea, Tonga, Vanuatu, and the Solomon Islands. First, the Papua-New Guinea study (Shearman, Ash, Mackey, Bryan, & Lokes, 2009) compares data from a land cover map for 1972 and satellite imagery for 2002 to show that deforestation was far higher than reported by FAO: overall, a net 15% of tropical forests were cleared and another 9% degraded by logging. An additional 13% of upper montane forests were cleared, with rates of clearing accelerating over time to 0.8 to 1.8%/year after 1990, and even higher at 1–3.4%/year in commercially accessible forests. The major driver besides logging was seen to be subsistence agriculture, linked to population growth.

A study on agricultural intensification and economic development on the coral atoll of Tongatapu in the Tonga Kingdom (Van der Velde, Green, Vanclooster, & Clothier, 2007) draws on economic and environmental data. Since 1987 Tonga has exported pumpkin squash to Japan, accounting recently for over 40% of all export earnings and about 2/3 of agricultural export earnings. Only 4% of the original forest remains, mostly in a national park and on uninhabited small islands. Increasing production and exports is putting increasing pressures on freshwater resources, while a big increase in use of chemicals in agriculture (mostly fertilizer, though soil moisture more than nutrients is the main productivity constraint) threatens water quality, with nitrates 5 times the WHO limit for drinking water (though most people use collected rainwater). In this type of small island, with a total population of only 100,000 living on a total land area under 800 km², 70,000 on Tongatapu, the main island, and 40,000 in the capital, there is very limited data collection capacity and therefore little data available, so the study pulls together what is available on the economy, which is environmentally vulnerable and dependent on squash exports and remittances from up to an estimated 100,000 Tongan workers living abroad. Remittances are several times the value of all exports and rising, while the second largest source of foreign exchange through the 1990s has declined considerably especially since 1999. Traditional agriculture has been mostly replaced by squash, so most food is increasingly imported, with a large negative trade balance being filled by remittances and foreign aid. Tonga is said to be exporting both its soil fertility and water quality via squash exports to Japan. Policy recommendations include less use of fertilizer, more coordinated government management of water, and promotion of more (rising from a low level) fish exports and tourism, with China being explored as a possible market, and seeking international environmental payments to protect its land and marine biodiversity.

In the case of Vanuatu, Simeoni and Lebot (2012) examined linkages between population, agriculture and forest cover, to assess constraints facing food production. Satellite data were digitalized in six layers (topography, soil potential, apparent residential dwelling density, etc.). Although Vanuatu is not densely populated overall (19 persons/km²), there is great variation across islands constituting the country, as well as in good land per household. In contrast to other islands, shifting cultivation is not seen as a serious threat to the environment. The main causes of deforestation are expansion of coconut plantations and permanent pastures, which contributes to increased pressures on land used for food production. The integration of layers of spatial data was found to be very useful for improving environmental planning

In the most recent study on the region on this topic here, Versteeg, Hansen, and Pouliot (2017) studied factors affecting deforestation, focusing on commercial tree planting in one province (Isabel) of the Solomons. Initially, programs to stimulate complementing traditional crops with commercial trees over 10 years, especially teak, were highly successful, but plantings then slowed and quality declined. The authors use mixed methods (household surveys and in-depth interviews) to collect data, and descriptive statistics and probit multiple regression to analyze the data, and conclude that finding more markets for teak, improving extension services, and incorporating more small-scale agro-forestry are keys to maintaining livelihoods and teak production.

Sub-Saharan Africa

It turns out that the only four studies that made the cut on the Africa region all pertain to one island, and all to the large one, Madagascar, beginning with one of the more significant publications on the topic of agricultural expansion and deforestation, by McConnell, Sweeney, and Mulley (2004), albeit dealing with a tiny part of the island country. The study area is 940 km² in the interior on an escarpment between the main port city on the Indian Ocean on the eastern coast and the capital city in the highlands, near the railroad connecting the two via a local road. The time reference is 1957–2000, with aerial photography and MSS/Landsat and ETM imagery used to measure changes over time in land use of pixels, primarily transforming from forest to agriculture, and to develop a logistic regression model to study the factors responsible. The only socio-economic data used was population density, based on data from population censuses, except for baseline data on residential structures observed in aerial photos, which were linked as well as possible to estimated cartographic boundaries (but later judged not very reliable --p. 182). Field visits were made during 1994–2000 to collect qualitative data from interviews with farmers and government conservation workers. Traditional land use has been mainly slash-and-burn, rainfed rice cultivation, with fields cultivated for 1–2 seasons only, then abandoned to fallow for 3–10 years, with other subsistence crops also grown close to dwellings. Access to farmland is based on traditional lineage relations with the village founders, under a community land chief. Of 795 cells subject to conversion, 527 were converted to agriculture. Overall, farmers converted low-lying land near villages at the forest margin much more than more distant interior land at higher elevations. A logit regression model was used to determine the factors predicting the conversion, finding prior land use the strongest predictor, followed by closeness to villages, being along a body of water (lake, river) or along the forest edge. Interestingly, no effects of city size (gravity) or population density were observed, attributable possibly to poor data. There was also, surprisingly, no effect of slope steepness in this small area, in contrast to the strong slope effects observed elsewhere in Madagascar by Green and Sussman (1990). The national park was effectively protected from clearing.

Moving to the other studies on Madagascar, in chronological order, first, Bakoariniaina, Kusky, and Raharimahefa (2006) provide historical context first, noting that the dramatic deforestation of the island began with French colonization as late as 1896, with many Malagasy fleeing the French into the interior to clear land for slash-and-burn cultivation, though this was already the dominant practice along the coast. By 2005 over 90% of the forest was gone, with the steep and difficult to access forests in the humid east and northeast the main ones remaining. The island's largest lake, Alaotra, at 750 m elevation in the east central area, is in a large basin known for high soil fertility and productive rice fields. But in the previous 30 years, silt from erosion caused by land clearing had come to clog the rivers and streams that feed the lake, shrinking it to 60% of its original size by the 1960s, 40% by the 1980s, and 20% by 2000, with fieldwork in 2003 revealing a small remnant lake with marshes converted to rice fields. Sadly, crop productivity also fell to 40% of its earlier level, yet clearing and slash-and-burn rice cultivation continue to expand, as the population grows. Two studies getting into the policy process shed light on this distressing situation, with Kull et al. (2007) contrasting the situations in Costa Rica and Madagascar, finding globalization pressures to contain deforestation in the latter ineffective. Focusing on western Madagascar, finally, Scales (2012) describes how conflicts between environmental narratives and identity politics have failed to lead to better management of forests. While international environmentalists have pushed for designating more areas as protected, government policies have focused on helping local communities, which results in complex politics surrounding access to resources, involving multiple and diverse stakeholders. The policy focus has been to seek alternatives to traditional slash-and-burn, through agro-forestry, changes in crops and more tourism, to preserve biodiversity. However, this has failed to translate conservation into local values, as local populations see themselves as able to practice slash-and-burn responsibly according to local taboos and custom.

Summary

A quick summary of the literature reviewed here on all geographical areas finds agricultural expansion a dominant factor in deforestation and changes in land use on islands, confirmed by a number of studies based mainly on satellite imagery. In fact, the literature on this topic is relatively rich, with a number of strong studies, including Rudel et al., Tole, Verburg and Veldcamp, McConnell et al., Henley, Yackulic et al., Newman et al., and Chopin et al. However, what underlies the agricultural expansion (e.g., population increase, higher commodity prices, export markets, government policies) varies greatly across islands and has not been effectively investigated thus far, due mainly to the lack of incorporation of socio-economic factors in models of land use change, which may in turn be linked to poor or non-existent demographic, social and economic data in many if not most of the countries/islands, apart from satellite data on land use. This calls for improvements in the generation of such data at the macro level, from, e.g., agricultural censuses, and the design of multi-topic household surveys that collect current and retrospective data on households in rural areas, their land use and clearing in the past and future intentions, combined with their other current economic activities providing livelihoods and possibilities for diversification if not rural-urban migration, which is occurring widely in any case.

Topic 2. Economic Approaches and Models

For purposes of this chapter and book, it is first desirable to delineate what is meant here by "economic" in the context of land use change (LULCC): it refers to studies that focus on linkages between economic factors and LULCC. Such studies may be at either the macro or micro level. At the macro level, this could include research looking at changes in economic sectors that affect land use beyond urban areas, on effects of changes in income levels, the location of economic activity, the value of imports or exports or the trade balance, or in prices of commodities of major importance to the particular island economy. At the micro level, such research might be examining such factors at the local community or household level, based on data collected from agricultural surveys or household surveys that include rural populations. This section also includes some studies providing descriptive analyses of long-term changes in the economy related to land use, such as through effects on agriculture or forest cover. Both macroeconomic models and household economic models incorporating land use are thus fair game. Some specific models that may be a priori useful on this topic regarding land use include dual-economy or two-sector (urban, rural) models (see Fei & Ranis, 1964; Lewis, 1954), agricultural household models (Barnum & Squire, 1979), as well as standard economic models (e.g., input-output, general equilibrium, etc.). As above, we follow the same geographic breakdown, beginning with the Caribbean (used to refer to the Western Hemisphere in general).

Caribbean

There is less of relevance regarding economic analyses of LCLUC in this region than there was regarding agriculture. One study (Berke & Beatley, 1995) looked at Jamaica's efforts to protect a tropical forest reserve, on the value of providing an appropriate mix of cooperation and market competition to stimulate people acting in their own interests to accomplish socially equitable development, while protecting the environment for future generations. This was said to require building local relationships, developing local economic activities supportive of conservation with clear boundaries, and monitoring and enforcement, in the context of a national land use plan.

Among the other studies, one (Modeste, 1995) describes the major overall positive economic effects of tourism on the economies of several small Caribbean countries (Barbados, Antigua/Barbuda and Anguilla), which can said to be true of (or sought by!) many island states all over the planet, as will be covered more in the companion study on population, tourism, etc. But Modeste also observes there may be negative effects on the agricultural sector, depleting it of manpower and therefore of food production and increasing dependency on food imports, but these effects tend to be dwarfed by the positive economic effects.

The most significant studies, however, are again on Puerto Rico, the first one (Lugo, 2002) focusing on the management of tropical forests in the unusual situation of substantial net reforestation in the latter part of the twentieth century, as noted under topic 1 above. First, Lugo notes the far easier and common transformation in tropical islands of the more fertile lowland forests to other land uses, whether agriculture or urban, while upland areas in contrast remain in forest, or, in the case of Puerto Rico become reforested, with a different species composition but similar in carbon level. Good management is said to require understanding and application of the natural resilience of ecosystems, ecological engineering in infrastructure construction, and enforcement of zoning, in the context of an overall conservation vision for the country.

The most significant study on a Caribbean island under this topic for purposes of this meta review is by Crk, Uriarte, Corsi, and Flynn (2009), studying the "biophysical, socio-economic and landscape factors" leading to forest recovery, in contrast to deforestation as in most studies, though this is starting to change. The focus is on Puerto Rico in 1991–2000, for which Landsat images exist revealing not only the continuing process of expansion of the urban built up area but also reforestation, albeit with increasingly fragmented forest cover. The spatial data are used to create a land use transition matrix limited to studying land cover changes that could realistically represent forest recovery (viz., from agricultural or logging, not from urban). A random set of 15,000 pixels was selected, such that none is closer than 300 m to the nearest other sample pixel, leaving 11,092 sample points. They use a logistic regression model to examine the biophysical, socio-economic and landscape features affecting reforestation. 21 potential drivers are examined, including landscape drivers (% forest, % coffee, %pasture, %urban within 300 m of the point), so-called socio-economic factors (every one being geographic—Euclidean distance measures, viz.,to nearest highway, other road, urban patch, forest reserve), and biophysical (elevation, slope, aspect vis a vis the sun, annual temperature, annual precipitation, soil agricultural capacity). In the logistic model, the dependent variable is probability of conversion of pixel from agriculture to forest, with a set of fixed effects independent variables as listed plus a random effect to control for the municipality of the pixel. As environmental conditions are spatially correlated, spatial autocorrelation was controlled for. The best prediction model estimated shows the two dominant explanatory variables to be %forest cover within 100m and slope—not surprising—with minor effects of distance to road and aspect. Soils with the highest and lowest agricultural capacities were found to have low probabilities of being reforested, the former due to its high agricultural value and use, the latter for its low biological capacity to regrow trees. Overall, forest recovery was seen as slowing and threatened by continuing urbanization. Most importantly, the model used is an excellent example of what could be done to study either reforestation or deforestation of pixels or farm plots/parcels, but many more demographic and socio-economic variables should be investigated in this context, both to better sort out the effects of biophysical and landscape factors as well as for the potential of results to assist in developing recommendations for development and environmental policies.

The most recent study here on Puerto Rico (Vance, Eason, & Cabezas, 2015) uses an innovative approach to assess the sustainability of island ecology, based on information theory. Authors note the need for continuity in environmental stewardship so that core system functions can endure or adapt to the significant perturbations that tend to occur in island economies. While the rural population declined, the urban population grew by 140% in 1970–2009 in Puerto Rico. In this context, the authors use Fisher information theory to evaluate changes in system dynamics in connection with the dramatic regional shifts in socio-economic and environmental conditions, especially in 1981–99, and stabilizing after 2005. Fisher information is said to be helpful in identifying warning signals of critical changes and regime shifts.

Asia

Not surprisingly, given their far larger sizes and populations, many of the most insightful articles from the Asia island region are from Indonesia and the Philippines. Especially insightful is a descriptive study on Palawan Island in the Philippines by Eder (2006) summarizing lessons learned from fieldwork during 1995–2005. In Palawan, as throughout the Philippines, an extremely skewed land distribution system exists with a small number of wealthy farmers owning the vast majority of the lowlands: thus as the population has grown poor farmers with little land have had to migrate up the mountainsides to clear forests and engage generally in swidden (slash-and-burn) deforestation to provide for their families, often competing with indigenous populations for space (see also Cruz, 1999). Over time, this leads to more land clearing, and mostly sedentary farming as remaining forest areas are on steep slopes and/or are declared by the state to be protected areas (39% still forested in 1992). Eder shows that, in contrast to the negative situation reached in Cebu uplands, in Palawan household livelihood strategies have continued to adapt, based on much qualitative research plus interviews/surveys with some 200 households. Government and NFO agroforestry programs have encouraged "sustainable development" which has helped many sustain farming activities (producing for market as well as some subsistence foods), but given the strong aspirations to improve their lives, especially to provide better livelihoods and education for their children, leads many if not most to establish second dwellings in the lowlands both for someone to work (most often the mother) and the children to attend secondary schools. As for the farming, as soils are depleted or land is cleared on poorer quality soils, rice gives way to corn, sweet potato, bananas, and cassava, along with tree crops (coconut, cashew, mango, citrus), most commonly managed by the male household head, but there is also non-timber forest product (NTFP) collection, fishing, overseas contract labor and remittances received, besides wage work in lowland towns. Fallow intervals have been shrunk by decades of population growth and shrinking plot sizes, now rarely over 2–4 years, with farms only a few ha—insufficient in themselves to support large households. But tree crops require much less labor than annual crops and facilitate split upland-lowland households. NTFPs are collected by both migrants and indigenous (honey and beeswax, rattans, tree resins/copal), but reported to be becoming unsustainable. Eder also reports a tendency for the successful mango farmers to buy out the others, leading to incipient land concentration. Still, he concludes that overall, people, with high aspirations for improving their livelihoods and even more those of their children, are generally attaining more productive and sustainable livelihood systems, combining farming with diverse other economic activities.

An island study from the mouth of the river of the world's third largest city is next. An island in Shanghai harbor, Chongming, has been increasing in size due to both river silt from the Yangtze plus land reclamation and an economic boom. The authors (Shen, Abdoul, Zhu, Wang, & Gong, 2017) seek to develop methods for a rapid cost-effective assessment to ensure sustainable development. Landsat imagery spanning 34 years shows the island expanding by 0.9% per year between 1979 and 2013, with the percent covered by vegetation declining from 71 to 45% and the built-up part growing from 5 to 20%, much urban, though the population was constant. The dominant factor underlying this expansion, economic development, is said to not be able continue in its present form without degrading the ecology of the island.

Similarly, economic growth is viewed as driving land use change and sedimentation of the upper Citarum River basin in Java, Indonesia (Noda, Yoshida, Shirakawa, Surahman, & Oki, 2017). The changes in LU occurred around the metropolis of Bandung city and are "typical of Southeast Asia", with paddy fields replaced by urban sprawl in the lowlands and cash crops replacing forests in the uplands, leading to increased risk of flooding in the urban area.

Finally, perhaps the most relevant study in the region is an econometric time-series model analysis of changes in forest land on Hainan, a large island off the southeast coast of China, over the long period, 1957–94 (Zhang, Uusivuori, & Kuuluvainen, 2000; see also Xu et al. 2002). Population growth is viewed as the major driving force behind the loss of natural forests and replacement by plantation forests. Hainan was originally covered by rainforests, but human interventions of logging, expansion of rubber production to meet domestic demand, shifting cultivation, and residential/industrial expansion led to natural forests falling to 30% by the 1950s and 15% by 1975, with most of the remaining in the mountainous center.

Then forest cover gradually rose in 1975–95, providing support for the "forest transition" hypothesis, with urban growth and massive plantation forests for woodchip production for export being planted, and efforts to protect forests and water. Government policy and World Bank loans for eucalyptus were important factors. The study is viewed as examining the roles of economic and institutional/policy factors on the changes in forest cover types, taking into account factors that affect each type of forest cover separately (prices) as well as jointly (e.g., population growth). Thus the authors examine the effects of changes in major product prices and policies on the decline in the areas in both natural forest and degraded land and corresponding large increases in plantation forests. The two key commodity prices used are "hardwood" for rainforest and eucalyptus for plantation forests.

The authors use a formal economic model based on the assumption that all land users (both individual landholders and state enterprises) allocate land to different uses based on the profitability in each use, which depends on relative prices. They show that the maximization conditions require that the ratio of the marginal products of land in use j to that in use j must equal the ratio of the prices of j and i. The overall role of institutional factors is reflected in economic growth (in output per capita) while that of population is manifest in population density, looking at the 13 county agglomerations of Hainan over the 37 year time period (221 observations), controlling for proportions of county land under private farm and state ownership. Generalized least squares was used to estimate the determinants of the area in each type of forests as a function of gross economic output per capita, population density, prices, and an overall agricultural products index, with dummy variables controlling for county effects. The results show that the % land in forest is negatively related to population density, private land ownership, economic growth, and timber price, and is positively linked to prices of agricultural and tropical product prices. The latter suggests that the expansion of agriculture in this context had no impact on forest cover. Meanwhile, managed forest cover (for wood chips, etc.) was positively linked to its price and negatively to tropical product prices (competing for same land), and positively to economic output, population density, and land in private hands. Especially noteworthy are the powerful associations in the expected directions for population (density)—the strongest of all variables in both models.

Pacific Islands

Studies on these islands focus more on specific issues than *sui generis* conditions in specific islands. This includes the role of the private sector, of government price policies, of foreign aid, of oil, and especially of fishing on the local economy and livelihoods of islanders. In an early study, Cole (1993) promoted the private sector as important to stimulate to shake the economies of the island states of the South Pacific out of stagnation, which have generally persisted so despite substantial foreign aid. He saw the widespread efforts of governments to promote import substitution and self-sufficiency instead of looking outward via exports as inefficient and not contributing to entrepreneurial efforts. Bertram (1993) also raised the issue of foreign aid, questioning whether the SIDS should depend on it so much, whether it is sustainable, and whether the islanders feel "entitled" to it. He also raised the question of whether the dependence on remittances from labor migrants is healthy, as it depends on the receiving countries accepting the migrants. Read (2008) notes that the role of foreign direct investment in SIDS economies has been neglected in research, perhaps due to its relatively small size, though it is often a critical contributor to investment and growth. And Reddy, Groves, and Nagavarapu (2014), looking at the case of Kiribati, state that economic development policies may have important economic and ecological consequences beyond the economic sector they target, and that it is important to understand those consequences to align policies with conservation of natural resources. Thus, most households in Kiribati engage in both copra and fishing, so when the government increased the price of the major export commodity, copra, that attracted even more labor to that sector than before but indirectly affected fishing and the coral reef ecosystem. Based on household survey data, a 30% increase in the price of copra led to a 32% increase in copra labor, with the larger growers of coconut trees switching to copra, while those with small plots could not make the switch so engaged in more fishing, by 38% in fishing time, which was expected to lead to overfishing and eventually a 20% drop in fish stocks. They therefore recommend a systems approach to policy design.

One study noted a particular problem of the SIDS in the Pacific, that none of the 13 islands have any oil so have to import it and suffer when the world price rises (Jayaraman & Choong, 2009). In a study on Samoa, Solomons, Tonga and Vanuatu, they find oil prices, Gross Domestic Product, and international monetary reserves intimately linked, with causality from (higher) oil prices to (lower) reserves to lower economic growth.

A topic most common in the research literature on the SIDS is the actual or potential role of fishing in SIDS economies. Kronen, Vunisea, Magron, and McArdle (2010) describe the factors underlying the artisanal coastal fisheries in Pacific island countries and territories and their dependence on fishing for their subsistence. But overfishing is increasing, so it is crucial to develop other sources of income for their livelihoods. Their study of 17 islands and territories found that the growing dependence on fishing was not only linked to population growth but also to people's choices, affected by resource exploitation rates and polices at the national level. Countries which are growing economically and creating other opportunities for livelihoods are having more success at ensuring sustainable coastal fisheries communities, which is therefore a key policy issue. At a more local scale, Barclay and Kinch (2013) analyze one fisheries development project each in Papua New Guinea and the Solomon Islands. The dependence of the fisher populations through exports should be viewed as global capitalist markets penetrating through the lens of local cultures. Therefore, the design of fisheries development projects should take into account local conditions, including local village approaches to fishing, the national political-economic context, ongoing transnational development assistance, and resource management.

Jaunky (2011) notes the vital importance of the sector to SIDS, and studies the statistical relationship between the growth in the value of fish exports and the growth in GDP for 23 SIDS over the period 1989–2002, finding a strong positive causal linkage. Zivin and Damon (2012) examine the job creation of the fisheries sector, focusing on tuna (c.f., also Barclay, 2010; Read, 2006) Based on an examination of the costs and benefits of alternative development strategies, they demonstrate that investment in better fisheries *management* can effectively encourage economic development and job creation more than alternatives such as selling access permits or improving fish processing capacity.

Africa

There are few island states associated with Africa, in contrast to Latin America and Asia, so only two appropriate references with an economic focus were found in the time frame of this meta study. First, in the Seychelles islands in the Indian Ocean, Campling and Rosalie (2006) found domestic and international factors threatening social gains linked to economic progress resulting mainly from tourism. This is interpreted as reflecting the serious vulnerabilities of SIDS (Small Island Developing States—widely cited in the literature). The importance of not neglecting the social dimensions in studies focusing on the economic and environmental vulnerabilities of SIDS is emphasized. Sobhee (2004) on Mauritius, focusing on fisheries, notes how environmental degradation of fisheries may be unfortunately linked to government investment in human capital and therefore to increasing inequality. Thus the government invested in education to contribute to successful economic development (including fertility decline), but children of poor families in coastal areas have limited access to education (beyond primary), especially due to the high opportunity cost to poor families of losing the labor of the young who often help in fishing and even child care so others can fish for family survival. A side effect is the overexploitation of fish and declining catches, creating a vicious circle of entrapment in poverty, at the same time while most of the island progresses.

Other

An interesting study on Malta (Conrad & Cassar, 2014) studies how to decouple economic growth from environmental degradation (that is, achieve the former without the latter) on an island country dominated by tourism, as most are. They investigate the degree to which this has been achieved in four economic sectors: (a) energy, air quality, climate change, (b) water, (c) waste management, and (d) land use. Some decoupling has been achieved between economic growth and also population growth vis a vis a to c above, but land development continues to be tied to both, suggesting both provide significant pressures on land change. The study also emphasizes the need for methodologies that go beyond quantitative analyses of environment-ecology relationships to study more intensively smaller scale areas, but within the context of larger areas to better understand environmental impacts.

A study which uses an economic model is that of Polo and Valle (2008), using a general equilibrium model to examine the impact of a possible fall in tourism in the Balearic Islands, in the Mediterranean off the coast of Spain. The 1997 input-output model shows economic dependence of the country on services, especially tourism. The authors explore the effects of a 10% fall in tourism using the input-output model to trace its impacts throughout the economy. Its effects are then examined using a general linear model with a social accounting matrix they develop.

Summary

In summarizing this section, in spite of many insights about the importance of specific economic variables (*viz.*, fishing, foreign aid and investment, exports, changes in key commodity prices, and income growth), it is clear that far less has been learned about the linkages between economic change and land use than has been learned about agriculture-deforestation relationships, as reviewed in the previous section. Some of this is due to the new technology, for example, emerging in recent decades via more and better satellite imagery, facilitating viewing changes in agriculture and forest cover cheaply and conveniently from above, in contrast to the weak socio-economic data bases available for most islands at either the macro (e.g., GDP national accounts, economic census, even population censuses) or micro (household survey) levels. Of all the potentially useful economic models and prospects for illuminating econometric analyses of economic variables such as prices, wages and income growth, there have been very few quality published studies on island countries or territories that capture such economic factors. Only a few of the references reviewed here use a macro- or micro-economic model to investigate potential linkages, and none, for example, uses a two-sector (rural-urban) economic macro model, nor anything close to an agent-based model with economic variables at the micro level to study economic factors and land use change for island developing countries, though several offer good starting points (e.g., Crk on Puerto Rico; Eder on Philippines; Zhang et al. on Hainan; Polo & Valle on Balearic Is.).

Topic 3. Integrating Models of Land Use and Broader Socio-Economic Factors

This section is a bit of a residual, picking up papers and models from a huge literature that relates to land use and land cover change but without a specific focus on the most common forestry-agriculture transition nor on linkages with the economy.

Nevertheless, there is doubtless overlap with not only those two topics above but also with topics covered in the second (future) review study focusing on population, tourism, etc. Finally, I exclude studies that focus on the use of satellite imagery or other spatial data to measure LCLUC *unless* they go beyond measurement to explore linkages with socio-economic processes or have interesting modelling or methodological approaches.

Caribbean

The most relevant study here is one on St. Croix, in the US Virgin Islands (Oliver, Lehrter, & Fisher, 2011). It is also rare in its dealing with linkages between land-based human activities (leading to pollution) and their implications for the health of coral reefs. It uses a Land Development Index (LDI) calculated from data on land use, which was found in fieldwork data collection to be negatively linked to coral health (stony coral colony size and density, and taxa richness or variety). Increasingly impervious surfaces in the watershed (from more built-up areas) was negatively related to coral cover. Overall, the LDI index was found to be a useful landscape indicator of human impacts on corals, indicating a close link between human land use and marine ecosystems, and therefore useful for land use and conservation planning. On this linkage, see also Ramos-Scharrón, et al. 2015, on Puerto Rico; Walsh & Mena, 2016, on Galapagos).

Asia

Selected studies from Asia range from China to Southeast Asia—with a concentration on Indonesia, especially Sumatra—to the Maldive Islands in the Indian Ocean (which has not come up here until now). These will be reviewed in that order, from north to south and east to west. The one study on China here (Pan et al., 2016) is on the Zhoushan islands, off the coast near Hangzhou and Shanghai. The authors note the spectacular process of urbanization between 1980 and 2013, driven by economic growth and accelerating since 2000. They quantify the socio-economic drivers of urban expansion using multivariate regression, finding the major drivers to be secondary and tertiary industry promoted by government investment and tourism. New bridges connecting to the mainland were also important, with population growth having much less effect. But the economic growth was so rapid, inefficient land use occurred, so land reclamation was resorted to for further construction. The study shows the need for better land use planning based on scientific principles to achieve more sustainable urban development.

A paper on Langkawi island in Malaysia (Leman, Ramli, & Khirotdin, 2016) applies a GIS-based evaluation of environmentally sensitive areas (ESAs) to promote sustainable land use planning, similar to studies on Reunion Island reviewed below under Africa. The need for better planning results from the considerable environmental degradation of the island due to economic growth, tourism and over-exploitation of natural resources in recent decades, resulting in conflicts between protecting the environment and meeting the needs of development and tourism, with the goal of making Langkawi a world-class Geopark. The authors examine different approaches to using GIS, providing guidelines for use of zoning and other strategies in ESAs.

In a study on the high-biodiversity island of Mindoro in the Philippines, Wagner, Leonides, Yap, and Yap (2015) examine the negative effects of extensive deforestation on ecosystem services. They conducted studies on soil and vegetation for sites of varying land use intensities along with qualitative interviews to complement the quantitative data. A history of exploitation of uplands involving slash-and-burn agriculture by the indigenous (Mangyan) minority had led to loss of biodiversity and soil erosion, though even allowing the land used to fallow for only a few years resulted in rapid growth of secondary forests in the tropical conditions. The dominant population (Tagalog) lives in the more fertile lowlands, growing rice and coconut palms. Population growth of both populations and political power concentrated in the hands of the Tagalog hurt efforts at poverty alleviation, driving illegal logging and more intensification of agriculture, leading to further declines in forest cover and soil quality and endangering biodiversity and livelihoods. This is one of the studies most clearly tracing both environmental and development problems to rapid population growth. It happens that the Philippines is the only country in southeast Asia with a (high) total fertility rate above 3 still (UN World Population Wall Chart, 2017).

The three studies on Indonesia here all pertain to Sumatra (Konagaya, 1999; Kunz et al., 2017; Robertson, Nelson, & De Pinto, 2009). In a mostly theoretical paper, Konagaya draws on the classic von Thunen-Alonso theory to study aggregate land use allocation ratios to maximize rent (returns on land use) at a regional scale, but finds it not possible to specify all the factors affecting bid-rents since land use is affected in different ways by many factors which differ from one city to another. They accordingly develop a probabilistic multi-city (GTA) model which they apply to Sumatra, noting the model has a theoretical foundation and yields plausible parameters. Robertson et al. also investigate the use of models, noting that discrete choice models are difficult to specify and also involve estimation problems due to spatial effects. They investigate the consequences of alternative assumptions of spatial effects using Monte Carlo simulations of a binary choice model of land use in Sumatra for *predicting* land use (based on past experience) rather than developing a research model for parameter estimation. They note that elasticities cannot be reliably computed from non-linear relations between parameters and predicted probabilities, since they depend on the other variables included in the model and their parameters. In the end, they find few statistically significant factors for predicting land use, that the simple inclusion of spatially lagged explanatory variables provides the best model for prediction. Ironically, it is the most recent of the studies that looks at the earlier historical antecedents of land use change in Sumatra, albeit in only one village: Kunz et al. (2017) note that since the Dutch colonial times, the establishment of privatized land tenure rights for colonists favored clearing land for rapid expansion of mono-culture cash crops for export, at the expense of traditional indigenous population swidden agriculture on communal land. The current National Development Plan in Indonesia does not change this, continuing to see Sumatra as a source for resource extraction, to reduce poverty and accelerate development. The field research of Kunz et al in a single village notes how *de jure* regulations conflict with customary land tenure practice, creating a contested land tenure environment which compromises resource use and poverty reduction.

It is important to mention another publication, which, albeit focusing on methodology, includes an application to the Philippines, so it is included here. Moulds, Buytaert, and Mijic (2015) describe a new land use modeling software package which is open source and free, apply it to two islands that have been modelled elsewhere in the literature, and encourage others to build upon it to improve it. They provide the source code, which they note is often not available or hard to find in current models, to facilitate others' add-on modeling. The model attempts to capture the complexity of LCLUC better than other existing models, and makes available an object-oriented software package to more flexibly model changes in land use of pixels as the spatial units change over time, including showing both their stability and transitions. It builds on previous models, LandSHIFT and IMAGE, estimating the suitability of pixels to change depending on their biophysical and other characteristics and location. It incorporates (integrates) flexible logistic statistical estimation functions more flexible than in the CLUE-s economic model (which allows only a binary specification of the dependent variable, land use). It can incorporate decision rules on plausible transitions (e.g., forest to agriculture and the reverse, but *not* urban to forest), and includes two (only) key "socio-economic" variables, population density and distance to nearest road. Brief applications are made for Plum Island, Massachusetts, USA, and Sibuyan Island in the Philippines. Even with its limited socio-economic variables, the authors state that a big challenge in its use will still usually be to obtain plausible values for all the baseline input data. Also, its approach is inductive rather than deductive, so it is a simulation model, not a research model, but could be expanded in the future to incorporate deductive modelling components as well as additional variables.

Pacific Islands

The islands involved here range from Guam (US) to the Solomon Islands, Samoa and the Galapagos islands of Ecuador. I begin with the oldest of relevant studies in the reference period, for Samoa (Zann, 1999), which begins by noting that increasing population size and development has led to growing pressures on coastal environments and fisheries in many small Pacific islands. Samoa is no exception, with its population growing 5–6 times in the previous 150 years, degrading wetlands, lagoons and coral reefs, which was complicated by recent severe cyclones, leading in turn to depletion of fish stocks. In 1990 a research study was undertaken to assess the coastal environmental situation and fish stocks and identify potential policies. Data were collected in a national population census, a household survey, and boat and market surveys, and a foreign aid program was initiated by Australia. It was decided to devolve some powers from the national government to villages and fishermen in a culturally appropriate co-management model that was tested and then adopted. Villagers were trained to subsequently undertake their own surveys, to identify factors affecting their fishing, and to seek ways to address those factors, and to develop a co-management plan with the national government. By 1998, 26 villages had entered the co-management program and 20 fisheries reserves had been created. The procedures for establishing such policies are said to be replicable in other islands in the South Pacific. In another study with a cultural focus, Reenberg et al. (2008) studied the coevolution of driving forces and adaptive strategies in the main island (Belona) of the Solomon Islands. They examined climate events, population growth, agriculture, non-agricultural activities, transport and other infrastructure, migration, politics, etc. Satellite imagery and aerial photos showed agricultural land intensity to change little despite 50% population growth over 1966–2006. Interviews with 48 households found shifting cultivation and fishing continued to be the main staple livelihoods but increasingly supplemented by other non-agricultural work, including government employment. Cultural bonds have also helped people cope with external shocks and socio-economic stressors.

In one of a number of increasingly sophisticated models of human activities and land use change in the Galapagos Islands (Ecuador), Miller et al. (2010) show that agent-based models (ABMs) can be useful to examine the complexity including feedbacks in human-environment systems. A spatially explicit ABM model was developed to examine linkages over time between major land cover/land use changes and people's livelihood decisions on the largest but sparsely settled island in the archipelago, Isabella. The model was implemented using free NetLogo software, which requires little programming knowledge. Isabela is characterized by a single urban small port community, an adjoining agricultural area, and is mostly a large area of the Galapagos National Park. A serious problem of invasive species (guava) exists in the agricultural area, and is modelled based on field data collected earlier. Other data inputs came from public data and household surveys on people's livelihoods—in agriculture, fishing and tourism. The model was used to explore scenarios of different levels of agricultural subsidies for controlling guava to assess results for land cover dynamics and people's livelihoods. They conclude that ABMs can have considerable utility for conceptualizing population-environment systems, testing policy scenarios (with some guessing of parameters on effects), and also revealing key data shortcomings. A number of additional studies using ABMs have been created for the Galapagos, mostly for the island of San Cristobal, which has the capital and largest city and a much larger agricultural area than Isabella, and adding variables and parameters (some estimated, others assumed) to include agricultural production, imports and exports of food, tourism flows and linked infrastructure, land use tradeoffs between agriculture and forest/scrub land, water, health, etc. (see Mena et al. and Pizzitutti elsewhere in this volume).

The final Pacific island study is on Guam (Clay, Valdez, & Norr, 2015), a model of land use with economic zones and transportation between zones. The authors note that urban land use models are often used in policy analyses to explore long run impacts of land use and urban infrastructure projects. The Large Zone Economic Model version of the SE3M model of simulated urban land use and transport was originally applied to Guam. In this paper the authors explore its adaptability for studying land use changes between urban and rural areas over time in Oahu, Hawaii, and Puerto Rico, from data for the years 2000, 2002 and 2007, and then validated for 2010. They conclude that it is reasonably accurate, and that if certain economic parameters are found, it can be used to model the entire regional economies along with the associated population and economic sector employment distribution.

Africa

In contrast to the two topics above, there are quite a number of publications that made it through screening on this LCLUC topic on Africa. These include three on the islands along the Atlantic coast (the Azores and Canary Islands) and nearly a dozen on the Indian Ocean (Zanzibar, Reunion, Mauritius, and Madagascar). First, on the Canary Islands, Otto, Krusi, and Kienast (2007), stating theirs to be the first long-run study of LCLUC on the islands, assessing socio-economic changes over 1964–92. Using a GIS of five main land use types, they show the arid coastal area of Southern Tenerife to have been totally transformed by mass tourism and intensified agriculture, resulting in destruction of coastal scrub and the natural endemic vegetation and replaced by irrigated crops, especially tomatoes then later bananas. Some land came to be abandoned, but most became built up, in urban housing, hotels, etc. The natural vegetation in nature reserves became degraded due to recreation and dumping of wastes, but this was better than in unprotected areas, where 60% of natural vegetation was totally lost. The authors concluded that half of those losses could have been avoided with better environmental planning, including creating fewer but larger protected areas.

An intriguing economic interaction model of land use is developed by Silveira and Dentinho (2010) and applied to one small island, Corvo, in the Azores archipelago. It involves setting up urban and non-urban zones where populations living in each zone are estimated based on economic activities, size of area, soil class, and location, with the economic activities (urban and touristic, horticulture, arable farming, pasture, forest) depending on "bid-rents" for land in the five uses estimated from land productivity per ha and employment productivity per ha in the various types of economic activities (which depends on technology assumptions) and assumptions about consumption per capita. Biophysical conditions determine land productivities, based on soil class (12 classes), rainfall, temperature, and slope. A spreadsheet is used to test spatial interactions. They simulate changes from the sixteenth to twentieth centuries based on sets of technical parameters assumed fixed over the sixteenth through the nineteenth centuries and then changing in the twentieth, with employment requirements/ha many times smaller in the twentieth century and production many times larger for all types of land use except forest cover, and adding tourism. Consumption per capita of those engaged in different forms of land use was also taken to be essentially constant except for those with pastureland. The model iterates until supply and demand are made equal through adjustments in bid-rents for each activity in each zone, taking into account the area in each type of soil in each zone and its aptitude for different types of production. The model is thus a general equilibrium simulation model. While its structure is intriguing, it could be expanded (*my thoughts*), e.g., to include internal migration between zones and also to be an "open" economy model (incorporate external factors, imports and exports, and remittances, besides tourism). In any case, like all models, it requires many assumptions about technological and other parameters for Corvo in the two time periods, and presents major challenges for including/testing policies. But with modern computers and more data, even borrowing parameters from other agricultural studies and other islands, it could be a basis for developing a model of island land use and change over time incorporating more socio-economic parameters than found so far in existing models.

In the other study on the Azores, Calado, Bragagnolo, Silva, and Pereira (2014) note the extraordinary biodiversity and vulnerability common to islands, including the Azores. They develop a multi-scale approach for implementing a regional conservation policy involving a new management system for protected areas, considering spatial units from the scales of archipelago to island groups to individual islands, integrating quantitative and qualitative information and land use studies to identify key areas of concern and therefore challenges for implementing conservation policy at multiple levels. They underline the importance of developing arrangements for dealing with the scale mismatches between the scales at which challenges are identified and the scales at administrative levels where policy decisions are made.

Moving to the east coast of Africa, I begin with the elephant in the room, the huge island country of Madagascar, with two studies. Vagen (2006) uses remote sensing imagery from 1972 to 2001 to analyze the overall change in land use and rates of change of different types in the eastern highlands. He finds rates of deforestation rose dramatically from 1972–92 (52 ha/year) to 1992–99 (341 ha/year), much attributed to political strife. Overall, elevation and accessibility to roads were found major predictors, manifest in a large increase (65%) in intensive crop cultivation of 3400 ha, half from cultivation of prior savannah grassland. Increasing population pressures on land thus led to an expansion (extensification) of agriculture on marginal grasslands, and soon to declining soil fertility. Meanwhile, Whitehurst, Sexton, and Dollar (2009) studied land cover change in the western dry forests of Madagascar, noting that there had been more deforestation in the west than in the more studied humid east. Much of what is left of the western dry forest is in Kirindy Mite National Park, leading the authors to assist in developing a management plan, so they began by using Landsat imagery to measure land cover changes in and around the Park in a 5 km buffer, based on images from 1990, 2000 and 2006. They found lower deforestation (1/3 as high) and higher reforestation (twice the rate) inside the Park compared to the buffer area outside. The rate of deforestation rose slightly in 2000–2006.

They suggest expanding the Park area and simplifying its shape, increasing coop-eration with local populations to better manage the buffer zone, and improving monitoring of human intrusions into the Park.

North of Madagascar is the island of Zanzibar, a small island (actually 2) which has been the subject of a number of studies, perhaps attracting scholars as well as tourists due to its fascinating history. Kukkonen and Kayhko (2014) study changes in forest cover and land use on the principal (southern) island, Unguja, over three decades based on satellite images, forest change trajectories and land cover analy-sis. Deforestation was found to be increasing from very low levels in 1975–96 to 0.46% per year in 1996–2009. They compared changes under three distinct land tenure systems, with deforestation high in both community-forest regimes and agro-forest systems, but for very different reasons: in the former, it was due to expanding traditional slash and burn cultivation, while in the latter it was due to urban sprawl, moving up hillsides to consume agroforest lands. On the other hand, in the third type of land ownership regime, government forests, tree cover increased in the ear-lier period due to large tree planting programs, but since 1996 deforestation has been high even in these areas. Factors driving forest clearing are said to be popula-tion growth, in-migration, urbanization and tourism, all leading to increased demand for agricultural and forest products. Actions proposed include establishment of a protected areas network connected by forest corridors, protecting trees in both urban and agricultural land use planning, replanting cleared government planta-tions, and extending plantations to the small northern island.

There is also one study on Mauritius, east of Madagascar, by Welsch et al. (2014), modeling climate, energy, water and land use simultaneously (CLEWS), noting they are interrelated, in contrast to most research and planning that looks at them individually. It explicitly looks at the interlinkages, albeit with a focus on energy. Mauritius was selected due to its policy goals to reduce dependence on imports of fossil fuels while also altering agricultural land use, along with its problems of water stress and its diverse climate zones. Scenarios for 2030 were set up and ana-lyzed using CLEWS. Pathways to 2030 based on treating the energy sector sepa-rately vs. taking into account interlinkages using CLEWS highlight important dynamics missed in the first approach. In particular, the added value of CLEWS was evident when anticipated rainfall declines were considered, by showing where land use changes are likely.

Finally, there are a surprising number of publications under this topic on the small island of Reunion, east of Madagascar and still a department of France (in 2019). Three studies were led by Lagabrielle (Lagabrielle et al., 2009, 2010, 2011), the first focusing on mapping biodiversity for conservation planning by identifying biodiversity processes and promoting accordingly conservation corridors to connect areas of high biodiversity. The second extends this by integrating modelling biodi-versity into land use planning, involving participatory modeling with 24 researchers in various disciplines along with key stakeholders. This involved mapping land use and biodiversity, developing a conservation plan based on conservation principles derived from a spatial optimization tool (MARXAN), and then simulating coupled LU-conservation scenarios with an ABM. This led to identifying priority areas for conservation on the coast, where most LU changes are occurring. However, the authors note that they did not have adequate representation of stakeholders among the 24, lacking any from both agriculture and urban sectors! In the third and last study, the authors develop operational controls for integrating conservation and restoration with land use planning. They first integrate ecological and socio-economic factors to identify the best spatial options (based on conservation costs) for conserving/restoring biodiversity, while minimizing LU conflicts. The planning protocol involves site prioritization and participation of stakeholders, resulting in implications for spatial planning. As elsewhere, Reunion has experienced rapid urbanization and agricultural expansion, with 43% of its land area in a National Park but only half of this contributes to biological targets, while another 21% of its land is stated that it should be in new ecological corridors connecting marine, terrestrial and freshwater sites.

The other, very recent study on Reunion (Lestrelin et al., 2017) looks at the interface between spatial modeling and urban planning, to deal with increasing problems of high population densities and urban sprawl that combine to increase pressures on agricultural land and ecosystems. Spatial models are co-developed with urban development scenarios and presented to institutional leaders to promote dialogue about future urbanization patterns, to facilitate creating a collaborative research network. This led to questions about and therefore modifying models and scenarios to shift from purely statistical approaches to more integrated perspectives. The paper focuses on the value of the *process* of landscape modeling and simulation in mediating debates among stakeholders.

Other

The final studies in this section are on Mediterranean islands of France (Corsica) and Greece (Lesvos and Nisyros). For Corsica, Sanz et al. (2013) analyze two study areas of 60K and 44K hectares over a long time frame, from 1774 to 2000, in areas dominated by chestnut trees—common in the Mediterranean. Data are from land-cover documents and human population counts, noting that populations rose steadily over time from 1770 to 1890, then fell precipitously to one-fifth their former size by 2005, involving massive land abandonment. This in turn has led to considerable reforestation, but of shrubs rather than chestnut trees, though current efforts aim at their restoration. The authors conclude that it is important to take a long-term perspective to properly understand the resilience of natural systems and provision of ecosystem services.

In the first study on a Greek island, Petanidou, Kizos, and Soulakellis (2008) look at the abandonment over time of agricultural terraces on Nisyros Island. Population grew in the nineteenth to early twentieth century, leading to significant expansion of terraces with stone walls to capture rainfall, facilitating cultivation on 58% of the island. But the collapse of agriculture and the resulting emigration led to land abandonment. Recently livestock grazing has become the main form of land use, involving a much smaller population with large farms. The authors recommend preserving portions of the unique forms of terraces as a "living resource for sustainable management". Moving on to Lesvos island, Dikou, Papapanagiotou, and Troumbis (2011) use remote sensing and multivariate statistical methods to (1) quantify land cover types and configurations (patch density, diversity, fractal dimension and contagion) for five coastal watersheds of Kalloni Gulf, from 1945 at the end of World War II to 2003; (2) evaluate the relative importance of biophysical (slope, substrate, stream) and human (roads, population density) variables; and (3) describe processes of LCLUC transitions over the five time periods covered. While principal land cover types did not change, population growth and an expanded road network led to increased heterogeneity of the landscape (patchiness), complexity (fractal dimension) and contagion (patch disaggregation). There was also some agricultural land abandonment and ecological succession. Hypothetical regulatory scenarios which take into account these land use dynamics can better predict future evolution in LCLUC, providing a valuable tool for regional planning.

Summary

The diversity of modeling approaches is evident in this section, along with interrelationships between LCLUC and many other factors going beyond the most common forest-agricultural transition and economic factors to geology, ecology, culture, urban expansion, politics and policy simulations, and participation of stakeholders. Many of these overlap with other topics, including those above and in the separate meta study on population, tourism, etc. Some here take a long historical view, and show the importance of culture and the resilience of natural systems. Several of the more interesting analytical models include Konagaya (noting the distinction between inductive and predictive models) and Moulds et al.'s software for modeling land use and its determinants over time with a binary structure for pixel transitions, cited under Asia above; Miller et al. using an agent-based model on the Galapagos, under Pacific islands; Silveira and Dentinho's model incorporating a bid-rent economic model in a land use study on the Azores; Lagabrielle et al. using a spatial optimization model on Reunion island, also under Africa; and Kukkonen and Kayhko on land tenure and land use on Zanzibar.

Topic 4. Remaining Models and Methods

This was intended to capture miscellaneous methodologies and modeling approaches not covered in the three sections above nor in the separate population-tourism-urbanization-climate study in preparation. Thus, it is a bit of a residual of the residual from above. Indeed, a large number of publications that were initially organized under this land use modeling sub-topic on the computer search were also in one or more of the topics above (or in the other four topics), so have been pre-empted by those topics, leaving a shorter residual list covered here. Indeed, almost every item reviewed in this overall meta assessment of the literature linking LCLUC to socio-economic factors could be listed under more than one topic, so two rules apply. One is the early bird gets the work (items classified in topics 1–2 above or in 3 on modeling are reviewed above and not here). The second rule is that someone has to decide, and that has to be me, the author.

Caribbean

The relevant studies remaining for this topic on this region are for Jamaica and St. Lucia. In the first one on St. Lucia, White and Engelen (1994, 1997) present a model integrating regional spatial dynamics with a mathematical modeling approach called cellular automata (see also Engelen at end of this section under Other for more details). They apply the model to St. Lucia (ibid., 1997) Their approach links a Geographic Information System (GIS) to a non-spatial model and population changes to investigate the implications of climate change for land use. They conclude that cellular automata facilitates a realistic depiction of land use changes over time, as well as serving as a good way to introduce spatial-environmental dimensions into standard economic-demographic models, which are said to be otherwise unrestrained in predicting future changes. Also for St. Lucia, Begin et al. (2014) study processes of sedimentation from land use change in watersheds causing damage to coral reefs. The authors note the lack of good data on sediments as well as on how to quantitatively link it to delivery and damage downstream on coral reefs.

There are also two publications on Jamaica qualifying for this general modeling topic, by Timms, Hayes, and McCracken (2013) and Setegn, Melesse, Grey, and Webber (2015). Timms et al investigate the long-term process of deforestation followed recently by reforestation, therefore, as another example of the forest transition theory of Rudel and others, cited in section 1 above. They first note that this theory has been found empirically valid so far mostly for advanced, industrial societies where it occurred as an endogenous process linked to their development tied to urbanization. Remote sensing data for 1987–2011 for the Cockpit County indicate it is in this process of change due to changes in demographics and in the agricultural sector, which is said to be decimated by neo-liberal (exogenous, not endogenous) economic policies that led to both changes in crops and land abandonment. Setegn et al., in a chapter in a book focusing on hydrology and climate, study spatial-temporal variation in hydrological processes in Caribbean islands. They note the demand for safe water has become a crucial issue, especially in overcrowded urban centers, affected by both population growth and environmental degradation and linked to land use practices and climate change. This indicated the value of developing a hydrological model, which they do for the Great River basin in the county. Surface water accounts for 28% of stream flow and ground water for 18%, varying across sub-basins due to variations in land cover. The hydrological model developed is said to be useful for predicting watershed impacts of climate and land use changes, including where water supply shortages are likely. This in turn can be used to assess irrigation potential, runoffs, and flood and pollution control needs.

Asia

Four studies on Asia and Asian islands were initially referenced here, but three were then re-classified under the topics of population and climate, so are not further reviewed here, leaving only one item, on Sumatra. However, it is one of the more interesting studies in this chapter. Villamor, Le, Djanibekov, van Noordwijk, and Vlek (2014) develop and use an Agent-based Model (ABM) to evaluate the effects of a payments for ecosystem services (PES) policy compared to alternative policies on land use dynamics, carbon storage and rural livelihoods in lowland Jambi province of Sumatra (Indonesia). Following brief, useful reviews of the literature on land use, PES programs and ABMs, they describe the study area and data used. The main socio-economic data are from a very small survey of 95 households in three villages, covering household characteristics, preferences and behavior, including land-use choices under several different hypothetical policy scenarios. Cluster analysis was used to categorize households as better-off vs. poor, based on incomes coming primarily from rubber or rice. For each type of household, multinomial logistic regression was used to assess willingness to adopt PES. Plot data were collected with the aid of participatory mapping, and role-playing games were played to help refine decision rules for the ABM. The ABM model used was LB-LUDAS, designed for studying land use decisions at the forest margin, with households as the human agents and land pixels (30m × 30 m, from Landsat ETM) as the landscape or biophysical agents. In the former, human resources are manifest by age, education, household size and dependency ratio; natural resources by land size; financial capital by gross income; physical capital by location (access to markets and town); and policy access by participation in PES, other direct subsidy or neither. The biophysical variables were land cover/wetness, etc., and the model is run in annual increments for 20 years. Environmental factors that drive LU include whether it is a protected area, with other variables including market prices, the 3 policy options, and neighborhood land use. Household behaviors manifest in state variables then can change over time with experience, and each year are functions of existing output prices of rubber and rice, labor costs, and input costs. Separate Cobb-Douglas production functions for rice and rubber ($O = AL^{\alpha}K^{\beta}$) were used to predict yields based on explanatory variables derived from the household survey (labor input L in days/ha/yr; land area K, and A reflecting agricultural technology manifest in chemical inputs and a wetness index driven by topography), with α and β the elasticities of inputs. The base model runs agent choices as populations grow (taken to be at 1.1%/yr) and biophysical conditions change over time. This is compared with a policy involving initially giving subsidies allowing conversions from initial land use to altered ones with labor allocated optimally each year, and a PES policy involving implied eco-certification of rubber if rubber forests are protected vs. rice, to improve carbon stocks and biodiversity, and inducing less tree clearing. The paper traces through various sub-models, decision trees and feedback loops allowing farmers to learn and change preferences, but no actual equations are presented in the publication, limiting ability to appraise it. The overall conclusion of the authors from running the model is that the PES program did not improve livelihoods as much as the direct subsidy program, though it had environmental benefits—an unfortunate trade-off, but based on the parameters and behavioral functions estimated from a very small sample in one small study area. The ABM methodology is nevertheless most promising, but does require collecting detailed data from an adequate sample of land users, implying data collection costs, but with potentially significant rewards in understanding behavior and therefore in formulating policy (see also Miller et al. and others on the Galapagos, including in this volume—by Pizzitutti, Walsh & Mena, & Pizzitutti, as well as other publications cited therein).

Africa

Four studies on Africa survived the filters in the three sections above (and for the other four topics for the future companion study on population-tourism, etc., three on Madagascar and one on nearby Reunion. The first two studies on Madagascar are by the same team of investigators led by Agarwal (Agarwal, Gelfand, & Silander Jr., 2002; Agarwal, Silander Jr, Gelfand, Dewar, & Mickelson Jr., 2005), both using statistical regression models to study tropical deforestation. The first uses a two-stage process in which the first stage is a spatial model on population size, involving counts of people by administrative units, referred to as town counts; the second is on land use based on remotely sensed data, at a 1 km × 1 km pixel level, which is, as always, not physically aligned with the population data from administrative/census sources. Regressions were used to align the data, for a tropical wet forest area on the eastern coast. In the second study in 2005, the authors say they seek to establish causality regarding the determinants of tropical deforestation in Madagascar, that better explanatory models are needed. They propose a hierarchical model incorporating spatial association to first deal with the physical misalignment identified in 2002, first, imputing measures of socio-economic variables when there are gaps in the data. They then use five data layers for the same study region to fit a spatial hierarchical model, at considerable effort due to computational demands, and finally a Bayesian approach allowing the model user to choose the initial model formulation.

The other study on Madagascar is by Laney (2004) on a northern interior highland region, Andapa. Laney uses "human-environment theory" implied by the induced intensification thesis of technological change in agriculture to characterize alternative strategies of farmers in Andapa available for confronting rising population pressures on land. The model is said to indicate the consequences of the strategies for LCLUC at the village level, though not necessarily the most numerically common response of farmers (mostly being very small), and therefore said to reflect the ecological fallacy: relationships at the higher scale level do not necessarily reflect those at the individual level (Robinson, 1950).

Finally, for the island of Reunion, Vayssieres, Vigne, Alary, and Lecomte (2011) use participatory modelling of farms to study the effects of various policy scenarios on agricultural intensification and its sustainability. They see population growth, increasing food demand, and increasing densification of agriculture as creating pressures on the environment in most of the world's regions in coming decades. Integration of crops with livestock raising is seen as one possible solution for better intensification of farming systems. They develop a whole-farm model called GAMEDE, and use it with six (!) "representative" farmers on the island to investigate ex-ante differences in farm sustainability at various degrees of crop-livestock integration. The model looks at the implications of decision processes in the context of biophysical conditions for gross income from outputs, labor needs, and energy and nutrient flows on the farm. The model draws upon both quantitative and qualitative data, reflecting actual farms rather than modelers' assumptions of what constitutes a *synthetic* farm, as in much of the modeling literature. This is said to capture the knowledge of farmers about how farms are actually managed. The six different prototype farms take farm diversity into account in developing policy interventions. The reliability of extrapolations and recommendations for policy resulting from the farm-level simulation was verified by checking *ex post* the representativeness of the six types in the island and also comparing expert opinions about the quantitative results from the multivariate analysis. "Our research indicates that actual farms can also be typical" (author abstract).

Other

Some years ago, Zhu, Aspinall, and Healey (1996) presented what they called a *knowledge-based spatial support system for land-use planning model*, and applied it to an island (Islay) off the west coast of Scotland. Calling it by the acronym ILUDSS (Islay Land Use Decision Support System), it puts into a software package a pre-programmed system of programs embodying the steps to develop and implement land use planning, to ease the task of planners. It includes functions for the initial query (what is being planned), formulation of a land use model for evaluating land use potential, evaluation of the model through automatic integration of data from the database, and implementation with results. It is intended to be user-friendly and flexible, and provide planners with a tool to utilize knowledge and expertise put together by modelers, based on their own personal modeling issues and preferences in relation to various criteria, such as physical suitability, proximity to roads, protected areas or other borders, and specified minimum area for each land parcel. ILUDSS guides the user to specify land use options and evaluation factors; then automatically formulates models to assess land use potential; and evaluates models automatically. It also assesses the database available and displays it graphically;

retrieves metadata on the data and existing models in the package; provides information on the modeling process, functionality and use; and generates maps and reports. Model builders can use it as a structure to build on and adapt. Flow charts illustrate the sequential steps in this kind of "automatic modelling". The application to Islay incorporates GIS (18 models, including buffering, reclassifying, etc.), analytical (land use, in Islay for three options—farming, forest and peat cutting, the major exploitable resource there) and rule-based models. The LU model takes into account three evaluation factors stated above: physical suitability, proximity to roads and protected areas, and minimum area for economic use, all specified by the user. The model was used to identify appropriate areas for expanding peat cutting. This "decision support system" is said to provide an operationally flexible, interactive and learning platform. In its form in 1996, it was only for rural areas, supports only three forms of land use, and provides only physical suitability for specific forms of land use (or changes), which would require sets of economic parameters to be added to determine economic potential. The authors note the platform could be expanded to overcome these limitations, and could also be useful for determining the suitability of habitats for wildlife or tourist facilities development.

It is intriguing to compare this platform for a land use modeling decision support system with that of *cellular automata* CA) as laid out by Engelen, White, Uljee, and Drazan (1995). CA provides a very different dynamic modelling method that can integrate socio-economic and environmental (physical) factors at micro (pixel) or higher-level scales. The framework comprises both "macro" and micro levels, with the so-called *macro level* model comprising the data and relationships between socio-economic and environmental factors and land use, whether estimated statistically or just postulated (guessed!). The combined macro-micro model was applied to investigate the implications of climate change on a small Caribbean island characterized by tourism and small-scale subsistence agriculture (see above application, for St. Lucia) to keep it simple in scale and content. The natural sub-system models (hypothetically) impacts of climate change (temperature and precipitation) on beach erosion; the social subsystem models population growth impacts; and the economic subsystem uses a simplified 5-sector input-output model to translate climate-and-population induced changes in demand via assumptions about land-productivity into aggregate changes needed in agricultural output and tourism into land use needs over time. This indicates the aggregate areas (e.g., pixels in different forms of land use, such as beach, agriculture, etc.) needed to expand to meet rising needs.

Once the suitability of lands around the island for tourism and agriculture are determined (down to the pixel level) and the existing baseline use is measured, the *micro-level model* will mathematically determine *which areas in which locations will transition* to the new uses, which can include not only unused or forested areas but also existing areas transitioning from one land use to another. Cells (pixels) change from their initial state according to not only their suitability in that particular form of land use vs. the alternative uses, but according to the state (land use) of *neighboring* cells. If cells are square, each has four main neighbors, and four more at the corners, plus potentially other cells beyond that, defined collectively as its "*neighborhood*". A cell cannot just flip land use if surrounded by similar land use cells if other cells in its neighborhood do not change, but, on the other hand, is prone to change if its neighbors are also changing. A neighborhood includes cells with weights determining forces of other cells, which logically decline with distance from the cell under consideration. The actual island application was based on only 113 cells, for which both initial land use as well as hypothetical suitability for each of the other forms of land use considered were set up a priori, and applied to model the effects of a 2 degree rise in temperature and a 20 cm rise in sea level over 40 years. The micro model is intriguing and highly dependent on initial land use (as it should be), sizes of neighborhood and neighborhood effects, and transition rules, and realistically ensures most changes occur logically through accretion rather than randomly. However, based on land suitability measures, one could conceivably have some new non-contiguous areas coming into a new use (e.g., new beaches converted from coastal scrub, or agriculture from highland forests). For other studies using CA, see applications to Ecuador and Thailand in Messina and Walsh (2001) and Walsh et al. (2006).

Overall Summary on Modelling

CA models (Engelen et al., 1995, the ILUDSS model of Zhu et al. (1996) and ABM models (Villamor et al., 2014)—covered in this topic 4—as well as other models reviewed in this meta assessment under topics 1–3 provide frameworks for studying processes of land use change. Their main limitations in practice are the lack of reliable estimates of parameters linking land-use to the socio-economic and environmental factors affecting it, which in turn in most islands is in turn related to the lack of adequate data (which tends to be even more limited than in usually much larger non-island countries, which have more capacity for collecting data). But in addition, models are just models, mechanisms for recognizing (but simplifying) relationships between things such as land use and the many potential factors affecting it. They can be, are in the process of, and will continue to be themselves dynamic, undergoing significant improvements in their capacity to (more) realistically portray land use— explaining processes that have occurred in the past, which is usually very important for predicting future changes, and hence can be an invaluable aid to planners (including identifying key missing data). Depending on the country context—physical and socio-economic conditions, and main land use issues—one model may be more useful than another. But all are only as good—as useful—as the quantity and quality of the underlying data available, which also must be improved. Garbage in, garbage out. The enormous advances in satellite imagery and its ability to identify and measure well different forms of land use (via free or cheap satellite data, from Landsat, Spot, and other sources) must be contrasted with the far more limited advances in recent decades in socio-economic data for most islands covered in this meta review.

Conclusions: State of the Art, Strengths, Gaps and Implications

There is little to be gained by attempting to summarize what is in essence already an attempt at summarizing such a vast and disparate literature on land use on islands in developing countries and its relationships to socio-economic factors. First, the focus is on analytical and modeling approaches, and for the time period of 1990 to 2017, though some basically descriptive and historical studies crept in. This meta effort tries to assist the reader interested in a particular aspect of these linkages, but was found to involve too many references to adequately cover the subject of land use in one book chapter. Therefore, we intend to prepare a second study, hopefully to be published in a different forum, on the parts of this literature which will focus on population, urbanization, tourism and climate linkages with land use on islands.

As noted, I organized this meta-assessment under each of the four topics by geographic area, for the convenience of those looking for references on the literature on a particular region or country. This is hopefully useful for students and scholars interested in particular islands, and those beginning research on them. The map also indicates the geographic distribution, and may be taken to provide a rough indication of the distribution of quality research on the topic of land use by island and region, indicating the concentration on particular islands (but note I specifically excluded the literature on Hawaii, as well as much on Puerto Rico and the Galapagos islands, as they are covered elsewhere in this book). Thus they are not adequately represented in this meta-review nor on the map. Other limitations may be inferred from the procedures we employed in the filters used, described at the beginning of this chapter and more fully below in the Appendix.

To investigate linkages between land use and socio-economic factors, data are evidently needed on both. While there has been spectacular progress in the quantity and quality of data on land use available on islands (even taking into account limited coverage of some islands from satellites and recurrent cloud cover), there has been far less progress on socio-economic data, severely limiting the relationships that could be examined for most islands. For some, there are not even basic data on national economic accounts nor values of production by economic sector or agricultural crop. Some islands and island countries are so small that they lack the capacity and resources to collect these basic data. In such island countries, researchers working there and interested in land use would do well to help train local people to collect the data, including seeking assistance from bilateral aid sources or multilateral agencies, such as the United Nations Statistical Office or the World Bank. They could identify and train appropriate local residents to prepare (update) workable data on national economic accounts and outputs by economic sector as a baseline, including local market prices, via economic censuses and surveys; at the same time, the islands are likely to have some data already on imports and exports since this involves foreign exchange transactions going through banks. Apart from economic data, demographic data may well also be in short supply, though is much easier to create, drawing upon established procedures and questionnaires used for many decades for population censuses and labor force, health and other household surveys.

Once this is done and institutionalized, data series over time can be created, which it is crucial to have in order to advance knowledge of the linkages between socio-economic factors and land use, and how they change over time, to then link to measues of land use from remote sensing, to investigate factors affecting land use change over time and try to identify relationships.

The meta-review here does not seek to select particular models as "ideal" ones, though comments are made here and there on how to improve existing ones—which in some cases has probably already occurred since the country studies reviewed here were implemented. There are thus hidden and unhidden lessons here about model architectures, key sector linkages, and occasionally plausible parameters from existing studies. While part of our original goal was to seek out the "best" studies and identify parameters linking key socio-economic factors to land use change which could be useful in formulating models for *other* countries lacking the necessary data or parameters to provide an estimate for an important missing parameter, circumstances tend to differ too much from one island to another to justify such borrowing, though sometimes it can be a plausible starting point. Moreover, models also are too different to transfer a parameter estimated with one type of model in one country context to a different model in another context without a great deal of religion. In fact, the vast majority of studies reviewed here do not present clear parameters that could be considered for such a purpose. Evidently, for that purpose, there is a need to seek out a few of the better and more flexible models to apply in several of the countries with relatively extensive data sets to seek to estimate more generally useful parameters and to compare them across models for a country, and then across countries.

Hopefully this is at least useful food for grist, for the next generation, which will have access to even better data on land use and surely more socio-economic data than has been available up to the time of this assessment. I hope this review can stimulate such further work on data collection as well as on improved research on relationships between socio-economic factors and land use on islands. The small size and much less complexity of islands does augur well for the value of any such sustained efforts of modelers to produce something more elegant and comprehensible (e.g., for policy-makers) than is possible for larger, land-based more complex economies.

Appendix: Method in Madness? Procedures Used in Island Land Cover/Land Use Change Literature Search

Meta-analyses are performed and published increasingly as literatures on many areas of interest and scientific inquiry grow faster than even scholars in the particular area can keep up with, much less scholars entering the area, and can provide

useful summaries/updates and introductions to fields of research.[1] Such meta studies vary greatly in subject scope and time frames for the years of publications covered, and inevitably involve subjective considerations about the boundaries of the
field covered and the kinds and quality of documents accepted for inclusion. The
scope, detail and length of meta-analyses depends on the topic, library and other
technical resources available, the number of authors and research assistants and
their disciplines, and therefore ultimately on the monetary resources available.
Procedures for undertaking meta-analysis are described in various publications,
with an extensive application to the topic of climate impacts on human conflict with
references to methodology in Hsiang, Burke, and Miguel (2013). There are many
other examples in the literature, including Geist and Lambin (2002) on land use and
a recent excellent one on the effects of environmental factors on human migration
in Hoffmann, Dimitrova, Muttarak, Crespo Cuaresma, and Peisker (2019).
Hoffmann et al. pared down the literature they were examining ultimately to 32
published articles which had actual parameter estimates from statistical studies of
the effects of a wide range of measurable environmental factors (as well as "control" variables, in multivariate statistical models) which could be compared. While
that is a noble goal and one we initially dreamed of in the case of this book to seek
general parameters to consider using in modeling factors affecting LCLUC in
islands, that would have required an extremely narrow review of the scientific literature, as well as more time and monetary resources than were available. In addition,
there is much to be said instead for beginning with a very broad-based journey
through the literature to assess the state of the arts and identify gaps, both analytical
and geographical. The approach here, being a first assessment, attempts to do something akin to the latter.

   The first step was to consider the full range of factors that might affect LCLUC
on islands, both stand-alone islands (whether constituting States or not) and islands
considered as part of archipelagos or even countries. The project was funded by
NASA (PI, Stephen Walsh), and involved many actual and virtual workshop meetings of the project staff: faculty members mostly from UNC but also three other US
universities (see authors of first chapter of this book), together with graduate student
research assistants and library staff at the Carolina Population Center at UNC. These
workshop meetings led me to settle initially on 11 major potential topics for the
literature search, which would be independent of the extensive literature focused on
the use of satellite remote sensing to *measure* land use, covered in a separate computer and literature search (see Walsh et al., chapter 3, in this volume).

---

[1]A global survey of country practices in collecting economic data from economic censuses and
surveys, registers of business establishments, or other administrative source was carried out by the
United Nations Statistics Division using a brief questionnaire in 2006 (UN, 2007). Over half the
countries responded, with the response rates higher for developed and transition economies than
low-income countries, but among the latter higher for the small island countries of Oceania.
Among the latter, while only 30% had carried out an economic census, this was expected to accelerate, with 50% anticipating implementing one within the next five years. It would evidently be
useful to have updated data on this situation, but researchers interested in a particular country
could contact sources in the country or the UN Statistics Division.

The NASA project islands research team decided that the 11 topics most relevant fields for the computerized search are primarily from the fields of geography, economics, demography, ecology, and agriculture. Accordingly, the Carolina Population Center Director of Library Services (Lori Delaney) and the island research team decided to search eight separate literature databases: PubMed, Scopus, Web of Science, GEOBASE, GeoRef, PAIS, IBSS, and Environment Complete as most relevant for identifying potentially relevant research. It was decided that the search should not be open-ended in time scope but rather limited to publications that are fairly recent, taken to be from 1990 to the time of conducting the computer searches in 2017.

Upon perusing the computerized files resulting from these searches for the each of the 11 topics[2], even after eliminating by computer evident duplications of citations across databases, the files were still huge, in the thousands. It was thus decided to drop two of the original 11 topics *a priori* before proceeding, namely, "sustainability" and "ecosystem goods and services", believing that virtually anything relevant to land use in either would already be captured in the remaining topics. A third topic, "international trade," was also dropped due to finding from its 693 computer citations only around 20 with any potential link to land use, so these 20 were examined and re-classified among the other eight topics, mostly under economics. The results of these initial processes are found in column 1 of the table in the introductory chapter of this book by Walsh et al.

In the relatively smaller databases (population, urbanization, tourism and climate change), the topic was initially coupled (filtered) with "island" or "islands" to produce the initial data, whereas with the larger, less manageable databases, the topic term was also filtered with "land use" to narrow down the results to make the screening of the larger literatures more manageable. The actual search strings are available by database searched (Word documents in SharePoint).

The hundreds to thousands of references for each of the eight final topics were printed out on computer sheets, 1-3 lines per citation. This was done, for each topic separately, alphabetically by author, noting the full name of the first author, date of publication and journal name, if any; the full title, publisher (if a book or part of a book, government or international agency, etc.) and page numbers. The next step was to look through these long lists as quickly as possible, using subjective filters, to screen the vast literatures for what *might* be relevant for our purpose. Since our focus was on the "serious" literature that might have something useful on the substantive relationships between that topic and LCLUC and/or on methodologies for exploring the relationships, one filter was to drop all very short papers (under 4 pp.); a second was to limit the language to English; a third was to exclude studies

---

[2]This part of the appendix draws on an excellent project document compiled by Ms. Sommer Barnes, August 24, 2017. Barnes was the major library staff person who implemented the original computer searches based on key words (topics cross listed with islands as geographic identifiers and usually also including a term for land use or change in land use). Then she worked closely with Bilsborrow to implement several filters, as described in the text, and also came up with some of her own original filters, including references with "false positives" such as urban "heat islands", etc.

based on pre-modern times and ancient civilizations, including those from archeology, paleontology, and climate studies; a fourth was, for each topic, to identify journal names that were natural sciences or otherwise so specialized to not be likely to have anything we could use, such as chemistry journals in the topics of agriculture and climate, entomology and infectious disease journals; family planning and sexual reproduction journals in the topic of population, as well as journals on non-human populations or bird demographics (fertility, mortality, migration, or population growth); nor journals under "modeling" on physics, computer science software, or mathematics. For some of the eight topics, Bilsborrow went through a sample of the long list of citations by author, title and journal name initially, to get a feel for the kinds of titles and journals appearing, relevant vs. not relevant, and then worked with Barnes together (for the largest topics) to train her to apply his kinds of criteria to complete the initial screening for that topic, while taking care to minimize type I errors—to be sure to not reject papers at the outset that *might* be relevant. Barnes then could go through the rest of the citations on that topic, leading to a follow-up much-reduced print-out that Bilsborrow would skim for any more obvious items to drop. This process tended to filter out about four-fifths of the original computer-generated items. While this process involved inevitable subjectivity, we are confident that few relevant items were dropped in error.

Our search strategy also included some screening for type II errors, to identify and exclude "false positives" (e.g., "heat islands", "islands of fertility", "tourist impacts on local cultures"). A number of other topics were usually screened out if the focus of the article is not linked to land use on the island, such as use of pesticides to control invasive species, land use of specific endemic vs. invasive species, unless linked to broader causal factors, ecological models such as of an insect or bird population, localized studies of urban air or water pollution, studies on landslides or water management problems unless linked to growing populations and land use; design of tourist complexes, or surveys of tourist attitudes/satisfaction; and even things apparently closer to land use, such as human impacts on coral reefs, if not linked specifically to land use or the impact of damage on tourism, for example.

The above process usually left us with between 100 and 500 journal articles *per topic* from among the thousands of titles. These were then printed out using EndNote with complete abstracts when available (as was the case for over 90%), which were then reviewed more closely, with Bilsborrow initially skimming the abstracts, leading to eliminating perhaps another 50–70% (exact records were not kept) that appeared clearly irrelevant. The rest were then classified at the same time into tier 1 papers (for which full papers were sought) and tier 2 titles (for which the abstract was expected to be sufficient to cite (or not) in the actual meta literature review on that topic).

A master list of all the tier 1 and tier 2 items was prepared and stored on the project drop box as an excel file and shared with all project staff and with NASA. This file of initially nearly 1000 entries has a single line for each final item retained, organized by topic, with the references listed alphabetically by tier 1 followed alphabetically again by tier 2, for each topic. For each reference, it shows the full name of the first author, full title, journal name, year published, and island/archipelago/country (if not in the title). However, in the process of writing the text of the meta evaluation here and reading or skimming the articles and abstracts, a number of items were reclassified from one topic to another and some from tier 1 to 2 and vice versa (including shifting items from the four topics covered in this chapter to topics covered in the second, future meta study mentioned above on population, tourism, etc.). The excel file was then revised to incoporate these changes in topic classification, without relocating items from the initial topic classification in the excel file (nor updating the numbers in table in the initial chapter of Walsh et al. showing citations by topic and tier. The entire revised and pared down excel list of over 700 items, in the two tier 1 and tier 2 classifications for all eight topics, is available from the author or the UNC Galapagos Science Center on request.

Acknowledgements I am grateful for excellent bibliographic assistance from Sommer Barnes and Lori Delaney of the Carolina Population Center library; to Brian Frizzelle of CPC for the beautiful informative map; and doctoral students in Geography at UNC, Sara Schmidt & Francisco Laso, for help in organizing and obtaining documents.

References

Agarwal, D. K., Gelfand, A. E., & Silander, J. A., Jr. (2002). Investigating tropical deforestation using two-stage spatially misaligned regression models. *Journal of Agricultural, Biological, and Environmental Statistics, 7*(3), 420–439.

Agarwal, D. K., Silander, J. A., Jr., Gelfand, A. E., Dewar, R. E., & Mickelson, J. G., Jr. (2005). Tropical deforestation in Madagascar: Analysis using hierarchical, spatially explicit, Bayesian regression models. *Ecological Modelling, 185*(1), 105–131.

Bakoariniaina, L. N., Kusky, T., & Raharimahefa, T. (2006). Disappearing Lake Alaotra: Monitoring catastrophic erosion, waterway silting, and land degradation hazards in Madagascar using Landsat imagery. *Journal of African Earth Sciences, 44*(2), 241–252.

Barclay, K. (2010). Impacts of tuna industries on coastal communities in Pacific Island countries. *Marine Policy., 34*(3), 406–413.

Barclay, K., & Kinch, J. (2013). Local capitalisms and sustainability in coastal fisheries: Cases from Papua New Guinea and Solomon Islands. *Research in economic anthropology, 33*, 107–138.

Barnum, H. N., & Squire, L. (1979). An econometric application of the theory of the farm-household. *Journal of Development Economics, 6*(1), 79–102.

Begin, C., Brooks, G., Larson, R. A., Dragicevic, S., Scharron, C. E. R., & Cote, I. M. (2014). Increased sediment loads over coral reefs in Saint Lucia in relation to land use change in contributing watersheds. *Ocean & Coastal Management, 95*, 35–45.

Berke, P. R., & Beatley, T. (1995). Sustaining Jamaica's forests: The protected areas resource conservation project. *Environmental Management, 19*(4), 527–545.

Bertram, G. (1993). Sustainability, aid, and material welfare in small South Pacific Island economies, 1900–90. *World Development, 21*(2), 247.

Bilsborrow, R.E. & Carr, D. (2001). Population, agricultural land use, and the environment in developing countries, in *Tradeoffs or synergies? Agricultural intensification, economic development and the environment,* Lee, D.R, Barrett, C.B. (eds), CABI, Wallingford, UK, pp. 35–56.

Bilsborrow, R. E., & Geores, M. (1992). *Rural population dynamics and agricultural development: Issues and consequences observed in Latin America.* Ithaca, NY: Cornell University

Population and Development Program and Cornell International Institute for Food, Agriculture and Development.

Bilsborrow, R. E., & Geores, M. (1994). Population change and agricultural intensification in developing countries. In L. Arizpe et al. (Eds.), *Population and the environment: Rethinking the debate* (pp. 171–207). Washington, DC: Westview Press, for Social Science Research Council.

Blaikie, P., & Brookfield, H. (1987). *Land degradation and society*. London: Methuen.

Boserup, E. (1965). *The conditions of agricultural growth: The economics of agrarian change under population pressure*. Chicago: Aldine Publishing Company.

Brookfield, H. (1984). Intensification revisited. *Pacific Viewpoint, 25*(1), 15–44.

Brookfield, H. (2001). *Exploring agrodiversity*. New York: Columbia University Press.

Brown, K., & Pearce, D. (Eds.). (1994). *The causes of tropical deforestation: Economic and statistical analyses of factors giving rise to the loss of Tropical Forests*. London: University College of London Press.

Calado, H., Bragagnolo, C., Silva, S. F., & Pereira, M. (2014). A multi-scale analysis to support the implementation of a regional conservation policy in a small-island archipelago-the Azores, Portugal. *Journal of Coastal Research, 70*, 485–489.

Campling, L., & Rosalie, M. (2006). Sustaining social development in a small island developing state? The case of Seychelles. *Sustainable Development, 14*(2), 115–125.

Chopin, P., Blazy, J. M., & Doré, T. (2014). A new method to assess farming system evolution at the landscape scale. *Agronomy for Sustainable Development, 35*(1), 325–337.

Chopin, P., Doré, T., Guindé, L., & Blazy, J. M. (2015). MOSAICA: A multi-scale bioeconomic model for the design and ex ante assessment of cropping system mosaics. *Agricultural Systems, 140*, 26–39.

Clay, M., Valdez, A., & Norr, A. (2015). Transferability study of the Large Zone Economic Module (LZEM) of the SE3M model of land use and transportation. *Applied Geography, 59*, 22–30.

Cole, R. V. (1993). Economic development in the South Pacific promoting the private sector. *World Development, 21*(2), 233.

Conrad, E., & Cassar, L. F. (2014). Decoupling economic growth and environmental degradation: Reviewing Progress to date in the small island state of Malta. *Sustainability, 6*(10), 6729–6750.

Crk, T., Uriarte, M., Corsi, F., & Flynn, D. (2009). Forest recovery in a tropical landscape: What is the relative importance of biophysical, socioeconomic, and landscape variables? *Landscape Ecology, 24*(5), 629–642.

Cruz, M. C. (1999). Population pressure, economic stagnation, and deforestation in Costa Rica and the Philippines. In R. Bilsborrow & D. Hogan (Eds.), *Population and deforestation in the humid tropics* (pp. 99–121). Liege, Belgium: International Union for the Scientific Study of Population.

Day, M. (2010). Challenges to sustainability in the Caribbean karst. *Geologia Croatica, 63*(2), 149–154.

Dikou, A., Papapanagiotou, E., & Troumbis, A. (2011). Integrating landscape ecology and geoinformatics to decipher landscape dynamics for regional planning. *Environmental Management, 48*(3), 523–538.

Dong, J. W., Xiao, X. M., Sheldon, S., Biradar, C., & Xie, G. S. (2012). Mapping tropical forests and rubber plantations in complex landscapes by integrating PALSAR and MODIS imagery. *Isprs Journal of Photogrammetry and Remote Sensing, 74*, 20–33.

Eder, J. F. (2006). Land use and economic change in the post-frontier upland Philippines. *Land Degradation and Development, 17*(2), 149–158.

Engelen, G., White, R., Uljee, I., & Drazan, P. (1995). Using cellular automata for integrated modelling of socio-environmental systems. *Environmental Monitoring and Assessment, 34*(2), 203–214.

Fei, J., & Ranis, G. (1964). *Development of the labor surplus economy: Theory and policy*. Homewood, IL: Irwin.

Geist, H. J., & Lambin, E. F. (2002). Proximate causes and underlying driving forces of tropical deforestation. *Bioscience, 52*(2), 143–150.

Green, G. M., & Sussman, R. W. (1990). Deforestation history of the eastern rainforests of Madagascar from satellite images. *Science, 248*, 212–215.

Helmer, E. H. (2004). Forest conservation and land development in Puerto Rico. *Landscape Ecology, 19*(1), 29–40.

Henley, D. (2005). Agrarian change and diversity in the light of Brookfield, Boserup and Malthus: Historical illustrations from Sulawesi, Indonesia. *Asia Pacific Viewpoint, 46*(2), 153–172.

Hoffmann, R., Dimitrova, A., Muttarak, R., Crespo Cuaresma, J., & Peisker, J. (2019). *Quantifying the evidence on environmental migration: A Meta-Analysis on Country-Level Studies*. Presented at Annual Meeting of Population Association of America, Austin, TX, April 10–13.

Hsiang, S. M., Burke, M., & Miguel, E. (2013). Quantifying the influence of climate on human conflict. *Science, 341*, 1235367. 1–14.

Jaunky, V. C. (2011). Fish exports and economic growth: The Case of SIDS. *Coastal Management, 39*(4), 377–395.

Jayaraman, T. K., & Choong, C.-K. (2009). Growth and oil price: A study of causal relationships in small Pacific Island countries. *Energy Policy, 37*(6), 2182–2189.

Kaimowitz, D., & Angelsen, A. (1998). *Economic models of deforestation: A review*. Bogor, Indonesia: Center for International Forestry Research.

Keller, M., Bustamente, M., Gash, J., & Silva Dias, P. (Eds.). (2009). *Amazonia and global change* (Geophysical monograph 186). Washington DC: American Geophysical Union.

Konagaya, K. (1999). The generalized Thunen-Alonso model for land use change in Sumatra Island. *Geographical and Environmental Modelling, 3*(2), 145–162.

Kronen, M., Vunisea, A., Magron, F., & McArdle, B. (2010). Socio-economic drivers and indicators for artisanal coastal fisheries in Pacific island countries and territories and their use for fisheries management strategies. *Marine Policy, 34*(6), 1135–1143.

Kukkonen, M., & Kayhko, N. (2014). Spatio-temporal analysis of forest changes in contrasting land use regimes of Zanzibar, Tanzania. *Applied Geography, 55*, 193–202.

Kull, C. A., Ibrahim, C. K., & Meredith, T. C. (2007). Tropical forest transitions and globalization: Neo-liberalism, migration, tourism, and international conservation agendas. *Society & Natural Resources, 20*(8), 723–737.

Kunz, Y., Steinebach, S., Dittrich, C., Hauser-Schäublin, B., Rosyani, I., Soetarto, E., & Faust, H. (2017). 'The fridge in the forest': Historical trajectories of land tenure regulations fostering landscape transformation in Jambi Province, Sumatra, Indonesia. *Forest Policy and Economics, 81*, 1–9.

Lagabrielle, E., Botta, A., Dare, W., David, D., Aubert, S., & Fabricius, C. (2010). Modelling with stakeholders to integrate biodiversity into land-use planning Lessons learned in Reunion Island (Western Indian Ocean). *Environmental Modelling & Software, 25*(11), 1413–1427.

Lagabrielle, E., Rouget, M., Le Bourgeois, T., Payet, K., Durieux, L., Baret, S., … Strasberg, D. (2011). Integrating conservation, restoration and land-use planning in islands—An illustrative case study in Réunion Island (Western Indian Ocean). *Landscape & Urban Planning., 101*(2), 120–130.

Lagabrielle, E., Rouget, M., Payet, K., Wistebaar, N., Durieux, L., Baret, S., … Strasberg, D. (2009). Identifying and mapping biodiversity processes for conservation planning in islands: A case study in Reunion Island (Western Indian Ocean). *Biological Conservation, 142*(7), 1523–1535.

Lambin, E. F., et al. (2001). The causes of land-use and land-cover change: moving beyond the myths. *Global Environmental Change, 11*(4), 261–269.

Laney, R. M. (2004). A process-led approach to modeling land change in agricultural landscapes: A case study from Madagascar. *Agriculture, Ecosystems and Environment, 101*(2–3), 135–153.

Leman, N., Ramli, M. F., & Khirotdin, R. P. K. (2016). GIS-based integrated evaluation of environmentally sensitive areas (ESAs) for land use planning in Langkawi, Malaysia. *Ecological Indicators, 61*, 293–308.

Lestrelin, G., Augusseau, X., David, D., Bourgoin, J., Lagabrielle, E., Lo Seen, D., & Degenne, P. (2017). Collaborative landscape research in Reunion Island: Using spatial modelling and simulation to support territorial foresight and urban planning. *Applied Geography, 78*, 66–77.

Lewis, W. A. (1954). Economic development with unlimited supplies of labour. *The Manchester School of Economic and Social Studies, 22*(2), 139–191.

Lugo, A. E. (2002). Can we manage tropical landscapes? – an answer from the Caribbean perspective. *Landscape Ecology, 17*(7), 601–615.

Malthus, T. R. (1960). *On Population (First Essay on Population, 1798, and Second Essay on Population, 1803),* Modern Library, for Random House, New York.

Marín-Spiotta, E., Ostertag, R., & Silver, W. L. (2007). Long-term patterns in tropical reforestation: Plant community composition and aboveground biomass accumulation. *Ecological Applications, 17*(3), 828–839.

McConnell, W. J., Sweeney, S. P., & Mulley, B. (2004). Physical and social access to land: Spatio-temporal patterns of agricultural expansion in Madagascar. *Agriculture, Ecosystems and Environment, 101*(2–3), 171–184.

Messina, J. P., & Walsh, S. J. (2001). 2.5D Morphogenesis: Modeling Landuse and Landcover Dynamics in the Ecuadorian Amazon. *Plant Ecology, 156*(1), 75–88.

Miller, B. W., Breckheimer, I., McCleary, A. L., Guzman-Ramirez, L., Caplow, S. C., Jones-Smith, J. C., & Walsh, S. J. (2010). Using stylized agent-based models for population-environment research: A case study from the Galapagos Islands. *Population and Environment, 31*(6), 401–426.

Modeste, N. C. (1995). The impact of growth in the tourism sector on economic development: the experience of selected Caribbean countries. *Economia Internazionale, 48*, 375–385.

Moulds, S., Buytaert, W., & Mijic, A. (2015). An open and extensible framework for spatially explicit land use change modelling: The lulcc R package. *Geoscientific Model Development, 8*(10), 3215–3229.

Newman, M, E,, McLaren, K. P., & Wilson, B. S. (2014). Long-term socio-economic and spatial pattern drivers of land cover change in a Caribbean tropical moist forest, the Cockpit Country, Jamaica. *Agriculture Ecosystems & Environment, 186*, 185–200.

Noda, K., Yoshida, K., Shirakawa, H., Surahman, U., & Oki, K. (2017). Effect of land use change driven by economic growth on sedimentation in River Reach in Southeast Asia – A case study in upper Citarum River Basin. *Journal of Agricultural Meteorology, 73*(1), 22–30.

Oliver, L. M., Lehrter, J. C., & Fisher, W. S. (2011). Relating landscape development intensity to coral reef condition in the watersheds of St. Croix, US Virgin Islands. *Marine Ecology Progress Series, 427*, 293–302.

Otto, R., Krusi, B. O., & Kienast, F. (2007). Degradation of an arid coastal landscape in relation to land use changes in Southern Tenerife (Canary Islands). *Journal of Arid Environments, 70*(3), 527–539.

Pan, W. K., & Bilsborrow, R. E. (2005). A multilevel study of fragmentation of plots and land use in the Ecuadorian Amazon. *Global and Planetary Change, 47*, 232–252.

Pan, Y., Zhai, M., Lin, L., Lin, Y., Cai, J., Deng, J. S., & Wang, K. (2016). Characterizing the spatiotemporal evolutions and impact of rapid urbanization on island sustainable development. *Habitat International, 53*, 215–227.

Parés-Ramos, I. K., Gould, W. A., & Aide, T. M. (2008). Agricultural abandonment, suburban growth, and forest expansion in Puerto Rico between 1991 and 2000. *Ecology and Society, 13*(2), 1.

Petanidou, T., Kizos, T., & Soulakellis, N. (2008). Socioeconomic dimensions of changes in the agricultural landscape of the Mediterranean basin: A case study of the abandonment of cultivation terraces on Nisyros Island, Greece. *Environmental Management, 41*(2), 250–266.

Pirard, R., & Mayer, J. (2009). Complementary labor opportunities in Indonesian pulpwood plantations with implications for land use. *Agroforestry Systems, 76*(2), 499–511.

Polo, C., & Valle, E. (2008). A general equilibrium assessment of the impact of a fall in tourism under alternative closure rules: The case of the Balearic Islands. *International Regional Science Review, 31*(1), 3–34.

Ramos-Scharrón, C. E., Torres-Pulliza, D., & Hernández-Delgado, E. A. (2015). Watershed- and island wide-scale land cover changes in Puerto Rico (1930s–2004) and their potential effects on coral reef ecosystems. *Science of the Total Environment, 506–507*, 241–251.

Read, R. (2006). Sustainable natural resource use and economic development in small states: The tuna fisheries in Fiji and Samoa. *Sustainable development, 14*(2), 93–103.

Read, R. (2008). Foreign direct investment in small island developing states. *Journal of international development, 20*(4), 502–525.

Reenberg, Anette, Torben Birch-Thomsen, Ole Mertz, Bjarne Fog, and Sofus Christiansen. (2008). Adaptation of human coping strategies in a small island society in the SW pacific - 50 years of change in the coupled human - Environment system on Bellona, Solomon Islands. *Human Ecology. 36*(6), 807–819.

Reddy, S. M., Groves, T., & Nagavarapu, S. (2014). Consequences of a government-controlled agricultural price increase on fishing and the coral reef ecosystem in the republic of Kiribati. *PLoS One, 9*(5), e96817.

Robertson, R. D., Nelson, G. C., & De Pinto, A. (2009). Investigating the predictive capabilities of discrete choice models in the presence of spatial effects. *Papers in Regional Science, 88*(2), 367–388.

Robinson, W. (1950). Ecological correlations and the behavior of individuals. *American Sociological Review, 15*(3), 351–357.

Rudel, T., et al. (2002). A tropical forest transition? Agricultural change, out-migration, and secondary forests in the Ecuadorian Amazon. *Annals of the Association of American Geographers, 92*(1), 87–102.

Rudel, T. K., Perez-Lugo, M., & Zichal, H. (2000). When Fields Revert to Forest: Development and Spontaneous Reforestation in Post-War Puerto Rico. *Professional Geographer, 52*(3), 386–397.

Sanz, A. S. R., Fernandez, C., Mouillot, F., Ferrat, L., Istria, D., & Pasqualini, V. (2013). Long-term forest dynamics and land-use abandonment in the Mediterranean Mountains, Corsica, France. *Ecology and Society, 18*(2), 38.

Scales, I. R. (2012). Lost in translation: Conflicting views of deforestation, land use and identity in western Madagascar. *Geographical Journal, 178*(1), 67–79.

Setegn, S. G., Melesse, A. M., Grey, O., & Webber, D. (2015). Understanding the spatiotemporal variability of hydrological processes for integrating watershed management and environmental public health in the great river basin, Jamaica. In: *Sustainability of Integrated Water Resources Management: Water Governance, Climate and Ecohydrology.* pp. 533–561.

Shearman, P. L., Ash, J., Mackey, B., Bryan, J. E., & Lokes, B. (2009). Forest Conversion and Degradation in Papua New Guinea 1972–2002. *Biotropica, 41*(3), 379–390.

Shen, G. R., Abdoul, N. I., Zhu, Y., Wang, Z. J., & Gong, J. H. (2017). Remote sensing of urban growth and landscape pattern changes in response to the expansion of Chongming Island in Shanghai, China. *Geocarto International, 32*(5), 488–502.

Shively, G. (2001). Agriculutral change, rural labor markets, and forest clearing: An illustrative Cae form the Philippines. *Land Economics, 77*, 268–284.

Silveira, P., & Dentinho, T. (2010). Spatial interaction model of land use – An application to Corvo Island from the 16th, 19th and 20th centuries. *Computers, Environment and Urban Systems, 34*(2), 91–103.

Simeoni, P., & Lebot, V. (2012). Spatial representation of land use and population density: Integrated layers of data contribute to environmental planning in Vanuatu. *Human Ecology, 40*(4), 541–555.

Sobhee, S. K. 2004. Economic development, income inequality and environmental degradation of fisheries resources in Mauritius. *Environ Manage. 34*(1), 150–157.

Timms, B. F., Hayes, J., & McCracken, M. (2013). From deforestation to reforestation: Applying the forest transition to the cockpit country of Jamaica. *Area, 45*(1), 77–87.

Tole, L. (2001). Jamaica's disappearing forests: Physical and human aspects. *Environmental Management, 28*(4), 455–467.

Tole, L. (2002). Population and poverty in Jamaican deforestation: Integrating satellite and household census data. *GeoJournal, 57*(4), 251–271.

UN. (2007). *The results of the UNSD survey of country practices in economic census*. Presented at 38th Session of Statistical Commission, Feb. 28–March 2, item 3(g) of provisional agenda. United Nations Statistics Division, New York, pp. 1–15.

Vagen, T.-G. (2006). Remote sensing of complex land use change trajectories-a case study from the highlands of Madagascar. *Agriculture, Ecosystems and Environment, 115*(1–4), 219–228.

Van der Velde, M., Green, S. R., Vanclooster, M., & Clothier, B. E. (2007). Sustainable development in small island developing states: Agricultural intensification, economic development, and freshwater resources management on the coral atoll of Tongatapu. *Ecological Economics, 61*(2–3), 456–468.

Vance, L., Eason, T., & Cabezas, H. (2015). An information theory-based approach to assessing the sustainability and stability of an island system. *International Journal of Sustainable Development & World Ecology, 22*(1), 64–75.

Vayssieres, J., Vigne, M., Alary, V., & Lecomte, P. (2011). Integrated participatory modelling of actual farms to support policy making on sustainable intensification. *Agricultural Systems, 104*(2), 146–161.

Veldkamp, A., & Fresco, L. O. (1996). CLUE-CR: An integrated multi-scale model to simulate land use change scenarios in Costa Rica. *Ecological Modelling, 91*, 231–248.

Verburg, P. H., & Veldkamp, A. (2004). Projecting land use transitions at forest fringes in the Philippines at two spatial scales. *Landscape Ecology, 19*(1), 77–98.

Verburg, P. H., Soepboer, W., Limpiada, R., Espaldon, M. V. O., & Sharifa, M. (2002). Land use change modelling at the regional scale: The CLUE-s model. *Environmental Management, 30*, 391–405.

Versteeg, S., Hansen, C. P., & Pouliot, M. (2017). Factors influencing smallholder commercial tree planting in Isabel Province, the Solomon Islands. *Agroforestry Systems, 91*(2), 375–392.

Villamor, G. B., Le, Q. B., Djanibekov, U., van Noordwijk, M., & Vlek, P. L. G. (2014). Biodiversity in rubber agroforests, carbon emissions, and rural livelihoods: An agent-based model of land-use dynamics in lowland Sumatra. *Environmental Modelling and Software, 61*, 151–165.

Wagner, A., Leonides, D., Yap, T., & Yap, H. (2015). Drivers and consequences of land use patterns in a developing country rural community. *Agriculture, Ecosystems and Environment, 214*, 78–85.

Walker, R. (2004). Theorizing land cover and land use change: The case of tropical deforestation. *International Regional Science Review, 27*(3), 247–270.

Walsh, S., Entwisle, B., Rindfuss, R., & Page, P. (2006). Spatial simulation modeling of land use/land cover change scenarios in Northeastern Thailand: A cellular automata approach. *Journal of Land Use Science, 1*(1), 5–28.

Walsh, S. J., & Mena, C. F. (2016). Interactions of social, terrestrial, and marine sub-systems in the Galapagos Islands, Ecuador. *Proc National Academy of Sciences, 113*(51), 14536–14543.

Wang, S., Ouyang, Z., Zhang, C., Xu, W., & Xiao, Y. (2012). The dynamics of spatial and temporal changes to forested land and key factors driving change on Hainan Island. *Shengtai Xuebao/Acta Ecologica Sinica, 32*(23), 7364–7374.

Wang, W., Pechacek, P., Zhang, M. X., Xiao, N. W., Zhu, J. G., & Li, J. S. (2013). Effectiveness of nature reserve system for conserving tropical forests: A statistical evaluation of Hainan Island, China. *Plos One, 8*(2), e57561.

Welsch, M., Hermann, S., Howells, M., Rogner, H. H., Young, C., Ramma, I., … Muller, A. (2014). Adding value with CLEWS – Modelling the energy system and its interdependencies for Mauritius. *Applied Energy, 113*, 1434–1445.

White, R., & Engelen, G. (1994). Cellular dynamics and GIS: Modelling spatial complexity. *Geographical Systems, 1*(3), 237–253.

White, R., & Engelen, G. (1997). Cellular automata as the basis of integrated dynamic regional modelling. *Environment and Planning B: Planning and Design., 24*(2), 235–246.

Whitehurst, A., Sexton, J., & Dollar, L. (2009). Land cover change in western Madagascar's dry deciduous forests: A comparison of forest changes in and around Kirindy Mite National Park. *ORYX, 43*(2), 275–283.

Xu, W.-d. (1990). Some aspects of agricultural regionalization in Hainan Island, South China. *Japanese Journal of Human Geography, 42*(3), 195–219.

Xu, X.-L., Zeng, L., & Zhuang, D. F. (2002). Analysis on land-use change and socio-economic driving factors in Hainan Island during 50 years from 1950 to 1999. *Chinese Geographical Science, 12*(3), 193–198.

Yackulic, C. B., Fagan, M., Jain, M., Jina, A., Lim, Y., Marlier, M., ... Uriarte, M. (2011). Biophysical and socioeconomic factors associated with forest transitions at multiple spatial and temporal scales. *Ecology and Society, 16*(3), 15.

Zann, L. P. (1999). A new (old) approach to inshore resources management in Samoa. *Ocean and Coastal Management, 42*(6–7), 569–590.

Zhang, Y., Uusivuori, J., & Kuuluvainen, J. (2000). Econometric analysis of the causes of forest land use changes in Hainan, China. *Canadian Journal of Forestry Research., 30*, 1913–1921.

Zhu, X., Aspinall, R. J., & Healey, R. G. (1996). ILUDSS: A knowledge-based spatial decision support system for strategic land-use planning. *Computers and Electronics in Agriculture, 15*(4), 279–301.

Zivin, J. G., & Damon, M. (2012). Environmental policy and political realities: Fisheries management and job creation in the Pacific Islands. *The Journal of Environment & Development, 21*(2), 198–218.

Social-Ecological Drivers of Land Cover/Land Use Change on Islands: A Synthesis of the Patterns and Processes of Change

Stephen J. Walsh, Laura Brewington, Francisco Laso, Yang Shao, Richard E. Bilsborrow, Javier Arce Nazario, Hernando Mattei, Philip H. Page, Brian G. Frizzelle, and Francesco Pizzitutti

Introduction

Globalization in the twenty-first century is posing new challenges to humans and natural ecosystems (Carter, Walsh, Jacobson, & Miller, 2014; Sun, 2014). From climate change to increasingly mobile human populations to the global economy, the relationship between humans and their environment is being modified in ways that will have long-term impacts on ecological health, biodiversity, ecosystem goods and services, human activities, land cover/land use change (LCLUC), and system sustainability (Kerr, 2005; Uyarra et al., 2005; Zhang & Walsh, 2018). These changes and challenges are perhaps nowhere more evident than in island ecosystems (Zhao et al., 2004; Charles & Chris 2009; Miller, Carter, Walsh, & Peake, 2014; Mena, Quiroga, & Walsh, 2020). Influenced by rising ocean temperatures, extreme weather events, sea-level rise, tourism, population migration and development, and invasive species (MacDonald, Anderson, & Dietrich, 1997; Tye, 2001; Alongi, 2008; Walsh et al., 2008; Brewington, 2013; Walsh, 2018), islands

S. J. Walsh (✉)
Center for Galapagos Studies, University of North Carolina at Chapel Hill, Chapel Hill, NC, USA
e-mail: swalsh@email.unc.edu

F. Laso · R. E. Bilsborrow · J. Arce Nazario · P. H. Page · B. G. Frizzelle · F. Pizzitutti
University of North Carolina at Chapel Hill, Chapel Hill, NC, USA
e-mail: richard_bilsborrow@unc.edu

L. Brewington
East-West Center, Honolulu, HI, USA

Y. Shao
Virginia Tech University, Blacksburg, VA, USA

H. Mattei
University of Puerto Rico, San Juan, Puerto Rico

© Springer Nature Switzerland AG 2020
S. J. Walsh et al. (eds.), *Land Cover and Land Use Change on Islands*, Social and Ecological Interactions in the Galapagos Islands,
https://doi.org/10.1007/978-3-030-43973-6_3

represent both great vulnerabilities as well as scientific opportunities for studying the impacts and significance of global change on ecosystem processes and sustainability (Schwartzman, Moreira, & Nepstad, 2000; Mejía & Brandt, 2015; Johannes de Haan, Quiroga, Walsh, & Bettencourt, 2019).

Human migration and tourism have brought profound changes to the natural environment in places like the Hawaiian and Canary Islands, and beyond (Juan, Eduardo, & Vanessa, 2008). With the rise in global wealth, pressure on unique island ecosystems increases as more people seek to visit and experience these "special" places, thereby introducing new threats to island sustainability. While islands are often geographically remote, relatively small in size, irregular in shape, and varied in their morphological, ecological, human, and topographic settings, they are also fragile and highly sensitive to changes caused by natural and anthropogenic factors. Processes such as climate change, urbanization, agricultural extensification, deforestation, and population migration are increasingly associated with the "temporary" migration of tourists to islands (Walsh & Mena, 2013). These forces and factors of change are often manifested as drivers of LCLUC that occur across a range of space-time scales and whose patterns and trajectories can be assessed through satellite remote sensing (Benitez, Mena, & Zurita-Arthos, 2018; Rivas-Torres, Benitez, Rueda, Sevilla, & Mena, 2018).

While islands are fundamentally different in their geography, ecology, protection, and intensity of human use and development, they share similar concerns, although at different scales and within different socio-economic and political structures. Islands are increasingly burdened by tensions between population-environment interactions and their trajectories of change (Ernoul & Wardell-Johnson, 2013; Walsh, Engie, Page, & Frizzelle, 2019). This paper presents the findings of a meta-analysis of papers published between 1988–2018 referenced in selected databases. After reviewing 309 socio-economic papers and 406 remote sensing papers for global islands, we synthesize the diverse drivers of LCLUC on islands, map the number of reviewed papers by selected geographic regions, and contextualize the findings relative to defined threats to island ecosystems. We focused on island sustainability and the exogenous and endogenous processes that impact islands and their LCLUC patterns. The central questions that motivated our work are as follows:

- What are the space-time patterns of social-ecological change affecting island sustainability?
- How do local and global forces of change threaten island ecosystems?
- How can complex adaptive systems be used to examine islands and alternate trajectories of change?
- How can a unifying template be developed for the assessment of global island ecosystems?

What follows is an interpretation of findings from the meta-analysis, a description of the general importance of islands to society and ecosystem goods and services, challenges to island sustainability, and considerations for future work in distilling the primary processes of change on islands, leading to a generalized Dynamic Systems Model of islands and the processes of change (see the Pizzitutti, Brewington & Walsh chapter in this volume).

Challenges & Opportunities for Island Studies

Islands are microcosms of larger mainland systems, but their smaller size, crisp boundaries, restricted access, and often historic isolation make them more manageable to study and measure factors that threaten their social-ecological sustainability. Oceanic islands may have been isolated through time as a consequence of their geography, for example, but development pressures linked to the expanding human dimension and demands of burgeoning tourism markets produce distinct LCLUC patterns. Islands with high-value amenity resources and iconic species also attract large numbers of visitors and associated resident populations to support the expanding tourism market. Projected LCLUC patterns for islands for the year 2100 indicate that change on islands will exceed that of continental areas (Kier et al., 2009). Further, the Human Impact Index, a measure of the current threats to LCLUC and development, is significantly higher on islands compared to continents (Kier et al., 2009). Many of the contemporary LCLUC drivers associated with the human dimension are tourism-related, and island tourism is a main component of global tourism. Historically, deforestation, agricultural extensification, establishment and expansion of human settlements, enhancements to community infrastructure, and development of port and transport facilities associated with trade, fisheries, and commerce have supported an expanding human impact on islands (Seddon, Froyd, Leng, Milne, Willis, & Gilbert, 2011). All of these factors have had direct and indirect consequences on LCLUC patterns. On many islands around the world, migrating residents are "pulled" to islands to work in the burgeoning tourism industry, often "pushed" from nearby continental settings or other island groups as a result of poor economic conditions. Employment opportunities in global tourism continue to grow and the expanding urban infrastructure meant to attract, service, and sustain long-term development is affecting LCLUC and ecosystem goods and services in direct and indirect ways.

Sustainability

On islands, even on oceanic islands that have been seemingly isolated from the pressures associated with continental settings, humans have become the primary driver of change (Moser et al., 2018). From the expanding human dimension seen through tourism, population migration, agriculture, fisheries, and urban development, islands and their inhabitants are often incapable of managing such rapid and extensive changes, posing challenges to long-term sustainability. The unforeseen consequences of development, such as the introduction of alien and invasive species can result in ecosystem degradation or loss, fragmentation, and depletion of natural resources that generate revenue for local residents and attract tourists (Walsh & Mena, 2016). The expanding human dimension on islands can degrade biodiversity and endemism, jeopardizing the long-term ecosystem provisioning that supports the

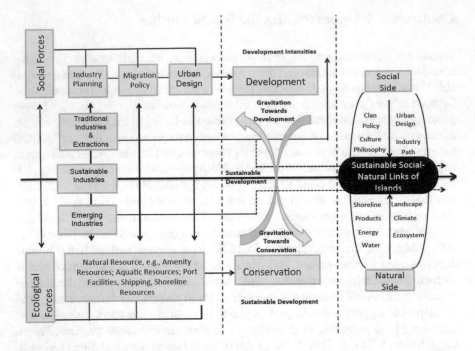

Fig. 1 Conceptual representation of an economic development and resource conservation gradient for islands shaped by social-ecological forces of change and characteristics of island ecosystems (Zhang & Walsh, 2018)

social, terrestrial, and marine subsystems and economic vitality. The expanding human dimension is not the only threat to island sustainability, however, as climate change and natural hazards further impact fragile and sensitive island ecosystems (Brewington, Frizzelle, Walsh, Mena, & Sampedro, 2014). Clearly observed in the Galapagos Island of Ecuador is the considerable challenge of balancing the needs of economic development and resource conservation (Walsh, Carter, Quiroga, & Mena, 2014). As resident and tourist populations continue to increase and exceed the capacity of local infrastructure to ensure environmental protection, the pressures on governance structures that engage a broad range of policy actors become more pronounced (Zupan et al., 2018).

Island sustainability may be conceptualized along a resource conservation—economic development gradient that represents a diversity of island forms and functions, shaped by exogenous and endogenous dynamics, and social and ecological processes of change. Whether supporting protected areas, iconic species, tourism sites, local communities, fisheries and agriculture, or urban and industrial development, islands are increasingly linked to the human dimension, mainland forces, and other island settings through the processes of globalization. Conservation and development efforts on islands require continuous monitoring of social-ecological interactions and the multi-scale drivers of change to implement adaptive management schemes and evaluate the desirability of outcomes. Figure 1 indicates the complex interplay between island social and ecological systems and their dynamics. Further, it points toward links between social-natural conditions and features related to island sustainability. The intent is to acknowledge the complex interactions between people, place, and the environment, and their impacts on island sustainability. Development and conservation are part of the sustainability discourse for islands and the changing vulnerability of people in altered settings imposed by forces of change.

Island sustainability is influenced by a host of processes acting across spatial scales that range from the molecular to the landscape, and over short- and long-term cycles. These space-time forces of change are often seen through the direct and indirect consequences of processes of change, such as (1) natural processes, including coastal erosion, ocean warming and acidification related to climate change, tectonics, short- and long-term climate variability such as the El Niño Southern Oscillation (ENSO), and natural hazards, (2) social processes, such as population migration and tourism, urban growth, and trade and economic development, and (3) integrated social-ecological processes, such as the introduction of alien species through the transportation of goods and materials, loss of ecosystem services through habitat degradation, over-exploitation of fisheries resources, and saturation of visitor sites, impacting amenity resources and iconic species.

Tourism as a Driver of Change

The causes and consequences of LCLUC on islands involve, at a minimum, individuals, households, and land parcels. Their combined influence is at the heart of evolving research in land change science.

Tourism is one of the most important global economic sectors and island tourism is a main component of world tourism. Today, approximately 1.2 billion tourists travel outside their borders, generating revenues in excess of $1.8 billion and contributing to an industry that accounts for over 14% of GDP (World Tourism Organization, 2017). As such, tourism has become an important and dynamic driver of change on islands (Currie & Falconer, 2014), especially those with high-quality, amenity resources and access to iconic species and landscapes.

As an example of the challenges that confront island ecosystems, beginning in the 1970s, the Galapagos Islands of Ecuador began to draw thousands of new residents who were attracted by lucrative opportunities linked to the islands' rich marine and terrestrial ecosystems and employment opportunities in construction, fisheries, and tourism (González, Montes, Rodríguez, & Tapia, 2008; Miller et al., 2010; Pizzitutti, Mena, & Walsh, 2014; Pizzitutti et al., 2017). Development of the tourism industry and a boom in fishing has more than tripled the local population in the past 15 years to over 30,000 residents. The number of tourists visiting the Islands has quadrupled over the same period and is now over 275,000 per year (2018). This expanding human imprint on the islands has contributed to (1) over-use of natural resources; (2) replacement of native and endemic species by invasive flora and

Fig. 2 Aerial view of the community of Puerto Ayora and the adjacent Galapagos National Park on Santa Cruz Island. Urban infilling has eliminated most open space within the urban zone due to the restricted area allocated for development

fauna; (3) extraction of marine resources at unprecedented rates; (4) expansion of tourism and associated development into increasingly fragile environments; and (5) a dramatic increase in human energy consumption and waste generation (Henderson, Dawson, & Whittaker, 2006; Brewington, 2013; Walsh, Page, Brewington, Bradley, & Mena, 2018). Recognizing these threats, in June 2007, UNESCO declared the Galapagos archipelago a World Heritage Site "in danger" as the Ecuadorian Government declared an "ecological emergency". UNESCO removed the Galapagos Islands from their list of World Heritage Sites "at risk" in July 2010, but much remains to be done to stem the flow of migrants to the islands and to reduce human impacts on the vulnerable ecology (Fig. 2).

A central challenge in the study of global island ecosystems is their remoteness and inaccessibility (Gil, 2003; Miller, Lieske, Walsh, & Carter, 2018). Islands many be connected directly or indirectly to other islands and archipelagos, or nearby continents. They are relatively small, occurring singularly or arrayed in archipelagos, irregular in shape, and varied in their ecological, economic, cultural, and topographic settings. Islands are also shaped over time through human and natural circumstances, impacted by episodic and continuous forces of change with historical and contemporary implications. Numerous factors serve to link islands to other islands and continents, with many factors having roots in historical and contemporary periods that involve economic, political, social, and biophysical forces (Pazmino, Serrao-Neumann, & Choy, 2018).

Meta-Analysis of Island Literature

The dominant outcome variable that we sought to understand regarding island systems is LCLUC. We searched eight databases—PubMed, Scopus, Web of Science, GeoBase, GeoRef, PAIS, IBSS, and Environment Complete—to identify papers published between 1988–2018 on islands and the social-ecological drivers of LCLUC. Initially, we searched the databases with a "broad brush" approach that yielded thousands of papers. Boolean operators effectively reduced the number of papers and provided a more targeted search whose outcome variable was LCLUC and the social-ecological drivers of change.

We separated our searches into dominant domains—socio-economic/LCLUC and remote sensing/LCLUC. For the socio-economic search, details on the screening process as the first step in the meta-analysis are provided by Bilsborrow in this volume. Literature results were classified into two tiers, Tier 1 included those that were considered to be of significant value due to their methodological content and Tier 2 included relevant but less-analytical papers. Table 1 summarizes the results for the socio-economic/LCLUC domain search that are reviewed elsewhere in this book by Bilsborrow.

We also conducted searches of the remote sensing literature using the same eight databases, but with a slightly different goal: to categorize papers by islands, archipelagos, and countries as well as by the types of remotely-sensed data, drivers, methods, and products. We identified 406 papers that were further defined by satellite type, classification approach, change-detection method, sensor fusion approach, pixel vs. object-based image analysis, image time-series analysis, and vegetation indices.

Three selected databases are briefly described below that provided us the greatest utility in meeting project goals. First, the United Nations Environmental Programme's (UNEP) "Global Distribution of Islands" (Depraetere & Dahl, 2007) dataset contains boundary files and global distributions for islands greater than 0.06 km² in size. This dataset, also known as the Island Biodiversity Programme of Work (IBPow) database, was developed in collaboration with the Institut de Recherche pour le Dévelopement (IRD) and has existed since 2005. The shapefile is

Table 1 Socio-economic/LCLUC search results for Tier 1 and Tier 2 categories

Dominant factor	Tier 1 papers	Tier 2 papers
LCLUC	48	160
Economic change, economic modeling	19	138
Other modelling, simulations	35	120
Urbanization	16	65
Agriculture, Deforestation, Reforest	23	125
Tourism	29	77
Population, Migration & Fertility	18	65
Climate change	8	11
Totals	196	761

based on the Global, Self-Consistent, Hierarchical, High-Resolution Shoreline (GSHHS) Database after Wessel and Smith (1996). Second, the "Threatened Island Biodiversity" database includes almost 2000 islands for nearly 1200 critically endangered and endangered terrestrial vertebrate species, collated from almost 1500 scientific literature sources, management documents, and databases, and from the contributions of more than 500 experts (Spatz et al., 2017). The dataset also contains information on vulnerable seabirds, the presence of invasive vertebrates, and important island characteristics, such as island size and human habitation that are often used in setting conservation priorities. Third, the "Bioclimatic and Physical Characterization" database provides information about the world's islands—standardized dataset to perform a comprehensive global environmental characterization for 17,883 of the world's marine islands greater than 1.0-km^2 (~98% of total island area) (Weigelt, Jetz, & Kreft, 2013). We used island area, mean temperature, mean precipitation, seasonality in temperature and precipitation, historic climate change, elevation, isolation, and past connectivity as key island characteristics and drivers of ecosystem processes from this database.

Global Island Databases and the Meta-Analysis

The unique biogeographical conditions of oceanic islands that give rise to their high endemism (geographic isolation, reduced surface area) are also the causes of high extinction (Whittaker & Fernández-Palacios, 2007; Moser et al., 2018). Islands are speciation and extinction hotspots that make them important sites for biodiversity conservation worldwide (Keitt et al., 2011). For example, endemic plant species in islands are more likely to be threatened, especially in small islands (Caujapé-Castells et al., 2010). In comparison to mainland biodiversity hotspots, islands or island groups are disproportionately vulnerable to invasive species (Reid et al., 2005; Kueffer et al., 2010; Bellard et al., 2014; van Kleunen et al., 2015). In addition, 80% of known species extinctions since 1500 have taken place in islands, despite islands representing less than 5% of the earth's landmass (Ricketts et al., 2005). Despite their relatively small surface area, the importance of islands transcends their limited geographical extension; marine animals are threatened with extinction because many of them, like seabirds, depend on island ecosystems for breeding (Spatz et al., 2014). We need homogenous data about islands around the world to understand the impacts that global change is having on island ecosystems, and how we can best conserve them and their inhabitants.

A literature review for available databases and an online search for the keywords "island database" and "global island database" revealed dozens of databases. These contain data about islands in a diversity of formats (CSV, KML, SHP, PDF, GeoTiff, and Web access), though most are either limited in geographic scope (are focused on a particular archipelago or region) or are not island-specific (including continental areas). Most databases define islands as landmasses smaller than Greenland, but their lower surface area limit is more variable, because it depends on the spatial resolution of the compiled data. Most databases also exclude freshwater islands. The IBPow database, Global Distribution of Islands, Global Island Database (version 1), is one of the first and most comprehensive efforts to compile homogenous global data specifically about oceanic islands. The database includes shorelines, bathymetry, topographic relief, place names, oceanic climatic data, and ecosystem classifications (UNEP-WCMC, Depraetere, & Dahl, 2010). This shapefile format archive (Wessel & Smith, 1996) converted island shorelines >0.06 km^2 into approximately 180,000 georeferenced polygons and assigned an "Island ID code" as a unique identifier. This code is useful to homogenize the dataset because island names can be variable across languages, local names, and spellings. In 2015, a second version of the Global Distribution of Islands database was created using OpenStreetMap data (based on Landsat products) to provide higher resolution geographic boundaries of approximately 460,000 islands. UNEP-WCMC has since depreciated the former, and the latter dataset includes only island boundaries. However, both datasets may be joined using their Island ID codes.

Subsequent efforts to collate physical and ecological data about global islands have built upon the UNEP-WCMC framework and individual islands can also be joined using common identifiers. Weigelt et al. (2013) made an essential contribution to our understanding of global island patterns through a physical characterization of 17,883 of the world's oceanic islands >1 km^2 (~98% of total island area), using WorldClim data. The authors considered key island characteristics and drivers of ecosystem processes, including area, temperature, precipitation, seasonality in temperature and precipitation, elevation, isolation, and past connectivity. These data may be joined with the Island Code field corresponding to IBPow Version 1 (2010).

Another contribution came from a partnership between conservation NGOs and universities that created the Threatened Island Biodiversity Database (TIB, http://tib.islandconservation.org/), a compilation of nearly 2000 islands and 1300 threatened vertebrates (as classified by the International Union for the Conservation of Nature, IUCN) that inhabit them. Similarly, they also created the Database of Island Invasive Species Eradications (DIISE, http://diise.islandconservation.org/), a compilation of all historical and current eradication projects on islands. The TIB and DIISE are the most comprehensive reviews of threatened and invasive vertebrates in island ecosystems, as well as of efforts to eradicate them (Jones et al., 2016; Keitt et al., 2011; Spatz, Newton, et al., 2014; Spatz, Zilliacus, et al., 2017).

Most recently, the Global Naturalized Alien Flora (GloNAF, https://glonaf.org/) was published and made available online for free with a Creative Commons license (van Kleunen et al., 2018). GloNAF currently lists the taxa of naturalized plants from 381 individual islands that comprise nearly one-third of all regions included in the entire database (van Kleunen et al., 2015). Linking these and other databases can help quantify vulnerability to natural and anthropogenic drivers of change on island ecosystems around the world (Pyšek et al., 2017).

The global island databases listed above compile their information at the scale of individual islands. However, most available literature aggregates islands into more general categories such as island groups, archipelagos, countries, or regions. This discrepancy presents a challenge when trying to link the insights from a qualitative

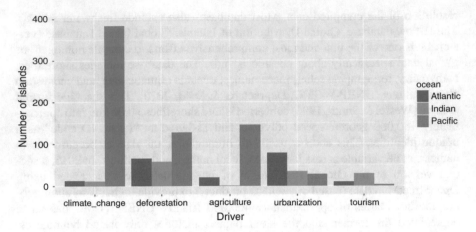

Fig. 3 Common anthropogenic drivers of LULC change on islands by ocean

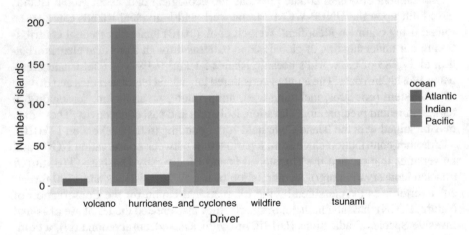

Fig. 4 Common natural drivers of LULC change on islands by ocean

meta-analysis with quantitative online databases. To overcome this hurdle, we manually linked islands that are part of more general island groups mentioned in the literature. For example, if an article was about the Galapagos, we marked all individual islands that belong to the archipelago to link the article with the specific Island ID codes. We did this for the 406 Tier-1 articles in our meta-analysis. We then used the "tidyverse" and "splitstackshape" packages in R (Version 3.4.2) to link the drivers of LCLUC from the literature review with individual islands.

The methodology above allowed us to ask questions at the scale of individual islands. For example, Figs. 3 and 4 suggest that islands in the Pacific Ocean may be better represented in the scientific literature, but they also tell us something about global drivers of change. For anthropogenic drivers of land use/land cover, most literature about Pacific Islands described the effects of climate change. Meanwhile,

Atlantic Ocean literature focused on urbanization. Most literature about islands in the Indian Ocean was concerned with tourism. For natural drivers of land cover/land use, most articles about Pacific Islands mentioned volcanoes, most about the Atlantic referenced hurricanes, and most Indian Ocean articles mentioned tsunamis. It should be noted that the ocean categorization for individual islands (based on IBPow Version 1) was likely based on political affiliations and might be different than what their geographic location might suggest.

Remote Sensing & Islands

All information from the meta-analysis was formalized in an excel spreadsheet that characterized each of the 406 papers. In addition, 34 island databases were identified, collected, and linked for subsequent analysis, and the meta-analysis was linked with the conflated island databases. Several graphics were generated to show the primary satellite systems used in one or more of the papers for a range of years. Figures 5, 6 and 7 show the number of papers that were identified in our meta-analysis that characterized LCLUC through, for instance, the use of Landsat, SPOT, and higher spatial resolution systems (i.e., GeoEye, Ikonos, QuickBird, WorldView-2).

In addition, the 406 remote sensing papers identified in our database searches were mapped according to islands for broad geographic regions. Figure 8 graphically represents the number of studies and papers identified for islands or archipelagos, globally. Figures 9, 10 and 11 further summarize studies and papers for Southeast Asia, the Galapagos, and the Hawaiian Islands, respectively.

In addition to identifying the sensor systems used to assess LCLUC on islands through our meta-analysis, as well as the number of referenced papers for them, we

Fig. 5 Landsat papers by sensor and year

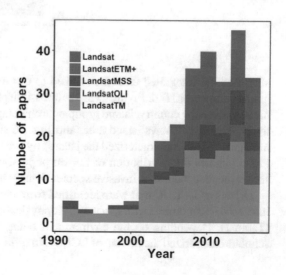

Fig. 6 SPOT papers by
sensor and year

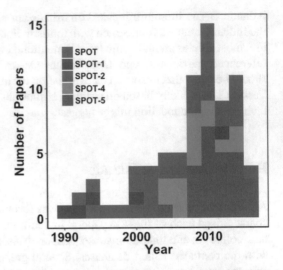

Fig. 7 High spatial
resolution (commercial)
satellite data by sensor
and year

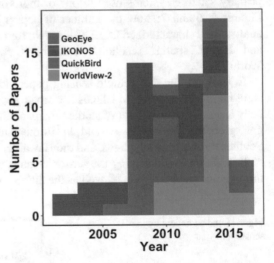

also used the integrated databases linked to our meta-analysis to summarize geographic characteristics. Figures 12, 13 and 14 display the number of islands in our meta-analysis by country, island group or archipelago, and geographic zone, respectively. Figure 15 shows island area, and Fig. 16 shows the distance to the nearest mainland. We also characterized the islands by ocean—Atlantic, Indian, and Pacific: Fig. 17 shows the distribution of human population density by ocean and Fig. 18 shows the distribution of invasive species density by ocean.

Drivers of LCLUC that were identified from the meta-analysis were synthesized into socio-economic factors, natural/biophysical factors, and geographic factors (Table 2). Depending on the environment being studied in the reviewed papers, scholars categorized the drivers of LCLUC in different ways. For instance, Agarwal,

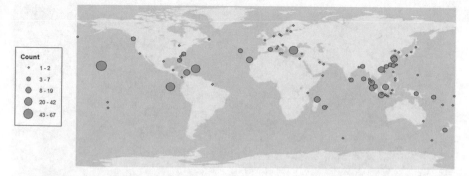

Fig. 8 Global islands—studies and papers

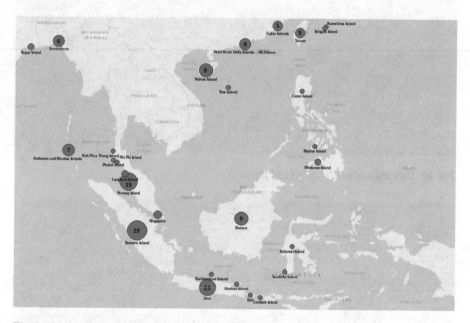

Fig. 9 Southeast Asia—studies and papers

Gelfand, and Silander (2002) identified tropical deforestation as a major LCLUC occurring on the east coast of Madagascar using defined classes of drivers—socio-economic (e.g., population growth & economic development), physical (e.g., topography or proximity to rivers and roads), government intervention (e.g., agricultural extensification and forest policies), and external (e.g., demand for exports or financing conditions). In the Galapagos Islands, Percy, Schmitt, Riveros-Iregui, and Mirus (2016) examined the potential of water shortages and reliance on hydro-climatic, anthropogenic, and pedo-hydrologic gradients across the archipelago to better understand the interactions between water and critical zone processes in tropical ecosystems.

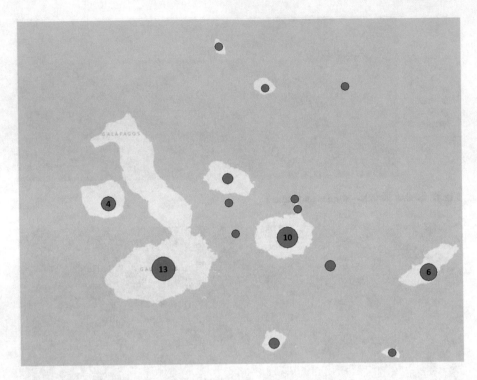

Fig. 10 Galapagos Islands—studies and papers

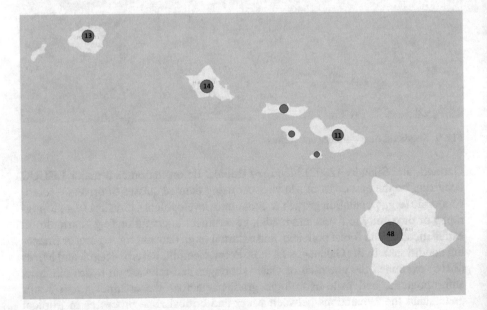

Fig. 11 Hawaiian Islands—studies and papers

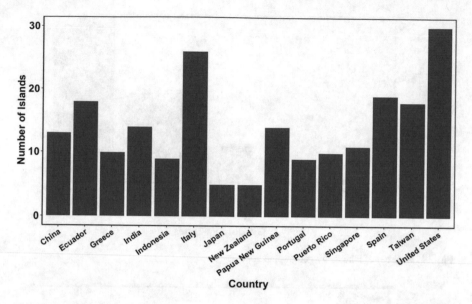

Fig. 12 Number of islands by country—only countries with greater than five islands represented in our study were summarized

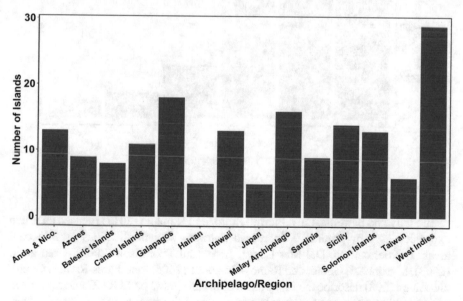

Fig. 13 Number of islands by island group or archipelago—only groups or archipelagos with greater than five islands were summarized

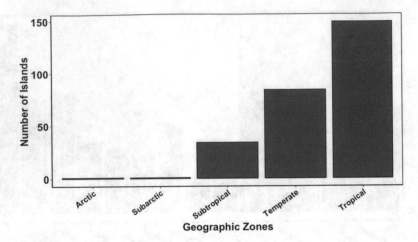

Fig. 14 Number of islands by geographic zone

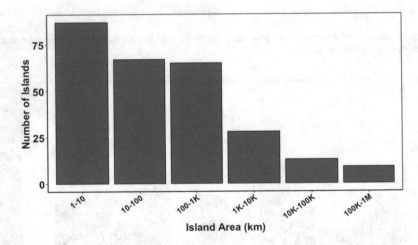

Fig. 15 Area of islands in our meta-analysis

In Sicily and Sardinia, Italy, Fiorini, Zullo, and Romano (2017) focused on urban expansion as a by-product of increased tourism, with implications on protected areas. In Puerto Rico, Del Mar Lopez, Aide, and Thomlison (2001), Grau et al. (2003), Martinuzzi, Gould, and Ramos-Gonzalez (2006), and Pares-Ramos, Gould, and Aide (2008) describe the importance of urban sprawl on LCLUC and associated agricultural abandonment as people migrate from the rural to urban areas for jobs, mostly in manufacturing. This land transition resulted in the reforestation of agricultural land and a change in the ecological function of abandoned farms through secondary forest succession. Most changes were associated with distance to roads, nature preserves, and urban areas as well as terrain settings of elevation, slope angle, and slope aspect. In Hawaii, several papers assessed coastal erosion, shoreline

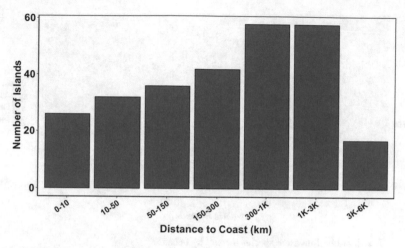

Fig. 16 Distance to the nearest mainland for islands in our meta-analysis

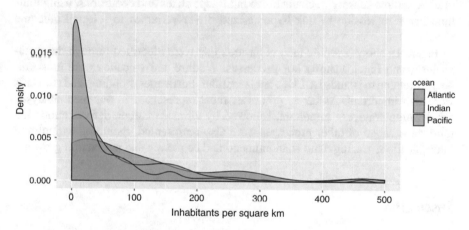

Fig. 17 Distribution of human population density by ocean

change, and climate change, particularly, sea-level rise (Ferrier et al., 2013; Anderson, Fletcher, Barbee, Frazer, & Romine, 2015), changing fire fuels and behavior influenced by invasive grasses (Ellsworth, Litton, Dale, & Miura, 2014) as well as changing fuel moisture in non-native tropical vegetation (Ellsworth, Dale, Litton, & Miura, 2017). Also, papers reported on Hawaiian dry forests as well as species richness, canopy density, canopy height, and basal area that were affected by a suite of biotic and abiotic characteristics (Pau, Gillespie, & Wolkovich, 2012).

In short, while there is likely a set of core processes of LCLUC in islands, there is considerable variability in islands relative to their geographic position, degree of isolation, morphology, and human-environment interactions. These differences shape the level of generalizability for analysis and modeling that extend beyond local contexts and their geographic site and situation. Most of our papers used

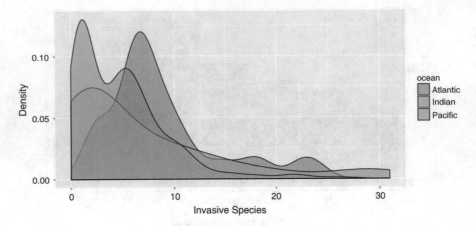

Fig. 18 Distribution of invasive species density by ocean

satellite remote sensing to characterize landscape states and conditions, using multiple linear regression to link hypothesized driver variables to selected outcome variables.

Figure 19 shows specific natural hazards that were identified through the meta-analysis, and Fig. 20 shows key processes. Wildfire and volcanos were most frequently related to islands LCLUC, but landslides, hurricanes, flooding, and tsunamis were also important. As far as processes most represented in our meta-analysis, urbanization was very important, followed by climate change, deforestation, and tourism. A range of other processes were also represented, including agricultural intensification, grazing, land abandonment, and economic and population growth.

Discussion

The meta-analysis highlighted several important issues surrounding LCLUC in global islands. For instance, Fig. 8 shows the locations of islands that were reviewed as part of the remote sensing-LCLUC assessment. Southeast Asia, Hawaii, and the Galapagos Islands of Ecuador were well represented in the papers that were reviewed as part of the meta-analysis. Figures 3 and 4 classify islands by ocean: most papers reported on Pacific Islands, where climate change, agriculture, and deforestation were the most frequently cited anthropogenic drivers of land change. Surprisingly, urbanization was not as important as other drivers, except for Atlantic islands. Deforestation was the most frequently referenced anthropogenic driver of LCLUC for islands in the Indian Ocean.

Figures 5–7 show the number of papers published between 1988 and 2018 for specific satellite systems. Landsat data were most commonly used to characterize LCLUC on islands, although SPOT data and higher-spatial resolution, commercial satellite data, i.e., GeoEye, Ikonos, Quickbird, and Worldview-2, were also well

Table 2 Key drivers of island LCLUC that were synthesized from the socio-economic and remote sensing meta-analysis

Socio-economic factors	Biophysical factors	Geographic factors
Tourism	Invasive species	Land degradation
Human occupation pattern	Protected area status	Island types: Continental, Oceanic
Rural development	Runoff: Erosion & Deposition	LCLUC patterns
Urban infrastructure	Land manage programs	Geographic position
Population growth	Volcanic deformation	Geographic context
Population density	Topographic position	Geographic isolation
Population migration	Natural hazards	Geographic connectivity
Community structure	Drainage patterns	Political boundaries: Community, Urban, Protected areas, Farm
Urbanization & Urban sprawl	Land productivity	Commercial forests
Household demographics	Soil fertility	Grazing, Agriculture agroforestry
Poverty & Social inequalities	Herbivory	Social-ecological barriers: Fences, Lava Flows, Fire, Reservoirs
Number of households	Freshwater availability	Land abandonment
Household livelihoods	Precipitation patterns	Settlement history
Employment patterns	Native, Endemic species	Geographic accessibility LCLU types
Standard of living	Iconic species and landscapes	Fishery zones
Technological change	Mangroves, Fisheries	Infrastructure zones
Land tenure	Coastal protection	Road type & Density
Credit & Capital	Species richness & Turnover	
International development	Climate change & Sea-level rise	
Colonial policies	ENSO events	
Globalization	Ecosystem services	
Public policy		

represented in the meta-analysis. The integration of higher spatial resolution systems was likely due to varying local island characteristics and conditions that require such data to assess landscape features, including pocket beaches, smaller agricultural plots, and patterns of rural, low-density settlements or rural-urban transitions. Figure 8 uses graduated circles to represent global patterns of islands and the number of papers within our meta-analysis.

By linking the meta-analysis to several island databases, we were able to synthesize the countries with islands that were most represented in our meta-analysis. The United States, Italy, Ecuador, Spain, and Taiwan contained the greatest number of islands that were reported (Fig. 12). Only countries with greater than five islands were represented. Similarly, we also grouped islands by archipelago, in which the

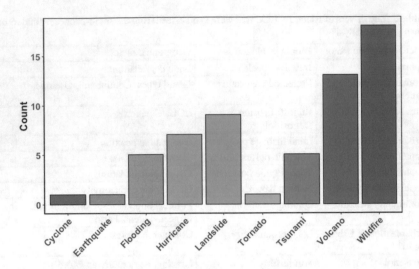

Fig. 19 Papers identified in our meta-analysis by natural hazard type

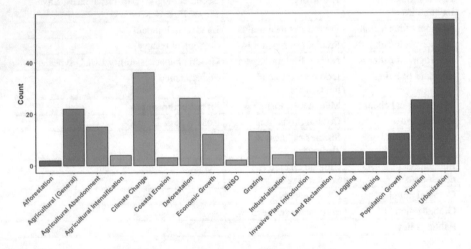

Fig. 20 Papers identified in our meta-analysis by processes of change

West Indies was most highly represented, followed by the Galapagos Islands, and the Malay Archipelago (Fig. 13). Further, we summarized island size (Fig. 14), distance to the coast/nearest continent (Fig. 15), and geographic zone (Fig. 14). Figure 15 shows that the greatest number of islands identified in our meta-analysis had an area of 1–10-km^2, followed by islands with areas of 10–100-km^2 and 100–1000-km^2. Fewer islands within our study were relatively close to a continent (Fig. 16), as most of the islands represented in the meta-analysis were 300–1000-km and 1000–3000 km away. As indicated in Fig. 14, most of the islands were tropical islands, followed by temperate and sub-tropical. Lastly, we evaluated the distribution of human population density (Fig. 17) and invasive species density (Fig. 18).

For papers within our meta-analysis, we reported the associated multi-dimensional drivers of LCLUC (Table 2). Where statistical models were used, we identified the most important variables and their statistical significance. Findings reported in the papers were context-relevant and place-specific; therefore, the scale and time periods of the process being studied were not always clear. Therefore, Table 2 is an unweighted listing of the most or least important drivers of LCLUC for certain types of islands (e.g., continental or oceanic), islands having a certain morphology (e.g., volcanic or coral), positioned in various settings (e.g., regional geography and ocean), human dimension (e.g., historical or contemporary context, rural or urban form, and cultural or political structures), environmental dimension (e.g., ecosystem type, spatial structure, and species richness), and the nature of isolation or connectedness as a consequence of trade, colonial legacies, and culture.

Conclusions

This descriptive study synthesized the literature on islands to (1) understand the social-ecological drivers of LCLUC, (2) identify the remote sensing satellite systems used to map and monitor islands, and (3) assess the methods used to characterize LCLUC on islands by an array of scholars and for diverse islands types. In general, our project objectives were achieved through the information collected, analyses conducted, and pattern-process relationships categorized by linking the papers to island databases that formed the foundation of a Dynamic Systems Model for islands (see the chapter by Pizzitutti, Brewington & Walsh in this volumne). The core model was initially developed and rooted in the Galapagos Islands of Ecuador and will be generalized to Hawaii and Puerto Rico as test cases. Our preliminary work suggests that the core model will be augmented by deriving modules that are sensitive to important social, ecological, geographical, morphological factors, such as tourism, trade, human migration, and agricultural abandonment.

Our work was framed within the concepts of complexity theory (Malanson & Walsh, 2015; Walsh & Mena, 2016) and was applied to understand the social-ecological threats to island ecosystems and their sustainability. Island biocomplexity combines complex, adaptive systems with a new island ecology that incorporates human-induced change on the environment, specifically LCLUC. It encompasses the complex interactions within and among ecological systems, the physical systems on which they depend, and the human systems with which they interact. Island ecosystems are complex adaptive systems because their macroscopic properties emerge from the interactions among the individual components of the ecosystem. Global change, including the forces associated with tourism, migration, and LCLUC, exert exogenous pressure on island ecosystems, with their own spatially contingent and endogenous dynamics.

Other papers developed by members of the project team explore related matters and point toward the meta-analysis described here. Forthcoming work will address the complex issues associated with socio-economic and ecological drivers of LCLUC for a diversity of island types, site locations, and degrees of human-environment interactions that may confound generalizability. The meta-analysis, however, moves us closer to recognizing the core issues important to islands, while retaining the local context and place-based conditions that serve to mediate social-ecological relationships. Results from this work highlight the potential to develop a set of unifying, multi-scale drivers of LCLUC on islands that have applicability to a general model for understanding island ecosystems worldwide.

References

Alongi, D. M. (2008). Mangrove forests: Resilience, protection from tsunamis, and response to global climate change. *Estuarine Coastal and Shelf Science, 76*, 1–13.

Agarwal, D. K., Gelfand, A. E., & Silander, J. A., Jr. (2002). Investigating tropical deforestation using two-stage spatially misaligned regression models. *Journal of Agricultural, Biological, and Environmental Statistics, 7*, 420. https://doi.org/10.1198/108571102348

Anderson, T. R., Fletcher, C. H., Barbee, M. M., Frazer, L. N., & Romine, B. M. (2015). Doubling of coastal erosion under rising sea level by mid-century in Hawaii. *Natural Hazards, 78*(1), 75–103.

Bellard, C., Leclerc, C., Leroy, B., Bakkenes, M., Veloz, S., Thuiller, W., & Courchamp, F. (2014). Vulnerability of biodiversity hotspots to global change. *Global Ecology and Biogeography, 23*(12), 1376–1386.

Benitez, F., Mena, C. F., & Zurita-Arthos, L. (2018). Urban land cover change in ecologically fragile environments: The case of the Galapagos Islands. *Land, 7*(1), 21. https://doi.org/10.3390/land7010021

Brewington, L. (2013). The double bind of tourism in Galapagos society. In S. J. Walsh & C. F. Mena (Eds.), *Science and conservation in the Galapagos Islands: Frameworks and perspectives* (pp. 105–125). New York: Springer Nature.

Brewington, L., Frizzelle, B. G., Walsh, S. J., Mena, C. F., & Sampedro, C. (2014). Remote sensing of the marine environment: Challenges and opportunities in the Galapagos Islands. In J. Denkinger & L. Vinueza (Eds.), *Galapagos marine reserve: A social-ecological system* (pp. 109–136). Springer.

Carter, R. W., Walsh, S. J., Jacobson, C., & Miller, M. L. (2014). Socio-economic challenges in managing iconic national parks. *The George Wright Forum*, A special issue on "Global change and the world's iconic protected areas" (P. Eagles & P. A. Taylor), *31*(3), 245–255.

Caujapé-Castells, J., Tye, A., Crawford, D. J., Santos-Guerra, A., Sakai, A., Beaver, K., & Jardim, R. (2010). Conservation of oceanic island floras: Present and future global challenges. *Perspectives in Plant Ecology, Evolution and Systematics, 12*(2), 107–129.

Charles, C. L., & Chris, C. (2009). Beyond sustainability: Optimizing island tourism development. *International Journal of Tourism, 11*, 89–103.

Currie, C., & Falconer, P. (2014). Maintaining sustainable island destinations in Scotland: The role of the transport–tourism relationship. *Journal of Destination Marketing & Management, 3*(3), 162–172.

Del Mar Lopez, T., Aide, T. M., & Thomlison, J. R. (2001). Urban expansion and the loss of prime agricultural lands in Puerto Rico. *Ambio, 30*(1), 49–54.

Depraetere, C., & Dahl, A. L. (2007). Island locations and classifications. In G. Baldacchino (Ed.), *A world of islands. An island studies reader* (pp. 57–105). Luqa, Malta: Agenda Academic Publishers.

Ellsworth, L. M., Dale, A. P., Litton, C. M., & Miura, T. (2017). Improved fuel moisture prediction in non-native tropical *Megathyrsus maximus* grasslands using Moderate-Resolution Imaging Spectroradiometer (MODIS)-derived vegetation indices. *International Journal of Wildland Fire, 25*(5), 384–392.

Ellsworth, L. M., Litton, C. M., Dale, A. P., & Miura, T. (2014). Invasive grasses change landscape structure and fire behavior in Hawaii. *Applied Vegetation Science.* https://doi.org/10.1111/avsc.12110

Ernoul, L., & Wardell-Johnson, A. (2013). Governance in integrated coastal zone management: A social networks analysis of cross-scale collaboration. *Environmental Conservation.* https://doi.org/10.1017/S0376892913000106

Ferrier, K. L., Taylor-Perron, L., Mukhopadhyay, S., Rosener, M., Stock, J. D., Huppert, K. L., & Slosberg, M. (2013). Covariation of climate and long-term erosion rates across a steep rainfall gradient on the Hawaiian Island of Kauai. *Bulletin of the Geological Society of America, 125*(7–8), 1146–1163.

Fiorini, L., Zullo, F., & Romano, B. (2017). Urban development of the coastal system of the Italian largest islands: Sicily and Sardinia. *Ocean & Coastal Management, 143*, 184–194.

Gil, S. M. (2003). Tourism development in the Canary Islands. *Annals of Tourism Research, 30*(3), 744–747.

González, J. A., Montes, C., Rodríguez, J., & Tapia, W. (2008). Rethinking the Galapagos Islands as a complex social-ecological system: Implications for conservation and management. *Ecology and Society, 13*(2) [online]. Retrieved from http://www.ecologyandsociety.org/vol13/iss2/art13/.

Grau, H. R., Aide, T. M., Zimmerman, J. K., Tomlison, J. R., Helmer, E., & Zou, X. (2003). The ecological consequences of socio-economic and land-use changes in post-agriculture Puerto Rico. *BioScience, 53*(12), 1159–1168.

Henderson, S., Dawson, T. P., & Whittaker, R. J. (2006). Progress in invasive plant research. *Progress in Physical Geography, 30*(1), 25–46.

Johannes de Haan, F., Quiroga, D., Walsh, S. J., & Bettencourt, L. (2019). Scales and transformative change—Transitions in the Galapagos. In *Urban Galapagos—Transitions to sustainability in complex adaptive systems* (T. Kvan & J. A. Karakiewicz, Guest Editors), Social and Ecological Interactions in the Galapagos Islands (S. J. Walsh & C. F. Mena, Series Editors) (pp. 43–58). Springer Nature.

Jones, H. P., Holmes, N. D., Butchartd, S. H. M., Tershy, B. R., Kappes, P. J., Corkery, I., & Croll, D. A. (2016). Invasive mammal eradication on islands results in substantial conservation gains. *Proceedings of the National Academy of Sciences, 113*(15), 4033–4038.

Juan, R. O. R., Eduardo, P. L., & Vanessa, Y. E. (2008). The sustainability of island destinations: Tourism area life cycle and teleological perspectives, the case of Tenerife. *Tourism Management, 29*, 53–65.

Keitt, B., Campbell, K., Saunders, A., Clout, M., Wang, Y., Heinz, R., & Tershy, B. (2011). The global islands invasive vertebrate eradication database: A tool to improve and facilitate restoration of island ecosystems. *Island Invasives Eradication and Management*, July 2015, 74–77.

Kerr, S. A. (2005). What is small island sustainable development about? *Ocean and Coastal Management, 48*(7–8), 503–524.

Kier, G., Kreft, H., Lee, T. M., Ibisch, P. L., Nowicki, C., Mutke, J., & Barthlott, W. (2009). A global assessment of endemism and species richness across island and mainland regions. *Proceedings of the National Academy of Sciences, 106*(23), 9322–9327.

Kueffer, C., Daehler, C. C., Torres-Santana, C. W., Lavergne, C., Meyer, J.-Y., Otto, R., & Silva, L. (2010). A global comparison of plant invasions on oceanic islands. *Perspectives in Plant Ecology, Evolution and Systematics, 12*(2), 145–161.

MacDonald, L. H., Anderson, D. M., & Dietrich, W. E. (1997). Paradise threatened: Land use and erosion on St. John, US Virgin Islands. *Environmental Management, 21*(6), 851–863.

Malanson, G. P., & Walsh, S. J. (2015). ABM: Individuals interacting in space (invited commentary). *Journal of Applied Geography, 56*, 95–98.

Martinuzzi, S., Gould, W. A., & Ramos-Gonzalez, O. M. (2006). Land development, land use and urban sprawl in Puerto Rico integrating remote sensing and population census data. *Landscape and Urban Planning, 79*(3–4), 288–297.

Mejía, C. V., & Brandt, S. (2015). Managing tourism in the Galapagos Islands through price incentives: A choice experiment approach. *Ecological Economics, 117*, 1–11.

Mena, C. F., Quiroga, D., & Walsh, S. J. (2020). Threats to sustainability in the Galapagos Islands: A social-ecological perspective. In F. O. Sarmiento & L. M. Frolich (Eds.), *International handbook of geography and sustainability*. Cheltenham, UK: Edward Elgar Publishing.

Miller, B. W., Breckheimer, I., McCleary, A. L., Guzman-Ramirez, L., Caplow, S. C., & Walsh, S. J. (2010). Using stylized agent-based models for population-environment research: A case from the Galapagos Islands. *Population and Environment, 31*(6), 401–426.

Miller, M. L., Carter, R. W., Walsh, S. J., & Peake, S. (2014). A conceptual model for studying global change, tourism and the sustainability of iconic national parks. *The George Wright Forum*, A special issue on "Global change and the world's iconic protected areas" (P. Eagles & P. A. Taylor), *31*(3), 256–269.

Miller, M. L., Lieske, S. N., Walsh, S. J., & Carter, R. W. (2018). Understanding the interaction between a protected destination system and conservation tourism through remote sensing. In *Comprehensive remote sensing, applications for societal benefits* (S. J. Walsh, Book Editor; S. Liang, Organizing Editor) (pp. 123–143). London: Elsevier.

Moser, D., Lenzner, B., Weigelt, P., Dawson, W., Kreft, H., Pergld, J., & Essla, F. (2018). Remoteness promotes biological invasions on islands worldwide. *Proceedings of the National Academy of Sciences, 115*(37), 9270–9275.

Pares-Ramos, I. K., Gould, W. A., & Aide, T. M. (2008). Agricultural abandonment, suburban growth, and forest expansion in Puerto Rico between 1991 and 2000. *Ecology & Society, 13*(2), 1.

Pau, S., Gillespie, T. W., & Wolkovich, E. M. (2012). Dissecting NDVI-species richness relationships in Hawaiian dry forests. *Journal of Biogeography*. https://doi.org/10.1111/j.1365-2699.2012.02731.x

Pazmino, A., Serrao-Neumann, S., & Choy, D. L. (2018). Towards comprehensive policy integration for the sustainability of small islands: A landscape-scale planning approach for the Galapagos Islands. *Sustainability, 10*, 1228. https://doi.org/10.3390/su10041228

Percy, M. S., Schmitt, S. R., Riveros-Iregui, D. A., & Mirus, B. B. (2016). The Galapagos archipelago: A natural laboratory to examine sharp hydroclimatic, geologic, and anthropogenic gradients. *WIREs Water*. https://doi.org/10.1002/wat2.1145

Pizzitutti, F., Mena, C. F., & Walsh, S. J. (2014). Modeling tourism in the Galapagos Islands: An agent-based model approach. *Journal of Artificial Societies and Social Simulation, 17*(1), 1–25.

Pizzitutti, F., Walsh, S. J., Rindfuss, R. R., Reck, G., Quiroga, D., Tippett, R., & Mena, C. F. (2017). Scenario planning for tourism management: A participatory and system dynamics model applied to the Galapagos Islands of Ecuador. *Journal of Sustainable Tourism, 25*(8), 1117–1137.

Pyšek, P., Pergl, J., Essl, F., Lenzner, B., Dawson, W., Kreft, H., & Van Kleunen, M. (2017). Naturalized alien flora of the world: Species diversity, taxonomic and phylogenetic patterns, geographic distribution and global hotspots of plant invasion. *Preslia, 89*(3), 203–274.

Reid, W. V., Mooney, H. A., Cropper, A., Capistrano, D., Carpenter, S. R., Chopra, K., & Zurek, M. B. (2005). *Ecosystems and human well-being: Synthesis. Millennium Ecosystem Assessment*. Washington, DC: Island Press.

Ricketts, T. H., Dinerstein, E., Boucher, T., Brooks, T. M., Butchart, S. H. M., Hoffmann, M., & Wikramanayake, E. (2005). Pinpointing and preventing imminent extinctions. *Proceedings of the National Academy of Sciences, 102*(51), 18497–18501.

Rivas-Torres, G., Benitez, F. L., Rueda, D., Sevilla, C., & Mena, C. F. (2018). A methodology for mapping native and invasive vegetation coverage in archipelagos: An example from the Galapagos Islands. *Progress in Physical Geography, 42*(1), 83–111.

Schwartzman, S., Moreira, A., & Nepstad, D. (2000). Rethinking tropical forest conservation: Perils in parks. *Conservation Biology, 14*(5), 1351–1357.

Seddon, A. W. R., Froyd, C. A., Leng, M. J., Milne, G. A., Willis, K. J., & Gilbert, J. A. (2011). Ecosystem, resilience and threshold response in the Galapagos coastal zone. *PLoS One, 6*(7), e22376.

Spatz, D. R., Newton, K. M., Heinz, R., Tershy, B., Holmes, N. D., Butchart, S. H. M., & Croll, D. A. (2014). The biogeography of globally threatened seabirds and island conservation opportunities. *Conservation Biology, 28*(5), 1282–1290.

Spatz, D. R., Zilliacus, K. M., Holmes, N. D., Butchart, S. H. M., Genovesi, P., Ceballos, G., & Croll, D. A. (2017). Globally threatened vertebrates on islands with invasive species. *Science Advances, 3*(10), e1603080.

Sun, Y. Y. (2014). A framework to account for the tourism carbon footprint at island destinations. *Tourism Management, 45*, 16–27.

Tye, A. (2001). Invasive plant problems and requirements for week risk assessment in the Galapagos Islands. *Weed Risk Assessment, 153*, 154–175.

UNEP-WCMC, Depraetere, C., & Dahl, A. (2010). Global distribution of islands IBPoW (2010) Global Island Database (version 1). UNEP World Conservation Monitoring Centre.

UN World Tourism Organization. (2017). *Annual Report 2017*. ISBN 978-92-844-1980-7 (electronic copy), 106p.

Uyarra, M. C., Cote, I. M., Gill, J. A., Tinch, R. R. T., Viner, D., & Watkinson, A. R. (2005). Island-specific preferences of tourists for environmental features: Implications of climate change for tourism-dependent states. *Environmental Conservation, 32*(1), 11–19.

van Kleunen, M., Dawson, W., Essl, F., Pergl, J., Winter, M., Weber, E., Kreft, H., Weigelt, P., Kartesz, J., Nishino, M., Antonova, L. A., Barcelona, J. F., Cabezas, F. J., Cárdenas, D., Cárdenas-Toro, J., Castaño, N., Chacón, E., Chatelain, C., Ebel, A. L., Figueiredo, E., Fuentes, N., Groom, Q. J., Henderson, L., Inderjit, Kupriyanov, A., Masciadri, S., Meerman, J., Morozova, O., Moser, D., Nickrent, D. L., Patzelt, A., Pelser, P. B., Baptiste, M. P., Poopath, M., Schulze, K., Seebens, H., Shu, W. S., Thomas, J., Velayos, M., Wieringa, J. J., & Pyšek, P. (2015). Global exchange and accumulation of non-native plants. *Nature 525*(7567), 100–103.

van Kleunen, M., Pyšek, P., Dawson, W., Essl, F., Kreft, H., Pergl, J., Weigelt, P., Stein, A., Dullinger, S., König, C., Lenzner, B., Maurel, N., Moser, D., Seebens, H., Kartesz, J., Nishino, M., Aleksanyan, A., Ansong, M., Antonova, L. A., Barcelona, J. F., Breckle, S. W., Brundu, G., Cabezas, F. J., Cárdenas, D., Cárdenas-Toro, J., Castaño, N., Chacón E.,, Chatelain, C., Conn, B., de Sá Dechoum, M., Dufour-Dror, J-M., Ebel, A. L., Figueiredo, E., Fragman-Sapir, O., Fuentes, N., Groom, Q. J., Henderson, L., Inderjit, Jogan, N., Krestov, P., Kupriyanov, A., Masciadri, S., Meerman, J., Morozova, O., Nickrent, D., Nowak, A., Patzelt, A., Pelser, P. B., Shu, W. S., Thomas, J., Uludag, A., Velayos, M., Verkhosina, A., Villaseñor, J. L., Weber, E., Wieringa, J. J., Yazlık, A., Zeddam, A., Zykova, E., & Winter, M. (2018). The Global Naturalized Alien Flora (GloNAF) database. *Ecology 100*(1).

Walsh, S. J. (2018). Multi-scale remote sensing of introduced and invasive species: An overview of approaches and perspectives. In *Understanding invasive species in the Galapagos Islands: From the molecular to the landscape* (M. L Torres & C. F. Mena, Guest Editors), Social and Ecological Interactions in the Galapagos Islands (S. J. Walsh & C. F. Mena, Series Editors) (pp. 143–154). Springer Nature.

Walsh, S. J., Carter, R. W., Quiroga, D., & Mena, C. F. (2014). Examining the vulnerability of iconic national parks through modeling global change and social and ecological threats. *The George Wright Forum*, A special issue on "Global change and the world's iconic protected areas" (P. Eagles & P. A. Taylor), *31*(3), 311–323.

Walsh, S. J., Engie, K., Page, P. H., & Frizzelle, B. G. (2019). Demographics of change: Modelling the transition of fishers to tourism in the Galapagos Islands. In *Urban Galapagos—Transitions to sustainability in complex adaptive systems* (T. Kvan & J. A. Karakiewicz, Guest Editors), Social and Ecological Interactions in the Galapagos Islands (S. J. Walsh & C. F. Mena, Series Editors) (pp. 61–83). Springer Nature.

Walsh, S. J., McCleary, A. L., Mena, C. F., Shao, Y., Tuttle, J. P., Gonzalez, A., & Atkinson, R. (2008). QuickBird and Hyperion data analysis of an invasive plant species in the Galapagos Islands of Ecuador: Implications for control and land use management. *Remote Sensing of Environment, Earth Observation for Biodiversity and Ecology, 112*, 1927–1941.

Walsh, S. J., & Mena, C. F. (2013). Perspectives for the study of the Galapagos Islands: Complex systems and human-environment interactions. In S. J. Walsh & C. F. Mena (Eds.), *Science and conservation in the Galapagos Islands—Frameworks & perspectives* (pp. 49–67). Springer Nature.

Walsh, S. J., & Mena, C. F. (2016). Interactions of social, terrestrial, and marine sub-systems in the Galapagos Islands, Ecuador. Sackler Colloquium on coupled human and environmental systems (Social Sciences, Environmental Sciences, Sustainability Science). *Proceedings of the National Academy of Sciences, 113*(51), 14536–14543.

Walsh, S. J., Page, P. H., Brewington, L., Bradley, J. R., & Mena, C. F. (2018). Beach vulnerability in the Galapagos Islands: Fusion of world-view 2 imagery, 3-D laser scanner data & unmanned aerial systems. In *Comprehensive remote sensing, applications for societal benefits* (S. J. Walsh, Book Editor; S. Liang, Organizing Editor) (pp. 159–176). London: Elsevier.

Weigelt, P., Jetz, W., & Kreft, H. (2013). Bioclimatic and physical characterization of the world's islands. *Proceedings of the National Academy of Sciences, 110*(38), 15307–15312.

Wessel, P., & Smith, W. H. F. (1996). A global, self-consistent, hierarchical, high-resolution shoreline database. *Journal of Geophysical Research: Solid Earth, 101*(B4), 8741–8743.

Whittaker, R. J., & Fernández-Palacios, J. M. (2007). *Island biogeography* (2nd ed.). New York: Oxford University Press.

Zhang, H., & Walsh, S. J. (2018). Comparison of the Zhoushan Islands, China and the Galapagos Islands, Ecuador: Island sustainability and forces of change. In *Comprehensive remote sensing, applications for societal benefits* (S. J. Walsh, Book Editor; S. Liang, Organizing Editor) (pp. 306–329). London: Elsevier.

Zhao, B., Kreuter, U., Li, B., Ma, Z., Chen, J., & Nakagoshi, N. (2004). An ecosystem service value assessment of land-use change on Chongming Island, China. *Land Use Policy, 21*, 139–148.

Zupan, M., Fragkopoulou, E., Claudet, J., Erzini, K., Horta e Costa, B., & Goncalves, E. (2018). Marine partially protected areas: Drivers of ecological effectiveness. *Frontiers of Ecology & Environment, 17*(7), 1–7. https://doi.org/10.1002/fee.1934

Transitions and Drivers of Land Use/Land Cover Change in Hawai'i: A Case Study of Maui

Laura Brewington

Introduction

Islands worldwide have undergone dramatic changes in land cover on the order of decades, centuries, and even millennia. The drivers of change may be natural, through ecosystem processes and events, or human driven, as in the case of climate change, population growth, species invasions, and globalization (Pelling & Uitto, 2001; Kerr, 2005; Kier et al., 2009; Fordham & Brook, 2010; Kueffer et al., 2010; Hay, Forbes, & Mimura, 2013; Spatz et al., 2014). The US state of Hawai'i is one of the most remote archipelagos in the world, but has nevertheless experienced extraordinary changes over its human history, from settlement by ancient Polynesian explorers to discovery by the western world, volcanic and tectonic activity, land utilization for commercial agriculture, and finally, access by the lucrative tourism industry.

Recently, a meta-analysis of global island land use/land cover change literature was conducted to categorize the diverse drivers of change and identify similarities and differences among island sites (Walsh et al., this volume). The results showed that in spite of the large shifts in land use in Hawai'i since Polynesian settlement and European contact, limited research has attempted to quantify them spatially and temporally. Changing agricultural practices in the islands have been a central theme. Ladefoged and Graves (2008) recreated ancient dryland agriculture plots on Hawai'i Island using aerial photography, and other efforts have utilized ground surveys and GIS to assess pre-contact Hawaiian field systems (Kurashima & Kirch, 2011; Lincoln & Ladefoged, 2014). Bartholomew, Hawkins, and Lopez (2012) described

L. Brewington (✉)
East-West Center, Honolulu, HI, USA
e-mail: BrewingL@EastWestCenter.org

© Springer Nature Switzerland AG 2020

S. J. Walsh et al. (eds.), *Land Cover and Land Use Change on Islands*, Social and Ecological Interactions in the Galapagos Islands, https://doi.org/10.1007/978-3-030-43973-6_4

the pineapple boom and bust that occurred in Hawai'i between the late-1800s and the mid-1900s, which ultimately succumbed to international competition from the Philippines and Thailand. Suryanata (2002) tracked agricultural intensification, commercialization, and diversification in the state throughout the latter part of the 1900s, examining the intense competition of Hawai'i products in the context of global economic forces (e.g. boutique crops such as coffee, pineapple, and macadamia nut). Perroy, Melrose, and Cares (2016) updated the 1980 state-wide agricultural assessment using high-resolution WorldView imagery and found gains in the spatial cover of macadamia nut, coffee, and diversified agriculture. Other spatially-explicit research has focused on coastal change, such as erosion due to sea-level rise, storm surge, and various shoreline modifications (Anderson, Fletcher, Barbee, Frazer, & Romine, 2015; Fletcher, Mullane, & Richmond, 1997; Fletcher, Rooney, Barbee, Lim, & Richmond, 2003; Miller & Fletcher, 2003), development intensity and wetland loss (Margriter, Bruland, Kudray, & Lepczyk, 2014), and the invasion of mangrove forests (D'iorio, Jupiter, Cochran, & Potts, 2007). Benning, LaPointe, Atkinson, and Vitousek (2002) and Fortini, Vorsino, Amidon, Paxton, and Jacobi (2015) have considered the interactions between a changing climate and endemic bird habitat in Hawai'i. The impacts of land cover change on Hawai'i's water resources (Brewington, Keener, Finucane, & Eaton, 2017; Ponette-Gonzalez et al., 2014), wildfire occurrence and intensity (Ellsworth, Litton, Dale, & Miura, 2014; Elmore, Asner, & Hughes, 2005), and forest resources (Asner et al., 2011, 2012; Morales, Miura, & Idol, 2008) have also been explored.

The above studies and more have provided invaluable spatial resources for local, regional, and state-level land management efforts. Nevertheless, archipelago- or island-wide land cover change assessments have remained elusive. A limited number of comprehensive land cover products do exist for Hawai'i, however, and this chapter summarizes them, selecting the island of Maui as a representative case for evaluating landscape transitions through time. First, a history of the drivers of land cover change on Maui is described, beginning with settlement by ancient Polynesian voyagers up to the present day. Then the available spatial data products for evaluating archipelago- and island-wide land cover are identified and modified for comparison. Changes in land cover over the past half-century are quantified and discussed in the context of Maui's recent trajectories. The chapter concludes with directions for future research linking the drivers of change with land cover products using newer high-resolution imagery and ancillary GIS data.

Background and Study Area

The island of Maui is located at 20.8°N, 156.5°W between the islands of Moloka'i, Lāna'i and Hawai'i in the Hawaiian archipelago (Fig. 1). Two basaltic shield volcanoes that were formed in the last 2 million years separate east and west Maui (Clague & Dalrymple, 1989), but west Maui is geologically older and has been exposed to extensive weathering that has created abundant beaches and reef systems,

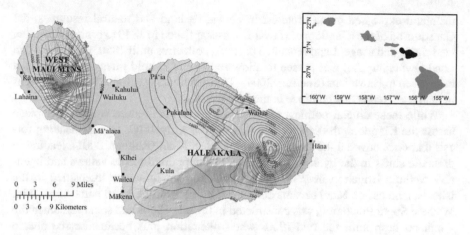

Fig. 1 The Hawaiian Islands and the island of Maui, showing topography, major geographic and population features, and mean annual rainfall contours (in mm/year)

compared to the younger cliffs and rocky shores of East Maui (Kirch, 2010). The unique and rugged topography of Maui generates variable rainfall, wind, and temperature patterns over short distances. The island experiences climatic extremes, from very high rainfall in the upper elevation windward (northeast-facing) slopes exceeding 1000 cm/year to leeward (southwest-facing) lower elevations receiving less than 60 cm/year (Giambelluca et al., 2013). Haleakalā, the large shield volcano that dominates East Maui, rises 3055 m above sea level and has alpine desert conditions at the summit, with occasional winter snowfall, while the low-lying isthmus in central Maui is characterized by hot and dry conditions with high winds (Juvik & Juvik, 1998).

Drivers of Land Cover Change on Maui

Maui has undergone dramatic changes in land use since human settlement. Prior to the nineteenth century, fishing, agriculture, and aquaculture were widely practiced by the island's indigenous Hawaiian populations (Cuddihy & Stone, 1990; Kirch, 1982). Ancient Hawaiian society was isolated from external influence for hundreds of years. Fire was extensively used across the landscape to clear forested areas and for agricultural practices. Irrigated wetland and rain-fed dryland agricultural production supported complex societies across the island and intensified from the 1200s until just before European contact in the 1700s (Kirch, 2010; Kirch et al., 2005; Kirch, Holson, & Baer, 2009). Following James Cook's arrival, the islands were integrated as the Kingdom of Hawai'i under King Kamehameha I, at which time Protestant missionaries were also expanding their influence throughout the Pacific Islands region. As descendants of the newcomers and authorities on the US mainland expressed greater interest in Hawai'i's land and natural resources, the Constitution of the Kingdom of Hawai'i was established in 1840 to protect Hawaiian land rights and usage. Land remained part of a collective until 1850, when the Alien Land Ownership Act was passed to allow foreigners to hold titles to land that had been given to Hawaiians (Stauffer, 2004). This paved the way for very rapid changes in land use, trends that continue in the islands today.

While these crucial political transformations in land tenure were taking place across the Kingdom, the sandalwood trade of the early 1800s was devastating forests that once covered the slopes of east and west Maui (Rhodes, 2001), leading to dramatic shifts in the hydrological regime that supported fertile valleys and intensive wetland irrigation near the coast. As American influence intensified in the islands, land use on Maui became dominated by agriculture. Maui's first sugar mill, Waihe'e Sugar Plantation, was constructed in the late 1800s and sugar remained the dominant crop until the mid-1900s when plantation jobs became scarce due to mechanization and the price of sugar began to fall (Maclennan, 2007). Central Maui sugarcane cultivation during this time covered around 36,000 ha. Declining rainfall in the latter half of the twentieth century stressed water resources for raising cattle and other livestock, and the greatest declines in agricultural land use began in the 1980s (Perroy et al., 2016). The last of Hawai'i's 14 commercial sugarcane plantations ceased operation on Maui in 2016, after 145 years (Tanji, 2016).

The decline of plantation-scale agriculture on Maui coincided with statehood and Hawai'i's post-WWII tourism boom, which was heavily concentrated in the Lahaina region of intensive coastal resort development (Blackford, 2001). By the 1960s, Maui had over 1300 hotel rooms, a number that almost tripled over the next 20 years. In light of concerns about watershed degradation and unchecked tourism development, the Hawai'i State Land Use Law was passed in the 1960s, which assigned all lands to conservation, agricultural, or urban districts (Hiatt, 1993). A rural district was later added to protect non-urban, inhabited areas from development activities otherwise permitted in the urban district. Streamflow diversions for agriculture and groundwater use for urbanization increased substantially throughout the twentieth century, however, as the state's populations of residents and visitors experienced exponential growth, from 220,000 residents and 17,000 annual visitors in the 1920s to 1.4 and 9.3 million today (HI Tourism Authority, 2017; US Census, 2010). Maui's unique geology and ecosystems, along with regional, domestic, and international accessibility, have made it a prime tourist destination with 2.74 million annual visitors (HI Tourism Authority, 2017) and it currently has the third largest resident population in the state (144,444; US Census, 2010).

Maui's coastal areas and shorelines include sandy beaches, dune ecosystems, cliffs, and low-growing vegetation that serve as popular recreation areas for island residents and visitors, as well as refugia for native and migratory shorebirds (HI Department of Land and Natural Resources, 2015). Some of the largest and earliest resort areas in Hawai'i were established on Maui (Cooper & Daws, 1985) but more recent residential and resort development has taken place along Maui's central south shore, near Kīhei. In lowland areas and along leeward slopes up to 2000 m, some native dry forest ecosystems persist in spite of extensive urban development and species introductions (HI Department of Land and Natural Resources, 2015; Juvik & Juvik, 1998). Mesic and wet forests dominated by native trees and shrubs are typically found in windward locations up to 2000 m in elevation (Crausbay, Frazier, Giambelluca, Longman, & Hotchkiss, 2014). In the mountains of east and west Maui, watershed partnerships manage higher elevation lands to protect native forestlands by removing introduced grasses, trees, and shrubs, and installing fences to exclude feral ungulates.

High-elevation sub-alpine and alpine areas (between 2000–3000 m) are found in east Maui near the summit of Haleakalā (3055 m) where the vegetation ranges from both wet and dry forests and shrubland at the lower end of the range, to arid and semi-barren habitats with highly specialized vegetation and grasslands (Leuschner & Schulte, 1991). The treeline around 2100 m is dominated by conifers that were introduced in early 1900s as part of an experimental forestry project. Areas near the summit contain aeolian deserts with grasses, lichens, and mosses (Juvik & Juvik, 1998). Haleakalā National Park was established in 1961 and protects a vast area of 135 km^2, which received over 1 million visitors in 2018 (NPS, 2019). A rapidly changing climate and high visitor traffic may increase invasive species establishment at the highest elevations, however, and the endemic, high-elevation Haleakalā silversword has experienced a 60% decline since 1990 (Krushelnycky et al., 2013). Warming air temperatures and long-term drying trends have impacted land use, natural resources, and ecosystem processes across the Hawaiian Islands (Keener et al., 2018). Quantifying and tracking the drivers of change therefore provide opportunities for targeted management and planning, to sustain Maui's population, economy, and unique environments into the future.

Methods

The purpose of this chapter was to evaluate how well the available land cover products for the Hawaiian Islands reflect the drivers of change on Maui; therefore, no new classification work or ground truth efforts were conducted. First, the existing land cover datasets for Hawai'i were identified. Only those that covered at least six of the eight main Hawaiian Islands (Kauai, O'ahu, Moloka'i, Maui, Lāna'i, and Hawai'i Island; the islands of Ni'ihau and Kaho'olawe are typically excluded from such products) were selected (Table 1). Land cover products were not required to be multi-temporal. The relevant metadata for each data product were collected, and if needed the dataset creators were contacted for clarification. The available products are briefly summarized below according to their spatial resolution, temporal coverage, type of land cover data, and intended use.

Table 1 Summary of existing land cover products for the main Hawaiian Islands

Data product	Timeframe	Resolution (m)	Description	Best for
Landfire Biophysical Settings	~1778	30	Represents vegetation that may have been dominant on the landscape prior to European contact, based on both the biophysical environment and an approximation of historical disturbance (i.e. wildfire)	Land cover estimation pre-contact
Land cover map	~1976	~200	Historical land cover data from 1970s and 1980s, created using aerial photography, land use maps, and surveys	Historical land use
NOAA C-CAP	1992, 2001, 2005, 2010	2.4–30	Standardized, raster-based inventories of land cover for the state of Hawai'i, updated every 5 years using Landsat, Quickbird, and WorldView imagery	State or island-level strategic planning
Landfire Existing Vegetation Type	2001, 2008, 2010, 2012, 2014	30	Distribution of plant community types that are currently present according to a nationally consistent set of ecological units, using Landsat imagery	State or island-level strategic planning
Hawai'i Gap Analysis Project	2005	30	Statewide land cover dataset created using automatic classifications of Landsat imagery with ancillary data to improve class distributions	State or island-level conservation planning
Carbon Assessment of Hawai'i	2014	30	Statewide land cover dataset prior mapping efforts (GAP) updated using very-high-resolution imagery	National carbon assessment; state or island-level strategic planning

Landfire BPS and EVT

As the name implies, the Landfire Project is designed to improve wildfire management, research, and development across the United States. Responsibility for the 30 m, Landsat-based gridded data products is shared between the US Department of Agriculture Forest Service (USFS) and the US Geological Survey (USGS), two of which are highlighted in this chapter: Landfire Biophysical Settings (BPS) and Existing Vegetation Type (EVT). The BPS layer represents natural vegetation communities that may have been dominant in Hawai'i prior to European contact and the arrival of British explorer James Cook in 1778 (Landfire, 2019a). The dataset creators attempted to incorporate what is known about indigenous influence on ecological processes and disturbance, such as fire (Rollins, 2009), and this layer has been particularly useful in mapping historic fire regimes and fuel loads. The Landfire EVT product contains distributions of plant community types according to a nationally consistent set of ecological units that are central to regional, multi-temporal assessments of habitat and dominant vegetation (Landfire, 2019b). Five EVT datasets are available for the main Hawaiian Islands from 2001 to 2014, and each temporal dataset is an update from the one before it according to enhancements in data records, imagery, and disturbance information. Landfire EVT data for Hawai'i contain multiple classes and subclasses that users may choose from, including specific categories for Hawaiian native and alien forest, shrubland, and grasslands, as well as aggregated types based on dominant species and national and agency-specific vegetation classification systems.

Hawai'i Land Cover Circa 1976

An early effort to map land cover in Hawai'i using aerial photographs from the 1970s and 1980s resulted in a relatively coarse spatial resolution (minimum polygon area of 200 m^2) map that represents ~1976 conditions (HI State GIS, 2019). This USGS-derived product was part of an effort to develop a consistent classification system for use with remotely-sensed data and generate land cover maps for the entire US. The two-level hierarchical system is described in Anderson, Hardy, Roach, and Witmer (1976). Level I classes are meant to be broad, inclusive categories (urban or built-up land, agricultural land, rangeland, forest land, water, wetland, and barren land), and Level II classes are more specified. Level II agricultural land classes include croplands, orchards, and feeding operations, for example. Neither class level distinguishes between native or non-native plant communities, or degree of development intensity.

NOAA C-CAP

The US National Oceanic and Atmospheric Administration's Coastal Change Analysis Program (NOAA C-CAP) derives nationally standardized inventories of land cover for US coastal areas to improve research and management efforts related to coastal and marine ecosystems (NOAA Office for Coastal Management, 2018). Medium (30 m) and high resolution (2.4 m) land cover data products for the main Hawaiian Islands are available from 1992 to 2010 in roughly five-year intervals. Land cover classes heavily represent coastal and wetland vegetation types but also incorporate categories for developed, cultivated, and forested land cover. NOAA C-CAP recommends that these data products be used as a screening tool for regional or site-specific management decisions, while small features and changes should be verified with a higher-resolution data source.

Hawai'i GAP and CAH

The USGS is responsible for the Gap Analysis Project (GAP), which provides national biodiversity range estimates to determine "gaps" in species or habitat-level protection. The Landsat-based products have been produced since the 1990s and are used for conservation purposes and management planning (Scott et al., 1993). The Hawai'i GAP dataset from 2005 adopts a classification system that was developed specifically for the islands, with 28 native vegetation classes and nine additional land use, non-native vegetation, or disturbed classes. Unlike the national GAP, the Hawai'i GAP dataset includes federally endangered plant species ranges (Gon III, 2006), which are useful for conservation planning but may not facilitate detailed assessments of other types of land cover—one example of this is the ambiguous 'uncharacterized forest' category. Closely related to Hawai'i GAP is a 2014 Carbon Assessment of Hawai'i (CAH) Land Cover Map that depicts the status of plant communities and disturbance in the main Hawaiian Islands, circa 2017 (Jacobi, Price, Fortini, Gon III, & Berkowitz, 2017a). The CAH was considered to be an update of the 2005 Hawai'i GAP dataset (Jacobi, personal communication, 2015). To form the land cover base for the CAH, USGS researchers modified the Hawai'i GAP land cover units using NOAA C-CAP land cover products, WorldView-2 imagery, and very high resolution Pictometry Online imagery. The resulting 30 m resolution data product, along with a companion CAH Habitat Status Map, serve as the basis for estimating current and future carbon stocks and fluxes as part of the national carbon assessment (Jacobi, Price, Fortini, Gon III, & Berkowitz, 2017b). The CAH land cover product differentiates between native and alien plant species but lacks the higher level of specificity from Hawai'i GAP, such as endangered plant species.

Multi-Temporal Comparison

For this assessment, the 1976 land cover map and the subsequent three contemporary products (NOAA C-CAP, Landfire EVT, and Hawai'i GAP/CAH) were selected for comparison. The 1976 historical land cover map was converted to a raster file and resampled from 200 to 30 m resolution, for change detection and comparison across all three multi-temporal products. The 2005 and 2010 NOAA C-CAP datasets have a 2.4 m resolution and were resampled to 30 m to match NOAA C-CAP datasets for 1992 and 2001, Landfire EVT, Hawai'i GAP, and the CAH land cover products. The selected products also contained a variety of land cover classes, some of which differed from one time period to the next. To represent and compare island-wide transitions through time, all original land cover classes were reclassified into eight main categories: agriculture, bare land, built-up, forest, grassland, shrubland, water body, and wetland. These categories were selected because they were relatively easy to consolidate across the various data products and would adequately reflect the major changes in land cover on Maui.

Following reclassification, area and relative (percent) land cover for each class were calculated for the final datasets. Comparable dates from the different land cover products were cross-tabulated against one another to derive Kappa indices that reflected the level of spatial agreement. Change trajectories and transition matrices were calculated between each consecutive time point and for the overall range of available dates. Spatial representations of change were interpreted according to the drivers of change described above.

Results

Reclassified Maui Land Cover

Table 2 shows how the original land cover classes from each dataset were reclassified into the eight main categories for analysis. Due to the diversity of original classes and the range of ecosystems each data product represents, reclassification was sometimes subject to visual comparison and interpretation. The Hawai'i GAP and CAH datasets, for example, included highly specific native or introduced vegetation classes; Hawai'i GAP contained 12 unique native forest categories that are summarized in the table below as 'native forest (various)'. Landfire EVT, on the other hand, made very limited distinctions between native and introduced plants ('Hawaiian rainforest/dry/mesic forest' and 'introduced upland vegetation-treed', for example). NOAA C-CAP did not distinguish between native and introduced vegetation at all. Due to its emphasis on Hawaiian ecosystem type and structure, the Hawai'i GAP dataset only contained two built-up categories. Landfire EVT was the lone dataset to include tree plantations as a unique land cover type, which was reclassified as 'forest' but other users may consider it as 'agriculture'. Hawai'i GAP and CAH also contained cliff vegetation categories, which at first did not appear to correspond directly to any of the redefined classes for this analysis. However, a general land cover variable in the CAH dataset listed these as 'native shrub', so they were reclassified as 'shrubland'. Classes that were reclassified to the 'wetland' category were especially diverse across products and included bogs, marshes, and aquatic beds.

Four of the reclassified maps were chosen for preliminary comparison: 1976 land cover, 2010 NOAA C-CAP, 2014 Landfire EVT, and 2014 CAH (Fig. 2). Table 3 contains the area and percent of each land cover class. Due to its coarse original resolution, the 1976 map contained much less spatial detail than the contemporary datasets. Land cover in 1976 was dominated by forest (36.1%), shrubland (23.6%), and agriculture (25.7%). Agricultural activities were concentrated in the central isthmus of the island, as well as west Maui near the coast. Commercial sugarcane and, to a lesser extent, pineapple production accounted for this extensive agricultural cover. Significant shrubland cover was found at the lower elevations of east Maui and was interspersed with grassland cover. Originally classified as 'shrub and

Table 2 Reclassification scheme for 1976, NOAA C-CAP, Landfire EVT, Hawai'i GAP, and CAH land cover products

New Class	Original class				
	1976 land cover	NOAA C-CAP	Landfire EVT	Hawai'i GAP	CAH
BU	Residential	Developed, low-intensity	Developed-low intensity	Low intensity developed	Low intensity developed
	Transportation, communications, and utilities	Open space developed	Developed-open space		Developed open space
		Impervious surface			
	Commercial and services	Developed, medium-intensity	Developed-medium intensity	High intensity developed	Medium intensity developed
	Industrial and commercial complexes	Developed, high-intensity	Developed-high intensity		High intensity developed
AG	Cropland and pasture	Cultivated land	Agriculture-cultivated crops and irrigated agriculture	Agriculture	Cultivated agriculture
	Orchards, groves, vineyards, nurseries and ornamental horticultural areas				
	Other agricultural land				
BL	Sandy areas other than beaches	Bare land	Barren	Very sparse vegetation to unvegetated	Very sparse vegetation to unvegetated
	Bare exposed rock				
	Strip mines, quarries, and gravel pits			Uncharacterized open-sparse vegetation	
	Transitional areas				
GL	Herbaceous rangeland	Grassland	Hawaiian dry/mesic grassland	Deschampsia grassland	Native mesic grassland
	Mixed rangeland	Pasture/hay	Introduced perennial grassland and forbland	Kikuyu grass grassland / pasture	Alien wet/mesic/dry grassland
				Alien grassland	

(continued)

Table 2 (continued)

	Original class				
New Class	1976 land cover	NOAA C-CAP	Landfire EVT	Hawai'i GAP	CAH
SL	Shrub and brush rangeland	Scrub shrub	Hawaiian dry/ mesic shrubland	Native shrubland (various)	Native wet/ mesic/dry shrubland
				Native dry/wet cliff vegetation	Native wet cliff community
		Palustrine scrub shrub wetland	Introduced upland vegetation-shrub	Alien shrubland, shrubs and grasses	Alien wet/ mesic/dry shrubland
				Uncharacterized shrubland	
F	Evergreen forest land	Palustrine forested wetland	Hawaiian rainforest/dry/ mesic forest	Native forest (various)	Native wet/ mesic/dry forest
		Evergreen	Introduced upland vegetation-treed	Alien forest	Alien wet/ mesic/dry forest
		Estuarine forested wetland	Managed tree plantation	Uncharacterized forest	
WL	Forested wetland	Palustrine emergent wetland	Inland marshes and prairies	Bog vegetation	Native bog community
	Nonforested wetland	Palustrine aquatic bed			
WB	Lakes	Open water	Open water	Water	Water
	Reservoirs				
	Bays and estuaries	Unconsolidated shore			

BU Built-up, *AG* Agriculture, *BL* Bare land, *GL* Grassland, *SL* Shrubland, *F* Forest, *WL* Wetland, *WB* Water body

brush rangeland', it can be assumed that this area was primarily used for ranching operations at that time. Built-up land accounted for only 3% of the total land cover island-wide in 1976.

Despite their similar timeframes, the three contemporary datasets (2010 NOAA C-CAP, 2014 Landfire EVT, and 2014 CAH) had multiple discrepancies in land cover. The NOAA C-CAP dataset had the lowest built-up cover (6.1%), compared to Landfire EVT (11%) and CAH (8.2%). The differences were most obvious in west Maui, where the NOAA C-CAP dataset showed grassland in areas that were built-up in the other two classifications, and in north central Maui, where shrubland was mapped instead of built-up cover. Particularly in west Maui, both the Landfire EVT and CAH datasets contained larger active crop production areas than NOAA

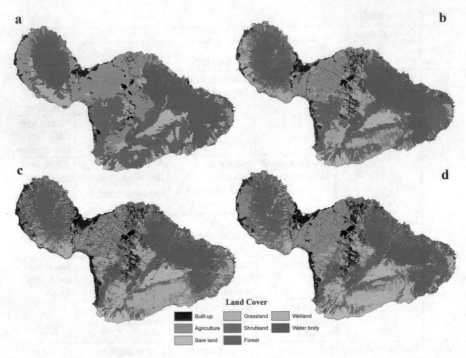

Fig. 2 Resampled and reclassified land cover maps for (**a**) 1976 land cover; (**b**) 2010 NOAA C-CAP; (**c**) 2014 Landfire EVT; and (**d**) 2014 CAH

Table 3 Area and percent of each land cover type for 1976 land cover, 2010 NOAA C-CAP, 2014 Landfire EVT, and 2014 CAH

Land cover	1976 area	%	2010 NOAA C-CAP area	%	2014 Landfire EVT area	%	2014 CAH area	%
BU	55.60	2.95	114.24	6.05	207.18	10.97	155.51	8.24
AG	485.11	25.73	181.69	9.63	201.93	10.70	265.92	14.09
BL	65.75	3.49	95.77	5.07	112.82	5.98	155.79	8.26
GL	149.40	7.92	317.78	16.83	441.74	23.40	320.84	17.00
SL	443.90	23.55	541.02	28.66	201.99	10.70	227.40	12.05
F	680.21	36.08	627.55	33.24	716.29	37.94	754.33	39.97
WL	2.52	0.13	3.33	0.18	0.95	0.05	3.58	0.19
WB	2.69	0.14	6.28	0.33	4.91	0.26	3.79	0.20

Area in km^2

C-CAP (201.93 and 265.92 km^2, versus 181.69 km^2), which in turn reflected more grassland. The original Landfire EVT classifications for this region were introduced shrub and trees, which were reclassified to shrubland and forest, respectively. The CAH agricultural category, meanwhile, extended into areas classified as shrubland

and/or grassland in the NOAA C-CAP and Landfire EVT datasets, such as the southern part of west Maui and into the upper elevations of east Maui.

There were obvious differences in reclassified shrubland cover, as well: NOAA C-CAP reflected over twice the amount of shrubland (28.7%) as Landfire EVT (10.7%) and CAH (12.1%), and this was especially noticeable in the area to the south and east of central Maui along the slopes of Haleakalā. NOAA C-CAP classified much of this region as 'scrub shrub', while the original Landfire EVT classes in the same area were either 'introduced perennial grassland and forbland' (reclassified to grassland) or 'introduced upland vegetation-treed' (reclassified to forest). In the CAH dataset, much of this area was categorized as 'kiawe dry forest and shrubland' in the detailed land cover class and as 'alien dry forest' in the generic land cover class, and was therefore reclassified to forest. In east Maui, too, there was a very large patch of shrubland in the NOAA C-CAP dataset that was not present in Landfire EVT or CAH. Low Kappa statistics reflected the low agreement between the three contemporary land cover classifications: the confusion matrix for 2010 NOAA C-CAP versus 2014 Landfire EVT yielded a Kappa coefficient of 0.47. For 2010 NOAA C-CAP versus 2014 CAH, Kappa was 0.52, and for 2014 Landfire EVT versus 2014 CAH, Kappa was 0.62.

Overall Land Cover Change (1976–2000s)

Change detection between the 1976 land cover map and the three contemporary land cover products reflected many of the demographic, economic, and environmental changes that have occurred on Maui since the 1970s (Fig. 3).

Changes to agriculture, grassland, and built-up cover spatially dominated the land cover transitions between 1976 and all three contemporary datasets (Fig. 4). All showed gains in built-up, bare land, grassland, and water body cover over the last 40 years. Consistent declines were recorded only for the agriculture class. In fact, hundreds of square kilometers of agriculture cover were lost between 1976 and the ~2010 era, the greatest of which were reflected by the 2010 NOAA C-CAP dataset ($303.4 km^2$, or 62.6%). Almost 50% of those losses were to new grassland cover, and the rest converted to forest, shrubland, or built-up areas (Table 4). These transitions were observed in west Maui, primarily, with new built-up areas appearing in the central isthmus (Fig. 4a). The NOAA C-CAP dataset revealed a 7.7% loss in forest cover where the other two current datasets both showed gains. A large new area of shrubland in east Maui and various patches in the mountains of west Maui were responsible for this forest loss. The forest-to-shrubland conversion also created a 21.9% increase in shrubland, compared to shrubland losses in both Landfire EVT (54.5%) and CAH (48.8%). Although the NOAA C-CAP dataset showed the least change in built-up cover, it still more than doubled in size, from $55.6 km^2$ in 1976 to $114.2 km^2$ in 2010.

The 2014 Landfire EVT dataset showed dramatic expansions in grassland cover ($292.4 km^2$, almost 200%) and built-up cover ($151.6 km^2$, 272.7%) compared to 1976. New grassland largely appeared in east Maui around the middle elevations of

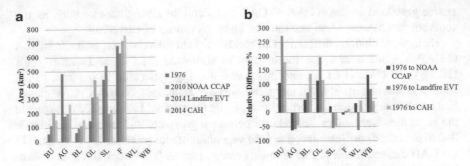

Fig. 3 (**a**) Area for each land cover class according to 1976, 2014 NOAA C-CAP, 2010 Landfire EVT, and 2010 CAH datasets, (**b**) relative percent change between 1976 and current land cover

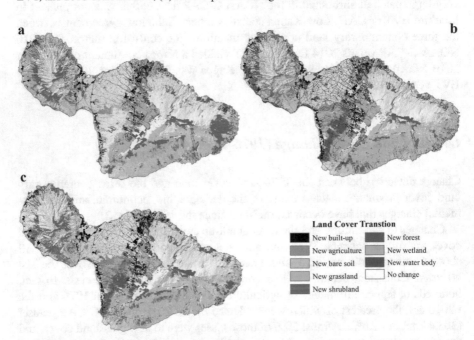

Fig. 4 Land cover changes between 1976 and the (**a**) 2010 NOAA C-CAP, (**b**) 2014 Landfire EVT, and (**c**) 2014 CAH datasets

Haleakalā and along the southern coast (Fig. 4b), a result of transitions out of agriculture (40.5%), shrubland (41.7%), and forest (42.5%; Table 5). The new built-up area was primarily converted from agriculture (32.9%) and appeared in patches throughout the island and especially along the coast.

The 2014 CAH dataset showed very large increases in built-up (99.9 km², 179.7%), bare land (90.0 km², 137%), and grassland (171.4 km², 114.8%) cover, due to transitions out of agriculture, shrubland, and forested areas (Table 6). Overall, according to this classification, forest actually increased by 11% and was concentrated in east Maui's middle to higher elevations (Fig. 4c). This was similar to changes mapped by

Table 4 Transition matrix from 1976 to 2010 NOAA C-CAP

1976	2010 NOAA C-CAP							
	BU	AG	BL	GL	SL	F	WL	WB
BU	36.98	1.03	1.03	2.94	3.40	9.54	0.05	0.54
AG	47.69	165.64	5.72	150.48	49.53	64.44	0.22	1.03
BL	3.50	0.80	37.95	5.15	15.09	2.96	0.01	0.20
GL	2.54	0.82	7.40	55.46	61.98	20.73	0.15	0.19
SL	12.76	11.78	37.20	74.62	259.48	47.37	0.05	0.31
F	10.27	1.26	5.02	28.75	150.60	481.44	1.75	0.79
WL	0.08		0.05	0.01	0.50	0.30	0.98	0.58
WB	0.11	0.35	0.08	0.26	0.19	0.27	0.12	1.30

Area in km^2

Table 5 Transition matrix from 1976 to 2014 Landfire EVT

1976	2014 Landfire EVT							
	BU	AG	BL	GL	SL	F	WL	WB
BU	46.59	1.06	0.43	2.61	1.11	3.40	0.00	0.30
AG	99.20	183.48	1.84	121.22	25.48	52.12		1.39
BL	4.83	0.56	40.32	4.37	12.39	3.12	0.00	0.05
GL	5.72	0.83	10.55	79.65	23.47	28.96	0.04	0.04
SL	28.60	12.79	47.61	156.28	68.53	129.57		0.19
F	20.36	2.89	10.24	77.46	70.21	497.46	0.91	0.30
WL	0.42		0.02		0.48	0.93		0.65
WB	0.29	0.31	0.03	0.14	0.07	0.34	0.00	1.54

Area in km^2

Table 6 Transition matrix from 1976 to 2014 CAH

1976	2014 CAH							
	BU	AG	BL	GL	SL	F	WL	WB
BU	45.11	1.74	1.41	1.96	0.72	4.31		0.22
AG	67.92	235.25	20.70	76.41	20.79	62.73		0.88
BL	4.87	1.52	39.39	5.56	10.03	4.26		0.01
GL	3.91	1.97	14.58	58.95	38.57	31.26	0.00	0.03
SL	18.46	18.27	54.55	130.24	87.13	134.74		0.14
F	14.20	6.15	22.38	46.98	70.08	516.21	3.58	0.20
WL	0.26	0.33	0.26	0.47	0.01	0.36		0.79
WB	0.16	0.65	0.03	0.07	0.01	0.26		1.52

Area in km^2

the Landfire EVT dataset, but differed from the forest losses observed by NOAA C-CAP. Bare land increased as a result of transitions out of shrubland, primarily, as well as agriculture and forest. A total of 219.2 km^2 of agricultural land was lost between 1976 and 2014, the least change of all three current datasets. The following sections describe the results for each multi-temporal dataset in more detail.

NOAA C-CAP Land Cover Change (1992–2010)

Four years were available for the NOAA C-CAP land cover data product (1992, 2001, 2005, and 2010) and the changes in area across the island of Maui, for each time period and overall, are shown in Fig. 5.

According to the NOAA C-CAP product, Maui land cover in 1992 was dominated by forest, shrubland, agriculture, and grassland. By 2001, however, agriculture cover had decreased by 27.5% and built-up land cover increased by 19.82%. Grassland and forest cover also increased during that period, while shrubland area was lost. The transition matrix (Table 7) showed that agricultural losses went largely to new built-up areas in central Maui and grassland areas in west Maui (Fig. 6a). Although they were small in total area, 86.13, 60.96, and 73.14% of the respective losses to bare land, grassland, and forest went to new built-up areas in the upcountry ranching areas of leeward eastern Maui, and new developments in west Maui. Overall, these three land cover classes increased slightly due to inputs from agriculture as well as shrubland, which lost around 6% of its 1992 cover. This was the only time period in which forest cover increased (only 5.35 km², less than 1% of total cover), which occurred mainly in the upper elevations of Haleakalā driven by losses

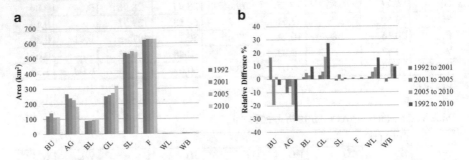

Fig. 5 (**a**) Area for each land cover class according to the NOAA C-CAP data product, (**b**) relative percent change for each time period and overall

Table 7 Transition matrix for the NOAA C-CAP data product, 1992–2001

1992	2001							
	BU	AG	BL	GL	SL	F	WL	WB
BU	119.25	0.01	0.00			0.01		
AG	11.04	234.58	1.32	10.48	5.31	2.11	0.04	0.06
BL	0.18	0.01	87.17	0.01	0.00	0.01	0.00	
GL	2.82	0.06	0.12	244.99	0.82	0.74		0.06
SL	4.69	2.78	0.18	1.11	522.87	4.01		0.01
F	1.11	0.01	0.02	0.13	0.25	620.72		0.00
WL						0.00	2.87	
WB	0.00	0.01		0.00	0.00	0.00		5.74

Area in km²

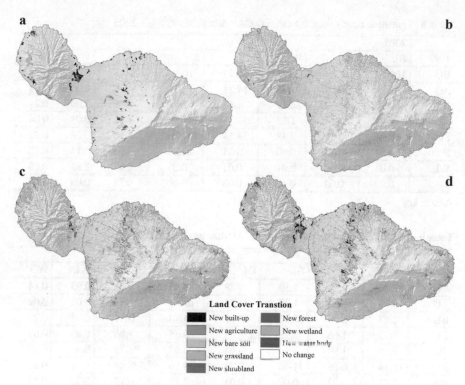

Fig. 6 NOAA C-CAP land cover changes between (**a**) 1992–2001; (**b**) 2001–2005; (**c**) 2005–2010; and (**d**) 1992–2010

to shrubland (31.4%) and grassland (16.1%). From 1992 to 2001, there were changes to 67.67 km² of the total land cover on Maui, or 3.58%.

Changes from 2001 to 2005 also largely impacted built-up land cover and agriculture, but in this period the built-up cover decreased by 26.89% to almost 7 km² less than the 1992 extent. The changes occurred in very small patches, however, that were likely due to errors resulting from resampling between the 30 m 2001 classification and the 2.4 m 2005 classification (Fig. 6b). These patchy shifts out of urban cover were distributed across the entire island and converted into forest, shrubland, agriculture, and grassland (Table 8). A conspicuous new urban area (previously grassland) appeared along Maui's south shore in the town of Kīhei, which was expanding at that time. The substantial increases to grassland (14.84%) area came from transitions out of agricultural land in west Maui, central Maui, and a large area on the northern coastline of east Maui. It is likely that this was a continuation of the transition out of sugarcane production. Meanwhile 12.9% of the losses to agriculture land cover in west and central Maui became shrubland, which increased by 18.57% island-wide. Total changes to land cover during this period affected 4.14% of Maui's total spatial area.

Table 8 Transition matrix for the NOAA C-CAP data product, 2001–2005

1992	2005							
	BU	AG	BL	GL	SL	F	WL	WB
BU	100.67	7.68	1.73	5.17	10.14	13.48	0.09	0.13
AG	2.22	215.98	1.36	14.31	2.77	0.77	0.00	0.04
BL	0.30	0.02	84.11	1.12	2.51	0.38	0.00	0.24
GL	1.97	0.25	1.53	232.73	15.30	4.88	0.01	0.04
SL	2.41	0.59	3.38	13.18	494.38	14.76	0.26	0.27
F	4.53	0.47	0.60	5.01	22.38	594.05	0.14	0.42
WL	0.01	0.00	0.01	0.01	0.26	0.07	2.53	0.03
8	0.09	0.02	0.47	0.03	0.09	0.13	0.04	4.74

Area in km²

Table 9 Transition matrix for the NOAA C-CAP data product, 2005–2010

2005	2010							
	BU	AG	BL	GL	SL	F	WL	WB
BU	83.00	2.54	1.50	9.28	6.82	8.88	0.05	0.14
AG	5.07	174.52	1.93	31.35	10.52	1.51	0.02	0.09
BL	2.27	0.05	81.92	3.67	3.80	0.51	0.03	0.75
GL	5.31	1.66	3.38	232.65	20.48	7.96	0.02	0.09
SL	7.26	1.99	4.60	30.94	470.63	31.73	0.43	0.20
F	11.14	0.89	1.46	9.78	28.16	576.48	0.30	0.25
WL	0.07	0.03	0.02	0.01	0.33	0.12	2.32	0.17
WB	0.11	0.01	0.60	0.07	0.25	0.31	0.15	4.17

Area in km²

Once again, landscape change between 2005 and 2010 was dominated by an increase in grassland cover (46.23%), most of which was the continued conversion out of agriculture (62% of lost agricultural land became grassland), leaving only Maui's central isthmus in sustained, commercial production (Table 9; Fig. 6c). Built-up land cover increased as the Kīhei continued to fill in along the southern shore and a new highway was developed to serve as an artery from the airport in Kahului. Urban areas also expanded near Wailuki in western central Maui, taking over land that had been in agriculture or grassland. Total shrubland cover gained areas along the slopes of Haleakalā due to the conversion out of agriculture, but overall declined by almost 7% from the previous period. Some of the added grassland cover was from bare land (33.1%) and shrubland (40.1%) losses. Although they made up the smallest total area on Maui, this period saw the largest changes in wetland and water body cover, which transitioned into forest and shrubland in some areas. Meanwhile, new wetlands and water bodies were converted from forest, shrubland, and bare soil along Maui's central north and southern shores. Island-wide changes totaled 5.5%.

Between 1992 and 2010, based on the NOAA C-CAP land cover product for Maui, island-wide agriculture cover declined by 31.43% as it transitioned primarily

Table 10 Transition matrix for the NOAA C-CAP data product, 1992–2010

	2010							
1992	BU	AG	BL	GL	SL	F	WL	WB
BU	77.56	5.56	1.70	10.58	9.74	13.80	0.10	0.21
AG	13.24	172.13	3.77	54.35	16.74	4.49	0.06	0.17
BL	0.96	0.05	80.03	1.83	3.24	0.42	0.03	0.58
GL	5.49	1.21	2.59	211.21	20.38	8.55	0.02	0.15
SL	8.23	1.82	5.13	30.40	460.70	28.63	0.45	0.23
F	8.57	0.88	1.64	9.32	29.70	571.36	0.36	0.35
WL	0.06	0.03	0.02	0.00	0.35	0.10	2.15	0.16
WB	0.11	0.00	0.59	0.07	0.16	0.16	0.16	4.05

Area in km^2

into grassland, built-up areas, and shrubland (Table 10). The greatest increases belonged to grassland (27.31%). Total built-up area actually declined by 4.22% overall, but substantial new developed areas appeared in coastal west Maui, central Maui and Kīhei, and new developments in upcountry along the leeward slopes of Haleakalā (Fig. 6d). Wetlands and water bodies increased by 15.98% and 9.13%, respectively, relatively recent transitions. Forest cover remained remarkably unchanged throughout the whole time period, with the only substantial gains occurring between 1992 and 2001, along leeward Haleakalā. The total change in Maui land cover from 199 2010 was 177.33 km^2, or 9.4%.

Landfire EVT Land Cover Change (2001–2014)

Landfire EVT land cover classifications for Hawai'i were available for 5 years (2001, 2008, 2010, 2012, and 2014). Due to the short time-steps between the 2008, 2010, and 2012 classifications, only transitions from 2001, 2008, and 2014 were analyzed for this chapter. Figure 7 summarizes the changes in area and relative percent, island-wide.

Compared to NOAA C-CAP, substantially fewer changes to land cover from 2001 to 2014 were represented by the Landfire EVT product and only affected the built-up, agriculture, grassland, shrubland, and forest land cover classes. No changes to bare soil, wetlands, or water bodies were mapped. From 2001 to 2008, 12.1% of built-up area was lost, the majority of which was converted to grassland (37.52%) and forest (31.13%; Table 11). The forest transitions were highly patchy, however, and total forest increases were only 2.92 km^2, or 0.41%. All changes in shrubland and forest land cover, on the other hand, converted to grassland, contributing an additional 5 km^2 to the total increase of 9.5 km^2. Most of these shifts occurred along the southern edge of west Maui and the leeward middle and upper elevations of Haleakalā (Fig. 8a). Total changes to land cover island-wide affected 30.57 km^2, or 1.62%.

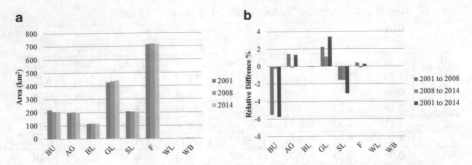

Fig. 7 (**a**) Area for each land cover class according to the Landfire EVT data product, (**b**) relative percent change for each time period and overall

Table 11 Transition matrix for the Landfire EVT data product, 2001–2008

	2008							
2001	BU	AG	BL	GL	SL	F	WL	WB
BU	207.72	2.87		4.53	0.91	3.75		
AG		199.26						
BL			112.82					
GL				427.30				
SL				4.14	204.31			
F				0.84		713.50		
WL							0.95	
WB								4.91

Area in km²

Fewer changes occurred between 2008 and 2014, according to the Landfire EVT data (Table 12). In fact, only grassland cover increased due to losses to agriculture, built-up area, shrubland, and forest. New grassland areas continued the expansion previously observed along southern west Maui, with a few large patches appearing in northwest Maui and along the southern coast of east Maui (Fig. 8b). No other transitions occurred during this period and land cover changes affected only 9.82 km² (0.52%) of the island.

Over the 14-year period from 2001 to 2014, changes in land cover were dominated by an almost 15% increase in grassland at the expense of built-up, agriculture, shrubland, and forest cover (Table 13). Of the relative losses in those categories, 100% of the transitions out of agriculture, shrubland, and forest were converted to grassland, while 40.4% of the built-up losses became grassland. Built-up area was also lost to forest (29.8%), agriculture (22.7%), and shrubland (7.12%). Maui land cover changes impacted a total of 38.1 km², 2% of island area.

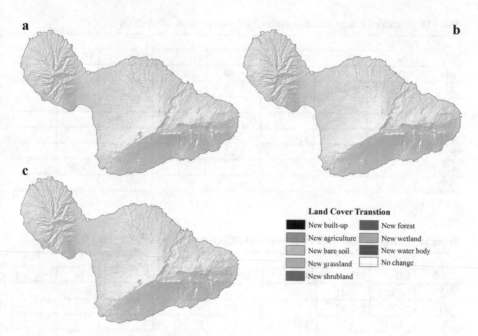

Fig. 8 Landfire EVT land cover changes at the end of each time period from (**a**) 2001–2008; (**b**) 2008–2014, and (**c**) 2001–2014

Table 12 Transition matrix for the Landfire EVT data product, 2008–2014

	2014							
2008	BU	AG	BL	GL	SL	F	WL	WB
BU	207.18			0.53				
AG		201.93		0.21				
BL			112.82					
GL				436.80				
SL				3.24	201.99			
F				0.97		716.29		
WL							0.95	
WB								4.91

Area in km²

Hawai'i GAP and CAH Land Cover Change (2005–2014)

The companion 2005 Hawai'i GAP and 2014 CAH datasets made up the third multi-temporal land cover product evaluated for this chapter. Changed detection revealed that only built-up and agriculture cover increased in area over the 10-year period, while all other classes declined (Fig. 9). Built-up cover increased dramatically, by 90.4 km² (138.8%). Those increases came from all of the other land cover types, namely transitions out of forest (26.1 km²), grassland (30.1 km²), and agriculture

Table 13 Transition matrix for the Landfire EVT data product, 2001–2014

2001	2014							
	BU	AG	BL	GL	SL	F	WL	WB
BU	207.18	2.86		5.09	0.90	3.75		
AG		199.07		0.20				
BL			112.82					
GL				427.30				
SL				7.36	201.09			
F				1.80		712.54		
WL							0.95	
WB								4.91

Area in km^2

Table 14 Transition matrix for the Hawai'i GAP/CAH data product, 2005–2014

2005	2014							
	BU	AG	BL	GL	SL	F	WL	WB
BU	63.70	1.41						
AG	19.86	242.43						
BL	7.47	1.26	155.79					
GL	30.08	4.76		320.84				
SL	7.79	0.85			227.40			
F	26.12	14.87				754.33		
WL	0.00						3.58	
WB	0.46	0.35						3.79

Area in km^2

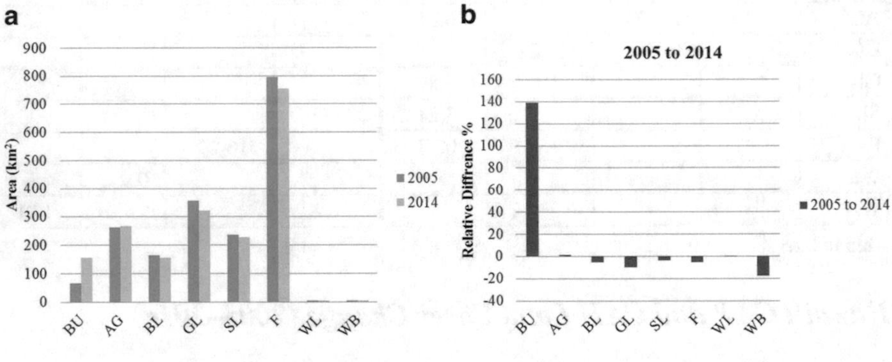

Fig. 9 (a) Area for each land cover class according to the Hawai'i GAP/CAH data product, (b) relative percent change for each time period and overall

(19.9 km^2; Table 14). New urban areas appeared along the previously developed coastal areas of Lahaina in west Maui, and Kahului/Kīhei in central Maui, as well as upcountry Maui along the slopes of Haleakalā (Fig. 10). Agriculture was the only

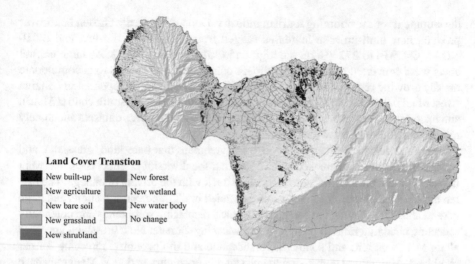

Land Cover Transtion

New built-up   New forest

New agriculture   New wetland

New bare soil   New water body

New grassland   No change

New shrubland

Fig. 10 Hawai'i GAP and CAH land cover changes from 2005–2014

other land cover class that increased in area (by 3.63 km², or 1.38%), and was the result of conversions out of forest (14.9 km²), grassland (4.76 km²), and built-up (1.41 km²) cover. New agricultural land cover appeared largely in central Maui where sugarcane has dominated the landscape for over a century, but small patches of grassland along the western and eastern coastlines also transitioned into agriculture, as well as the higher elevations of upcountry east Maui. Although small in size, a substantial percentage of water body cover was lost (17.6%), split almost evenly between the built-up and agriculture land cover classes. Water body areas that were claimed by development were found along central Maui's northern and southern shores, whereas water body conversions to agriculture mostly took place inland in central Maui. The total area that underwent land cover change on Maui from 2005 to 2014 was 188 km², or 10%.

Discussion

Maui has undergone dramatic changes in land cover over time, which was evident from the four decades of classification data that were analyzed for this chapter. The observed changes in agriculture and built-up land cover were most representative of the documented drivers of change. Consistent with the dramatic decline in sugarcane and pineapple production statewide at the end of the twentieth century, Maui's agricultural cover decreased by around 50% between 1976 and the most recent land cover classifications. Some new agricultural areas appeared in upcountry Maui (leeward Haleakalā) as a result of the transition toward boutique coffee, pineapple, and macadamia nut crops, as well as diversified agriculture for local food. Most of the contraction occurred in west Maui, however, which underwent significant urban development for the booming tourism industry. Depending on the chosen land cover product, new built-up area increases ranged from 105.5% on the low end (2010 NOAA C-CAP) to 272.7% on the high end (2014 Landfire EVT). As large coastal areas were converted to homes, apartment complexes, and resorts to accommodate rapidly growing resident and visitor populations, built-up areas appeared in Lahaina (west Maui), Wailuku and Kahului (north central Maui), Kīhei (south central Maui), and upcountry (leeward Haleakalā). All three current land cover datasets adequately represented these island-wide trends.

The three current products were also in agreement that bare land, grassland, and water body cover has increased since 1976, but the diversity of subclasses that went into each product made this more difficult to track with the drivers of change. The barren areas near the summit of Haleakalā expanded in all three classifications and large new patches spread along the southeastern coast, displacing shrublands and potentially reflecting changing rainfall patterns and a warming climate. New grassland appeared along Maui's eastern and southeastern shoreline and in upcountry, replacing former shrubland. Agricultural conversion to grassland in upcountry and west Maui may have been the consequence of fields going fallow, or conversion to ranching over time.

Some land cover transitions were difficult if not impossible to relate to the known drivers of change on Maui, however, and this highlighted the uncertainty inherent in attempting to compare multiple classification data products. While the NOAA C-CAP dataset showed an increase in shrubland cover from the 1970s, the Landfire EVT and CAH datasets showed a decline. The opposite was true for forest cover, with NOAA C-CAP showing losses in forest after 1976 and Landfire and CAH showing gains. NOAA C-CAP and CAH revealed increases in wetland cover, while Landfire EVT recorded a loss. How these shifts were recorded and why they show such divergent trends across datasets should be subjects of careful consideration by future users, especially if the focus is watershed conservation or wetland management. In addition, neither the 1976 classification nor the NOAA C-CAP datasets differentiated between native and introduced plant species. What constitutes a wetland also varied from one product to the next. Finally, it is likely that in the more complex ecosystems near the coastal areas, no single land cover class would be adequate even at the 30 m spatial resolution of all but the most current datasets.

Compared to transitions from the 1970s to present day, decadal and intra-decadal trends were far more difficult to distinguish using the available land cover classifications for Maui. Of the three available products, NOAA C-CAP data from 1992 to 2010 reflected the most change, which largely continued trends that began in the 1950s and 60s (agricultural declines and increases in built-up and grassland area). The 17% increases in built-up area that occurred between 1992 and 2001, meanwhile, had been reversed by 2005 (a 19% loss), which was in some agreement with transitions observed using the Landfire EVT data from 2001 to 2008 (a 5.5% loss). Generally speaking, however, the Landfire EVT products showed very few changes, including almost imperceptible increases in agriculture (<2%) that did not appear to reflect the losses to that sector over the last 20 years. The 2005 GAP and 2014 CAH datasets showed a 139% increase in built-up area over a 10-year period, a 1% increase in agriculture, and declines in all other land cover, including grassland.

The dramatic differences in landscape change represented by these three data products once again underscores the need to acknowledge and account for uncertainty, which could be aided by the use of very high resolution imagery and other geospatial data sources for validation.

Conclusions

In this assessment, the island of Maui was selected as a case study to describe the drivers of land cover change and assess the applicability of available land cover products for spatial analysis. Four products were selected to compare land cover change through time: a 1976 land cover classification for the state, NOAA's C-CAP land cover classifications (1992, 2001, 2005, and 2010), the Landfire EVT product (2000, 2008, and 2014), and the Hawai'i GAP and CAH datasets (2005 and 2014). Trajectories of change and transition matrices were calculated between the 1976 land cover dataset and the latest date for each multi-temporal product (e.g. 2010 NOAA C-CAP, 2014 Landfire EVT 2014, and 2014 CAH). Consecutive dates within the same multi-temporal product were also used for change detection.

The available multi-temporal land cover products for Hawai'i reflected very different trends through time on Maui. Considering the drivers of land cover change that were described above, NOAA C-CAP offers the most comprehensive spatial product for tracking change. It has a number of advantages over the Landfire EVT and Hawai'i GAP/CAH products. First, classifications from 2005 and onward have a higher spatial resolution (2.4 m) and are based on Quickbird or WorldView imagery, making them more applicable in diverse, rapidly-changing ecosystems like Hawai'i's coastal areas. Second, classes are standardized across time, with the exception of built-up land cover (open space and low-, medium-, and high-intensity development) that was simplified to 'impervious surface' beginning in 2005. Third, like Landfire EVT, it is updated on a semi-regular basis as new cloud-free imagery becomes available. Unlike Landfire EVT that updates only portions of the dataset at a time, however, classification is performed for the entire image mosaic at each time point using training data collected within 1–2 years of image capture. Designed with ecosystems management and change detection in mind, NOAA C-CAP is currently the only high-resolution, internally-consistent product that is updated at regular intervals.

Given the limited number of landscape-level land cover products for the main Hawaiian Islands and the discrepancies between them, however, this analysis also underscored the fact that ancillary spatial information is needed to better characterize land cover change and its drivers. The state of Hawai'i is fortunate to possess an extensive, high-resolution remote sensing image catalog that is free for most users. The Hawai'i Statewide GIS Program (planning.hawaii.gov/gis/) contains the efforts of multiple agencies to consolidate the available geospatial data and promote its use for planning, research, and management. This data can be combined with population, tourism, and economic census data to spatially render the drivers of land cover change, such as zoning district boundaries, development plans, rainfall and temperature data, critical habitats, ungulate fencing, watershed and conservation zones, wildfire footprints, important agricultural lands, and species habitats and distributions. As newer imagery and data become available, they will provide further opportunities to refine existing products, reduce uncertainty, and generate new information for tracking land cover change in Hawai'i.

References

Anderson, J. R., Hardy, E. E., Roach, J. T., & Witmer, R. E. (1976). A land use and land cover classification system for use with remote sensor data, Rep. No. 964. Land Cover Institute, Reston.

Anderson, T. R., Fletcher, C. H., Barbee, M. M., Frazer, L. N., & Romine, B. M. (2015). Doubling of coastal erosion under rising sea level by mid-century in Hawai'i. *Natural Hazards, 78,* 75–103.

Asner, G. P., Hughes, R. F., Mascaro, J., Uowolo, A. L., Knapp, D. E., Jacobson, J., … Clark, J. K. (2011). High-resolution carbon mapping on the million-hectare Island of Hawai'i. *Frontiers in Ecology and the Environment, 9,* 434–439.

Asner, G. P., Mascaro, J., Muller-Landau, H. C., Vieilledent, G., Vaudry, R., Rasamoelina, M., … van Breugel, M. (2012). A universal airborne LiDAR approach for tropical forest carbon mapping. *Oecologia, 168,* 1147–1160.

Bartholomew, D. P., Hawkins, R. A., & Lopez, J. A. (2012). Hawai'i pineapple: The rise and fall of an industry. *HortScience, 47,* 1390–1398.

Benning, T. L., LaPointe, D., Atkinson, C. T., & Vitousek, P. M. (2002). Interactions of climate change with biological invasions and land use in the Hawaiian Islands: Modeling the fate of endemic birds using a geographic information system. *Proceedings of the National Academy of Sciences of the United States of America, 99,* 14246–14249.

Blackford, M. (2001). *Fragile paradise: The impact of tourism on Maui, 1959–2000.* Lawrence: University of Kansas Press.

Brewington, L., Keener, V., Finucane, M., & Eaton, P. (2017). Participatory scenario planning for climate change adaptation using remote sensing and GIS. In S. J. Walsh (Ed.), *Remote sensing for societal benefits* (pp. 236–252). Amsterdam: Elsevier.

Clague, D., & Dalrymple, G. (1989). Tectonics, geochronology, and origin of the Hawaiian emperor volcanic chain. In E. Winterer, D. Hussong, & R. Decker (Eds.), *The eastern Pacific Ocean and Hawai'i* (pp. 188–217). Boulder: The Geological Society of America.

Cooper, G., & Daws, G. (1985). *Land and power in Hawai'i: The Democratic years.* Honolulu: University of Hawai'i Press.

Crausbay, S. D., Frazier, A. G., Giambelluca, T. W., Longman, R. J., & Hotchkiss, S. C. (2014). Moisture status during a strong El Niño explains a tropical montane cloud forest's upper limit. *Oecologia, 175,* 273–284.

Cuddihy, L., & Stone, C. (1990). *Alteration of native Hawaiian vegetation: Effects of humans, their activities and introductions.* Honolulu: University of Hawai'i Cooperative National Park Resources Studies Unit.

D'iorio, M., Jupiter, S. D., Cochran, S. A., & Potts, D. C. (2007). Optimizing remote sensing and GIS tools for mapping and managing the distribution of an invasive mangrove (Rhizophora mangle) on South Moloka'i, Hawai'i. *Marine Geodesy, 30,* 125–144.

Ellsworth, L. M., Litton, C. M., Dale, A. P., & Miura, T. (2014). Invasive grasses change landscape structure and fire behaviour in Hawai'i. *Applied Vegetation Science, 17,* 680–689.

Elmore, A. J., Asner, G. P., & Hughes, R. F. (2005). Satellite monitoring of vegetation phenology and fire fuel conditions in Hawaiian drylands. *Earth Interactions, 9,* 1–21.

Fletcher, C., Mullane, R., & Richmond, B. (1997). Beach loss along armored shorelines on Oahu, Hawaiian Islands. *Journal of Coastal Research, 13*, 209–215.

Fletcher, C., Rooney, J., Barbee, M., Lim, S. C., & Richmond, B. (2003). Mapping shoreline change using digital orthophotogrammetry on Maui, Hawai'i. *Journal of Coastal Research, Special Issue, 38*, 106–124.

Fordham, D. A., & Brook, B. W. (2010). Why tropical island endemics are acutely susceptible to global change. *Biodiversity and Conservation, 19*, 329–342.

Fortini, L. B., Vorsino, A. E., Amidon, F. A., Paxton, E. H., & Jacobi, J. D. (2015). Large-scale range collapse of Hawaiian forest birds under climate change and the need for 21st century conservation options. *PLoS One, 10*, e0140389.

Giambelluca, T., Chen, Q., Frazier, A., Price, J., Chen, Y., Chu, P., ... Delparte, D. (2013). Online rainfall atlas of Hawai'i. *Bulletin of the American Meteorological Society, 94*, 313–316.

Gon III, S. (2006). *The Hawai'i gap analysis project final report*. University of Hawai'i, Research Corporation of the University of Hawai'i, Honolulu.

Hawai'i Department of Land and Natural Resources. (2015). *Hawai'i state wildlife action plan (SWAP)*. Honolulu: H.T. Harvey and Associates.

Hawai'i State GIS. (2019). Land use land cover of main Hawaiian Islands as of 1976. Retrieved April 19, 2019, from https://geoportal.Hawai'i.gov/datasets/e00b356bcc9d4fabb6e07d6319a7b543_11

Hawai'i Tourism Authority. (2017). *2017 annual visitor research report*. Honolulu: Hawai'i Tourism Authority.

Hay, J. E., Forbes, D. L., & Mimura, N. (2013). Understanding and managing global change in small islands. *Sustainability Science, 8*, 303–308.

Hiatt, W. (1993). Hawai'i: Growth, government, and economy. *Journal of Urban Planning and Development, 119*, 97–115.

Jacobi, J. D., Price, J., Fortini, L. D., Gon III, S., & Berkowitz, P, (2017a). *Carbon assessment of Hawai'i*. US Geological Survey data release, https://doi.org/10.5066/F7DB80B9.

Jacobi, J. D., Price, J., Fortini, L. B., Gon, S., III, & Berkowitz, P. (2017b). Baseline land cover. In P. Selmants, C. P. Giardina, J. D. Jacobi, & Z. Zhu (Eds.), *Baseline and projected future carbon storage and carbon fluxes in ecosystems of Hawai'i* (US geological survey professional paper 1834) (pp. 9–20). Reston: US Geological Survey.

Juvik, S., & Juvik, J. O. (1998). *Atlas of Hawai'i* (3rd Rev. ed.). Honolulu: University of Hawai'i Press.

Keener, V. W., Helweg, D. A., Asam, S., Balwani, S., Burkett, M., Fletcher, C., ... Tribble, G. (2018). Ch. 27: Hawai'i and U.S. affiliated Pacific Islands. In *Impacts, risks, and adaptation in the United States: Fourth national climate assessment* (Vol. II, pp. 1242–1308). Washington, DC: US Global Change Research Program.

Kerr, S. A. (2005). What is small island sustainable development about? *Ocean & Coastal Management, 48*, 503–524.

Kier, G., Kreft, H., Lee, T. M., Jetz, W., Ibisch, P. L., Nowicki, C., ... Barthlott, W. (2009). A global assessment of endemism and species richness across island and mainland regions. *Proceedings of the National Academy of Sciences, 106*, 9322–9327.

Kirch, P. (1982). The impact of the prehistoric Polynesians of the Hawaiian ecosystem. *Pacific Science, 36*, 1–14.

Kirch, P. (2010). *How chiefs became kings: Divine kingship and the rise of archaic states in ancient Hawai'i*. Berkeley: University of California Press.

Kirch, P., Coil, J., Hartshorn, A., Jeraj, M., Vitousek, P. M., & Chadwick, O. (2005). Intensive dryland farming on the leeward slopes of Haleakala, Maui, Hawaiian Islands: Archaeological, archaeobotanical, and geochemical perspectives. *World Archaeology, 37*, 240–258.

Kirch, P. V., Holson, J., & Baer, A. (2009). Intensive dryland agriculture in Kaupō, Maui, Hawaiian Islands. *Asian Perspectives, 48*, 265–290.

Krushelnycky, P. D., Loope, L. L., Giambelluca, T. W., Starr, F., Starr, K., Drake, D. R., ... Robichaux, R. H. (2013). Climate-associated population declines reverse recovery and threaten future of an iconic high-elevation plant. *Global Change Biology, 19*, 911–922.

Kueffer, C., Daehler, C. C., Torres-Santana, C. W., Lavergne, C., Meyer, J.-Y., Otto, R., & Silva, L. (2010). A global comparison of plant invasions on oceanic islands. *Perspectives in Plant Ecology, Evolution and Systematics, 12*, 145–161.

Kurashima, N., & Kirch, P. (2011). Geospatial modeling of pre-contact Hawaiian production systems on Molokaʻi Island, Hawaiian Islands. *Journal of Archaeological Science, 38*, 3662–3674.

Ladefoged, T. N., & Graves, M. W. (2008). Variable development of dryland agriculture in Hawaiʻi: A fine-grained chronology from the Kohala field system, Hawaiʻi Island. *Current Anthropology, 49*, 771–802.

Landfire. (2019a). *Biophysical Settings Layer*. US Department of Agriculture and US Department of the Interior. Retrieved July 9, 2019, from https://www.landfire.gov/evt.php

Landfire. (2019b). *Existing Vegetation Type*. US Department of Agriculture and US Department of the Interior. Retrieved July 9, 2019, from https://www.landfire.gov/evt.php

Leuschner, C., & Schulte, M. (1991). Microclimatological investigations in the tropical alpine scrub of Maui, Hawaiʻi: Evidence for a drought-induced alpine timberline. *Pacific Science, 45*, 152–168.

Lincoln, N., & Ladefoged, T. (2014). Agroecology of pre-contact Hawaiian dryland farming: The spatial extent, yield and social impact of Hawaiian breadfruit groves in Kona, Hawaiʻi. *Journal of Archaeological Science, 49*, 192–202.

Maclennan, C. (2007). An introduction to WAI: Indigenous water, industrial water in Hawaiʻi. *Organization and Environment, 20*, 497–505.

Margriter, S. C., Bruland, G. L., Kudray, G. M., & Lepczyk, C. A. (2014). Using indicators of land-use development intensity to assess the condition of coastal wetlands in Hawaiʻi. *Landscape Ecology, 29*, 517–528.

Miller, T. L., & Fletcher, C. H. (2003). Waikiki: Historical analysis of an engineered shoreline. *Journal of Coastal Research, 19*, 1026–1043.

Morales, R. M., Miura, T., & Idol, T. (2008). An assessment of Hawaiian dry forest condition with fine resolution remote sensing. *Forest Ecology and Management, 255*, 2524–2532.

National Park Service. (2019). Haleakala National Park. National Park Service. Retrieved September 2, 2019, from https://www.nps.gov/hale/index.htm

NOAA Office for Coastal Management. (2018). *C-CAP land cover atlas. Coastal change analysis program (C-CAP) high-resolution land cover*. NOAA Office for Coastal Management, Charleston. Retrieved November, 19, 2018, from https://coast.noaa.gov/digitalcoast/tools/lca.html

Pelling, M., & Uitto, J. I. (2001). Small island developing states: Natural disaster vulnerability and global change. *Global Environmental Change Part B: Environmental Hazards, 3*, 49–62.

Perroy, R. L., Melrose, J., & Cares, S. (2016). The evolving agricultural landscape of post-plantation Hawaiʻi. *Applied Geography, 76*, 154–162.

Ponette-Gonzalez, A. G., Marin-Spiotta, E., Brauman, K. A., Farley, K. A., Weathers, K. C., & Young, K. R. (2014). Hydrologic connectivity in the high-elevation tropics: Heterogeneous responses to land change. *Bioscience, 64*, 92–104.

Rhodes, D. (2001). Changes in the sandalwood trade of Hawaiʻi. US National Park Service. Retrieved from https://www.nps.gov/parkhistory/online_books/kona/history5e.htm

Rollins, M. G. (2009). LANDFIRE: A nationally consistent vegetation, wildland fire, and fuel assessment. *International Journal of Wildland Fire, 18*, 235–249.

Scott, J. M., Davis, F., Csuti, B., Noss, R., Butterfield, B., Groves, C., … Wright, R. G. (1993). Gap analysis: A geographic approach to protection of biological diversity. *Wildlife Monographs, 123*, 3–41.

Spatz, D. R., Newton, K. M., Heinz, R., Tershy, B., Holmes, N. D., Butchart, S. H. M., & Croll, D. A. (2014). The biogeography of globally threatened seabirds and island conservation opportunities. *Conservation Biology, 28*, 1282–1290.

Stauffer, R. H. (2004). *Kahana: How the land was lost*. Honolulu: University of Hawaiʻi Press.

Suryanata, K. (2002). Diversified agriculture, land use, and agrofood networks in Hawaiʻi. *Economic Geography, 78*, 71–86.

Tanji, M. (2016). Operations winding down at HC&S. *The Maui News*, Wailuku.

US Census Bureau. (2010). *Census 2010*. US Census Bureau, Washington. Retrieved from https://www.census.gov/2010census/

Walsh, S. J., Brewington, L., Shao, Y., Laso, F., Bilsborrow, R. E., Nazario, J. A., … Pizzitutti, F. (this volume). Social-ecological drivers of land cover/land use change on islands: A synthesis of the patterns and processes of change. In *Land cover and land use change on islands: Social and ecological threats to sustainability*. Springer.

Threats of Climate Change in Small Oceanic Islands: The Case of Climate and Agriculture in the Galapagos Islands, Ecuador

Carlos F. Mena, Homero A. Paltán, Fatima L. Benitez, Carolina Sampedro, and Marilú Valverde

Introduction

Islands are among the most vulnerable to climate change and climate variability (Allen et al., 2014), due these are land masses that evolved in isolation surrounded by large bodies of water, sustaining fragile and complex socio-environmental systems, which have been shaped by the effects of ocean currents, ocean-borne storm systems, tidal fluctuations, and other oceanic phenomenon (Shea, 2003). Thus, characteristics such as the long coastlines relative to land areas, makes islands particularly susceptible to natural hazard (Campbell, 2020). Additionally, having a restricted size limit the access to natural and social resource (Connell, Lowitt, Saint Ville, & Hickey, 2019), which, added to big distance to the continent and the lack of connection to the social and economic dynamics that this entails, makes islands systems more sensitives to changes.

Specifically, Islands in the Pacific Ocean are particularly vulnerable as they sit in the heart-beat of El Niño Southern Oscillation (ENSO) (Barnett, 2011) considered the largest phenomenon of inter-annual variability of the global climate system, with a recurrence period, between 2 to 8 years (dos Santos Coelho & Ambrizzi, 1999). This phenomenon expose the pacific islands to extreme changing climatic conditions such as increment/decrement of sea surface temperature and salinity,

The original version of this chapter was revised. The correction to this chapter is available at https://doi.org/10.1007/978-3-030-43973-6_14

C. F. Mena (✉) · H. A. Paltán · F. L. Benitez · C. Sampedro · M. Valverde
Institute of Geography, Universidad San Francisco de Quito, Quito, Ecuador

Galapagos Science Center, Puerto Baquerizo Moreno, Ecuador
e-mail: cmena@usfq.edu.ec

increment/decrement of sea level and wave activity, increment/decrement of air temperature and amount of ultra violet radiation reaching the surface of the earth, change in the rainfall and evaporation patterns, with direct and indirect impacts on the socio-environmental systems of the islands (Santos, 2006). And these phenomenon is expected to increase in frequency and intensity because of climate change effects (Barnett, 2011).

Climate change impacts are not the same in every island, thus some islands may be more exposed than others and different islands will be exposed to different effects (Campbell, 2020). Some islands are affected dramatically by repeated and every time more intense natural hazards, while other islands experience more subtle climatic events, such as prolonged periods of drought and more extreme precipitation patterns, that could trigger crises in medium and long term (Barnett, 2011; Barnett et al., 2015). Although different impacts are expected from both, subtle climatic events like a dry spell, might over a period of years, set in motion a host of interrelated problems that is nearly as costly on many fronts as any individual disaster, for example a cyclone, that strikes in a matter of hours (Allen et al., 2014). However, the magnitude of the damage will depend greatly on the climate scenario and the adaptation potential of the socio-environmental system (Rosegrant & Cline, 2003). A specific stressor, across different islands is the influx of touristic flows that pressure infrastructure, emergency response procedures, and more importantly food supply systems.

An example of the potential impacts of climate change, tourism pressure and agricultural land degradation is the Galapagos Archipelago, which is considered one of the most iconic islands in the world. Its high endemism and ecological fragility make it a highly valued ecosystem for natural conservation purpose. Human-related pressures such as invasive species, community development, tourism growth and changes in the local food consumption (Sampedro, Pizzitutti, Quiroga, Walsh, & Mena, 2018) are constantly compromising these characteristics as well as the sustainable development in the islands.

Furthermore, the effects of the climate change are potentially devastating for the Islands in the Pacific region like Galapagos, increasing its vulnerability on diverse sectors such as agriculture, biodiversity, health, economy, and natural resources. Strong local human-related pressures would interact with climate change impacts and increase threats. The understanding of this interaction and the potential impacts of a changing climate on conservation and food security outcomes is key to understand the types of responses needed.

Determinants of the Climate in the Galapagos Islands

Given the Galapagos Islands location, ~1000 km off the coast of Ecuador, a complex interplay of ocean currents and winds influence their climate. These mechanisms are mainly governed by interactions of the Inter-Tropical Convergence Zone (ITCZ) as well as El Niño Southern Oscillation (ENSO) (Chavez & Brusca, 1991; Houvenaghel, 1974; Sachs & Ladd, 2010). Primary, the ITCZ migration influences the main bi-seasonal characteristics of currents and winds in the Island whereas ENSO regulates yearly-decadal fluctuations (Hamann, 1979, 1985; Hartten & Gage, 2000).

As such, from June to December when the ITCZ is northwards from the Galapagos, the southeast trade winds bring to the Islands cold air upwelled from southern regions of the Pacific (Koutavas & Lynch-Stieglitz, 2004; Trueman & d'Ozouville, 2010). This results in cool Sea Surface Temperatures (SST) and cooler conditions in the Islands. Yet, between January and May when the ITCZ shifts to the south, such winds are reduced bringing warm ocean currents from the north Pacific and hot conditions prevail. As such, the ITCZ dictates the regular characteristics of the Islands' hot/wet and cold/dry seasons. Precipitation during these seasons is also associated with altitude and topographic conditions in the islands. In the hot season, precipitation increases with altitude, whereas the cold season creates conditions which favor condensation resulting in drizzles and fogs above certain altitude thresholds while the lowlands remain dry (Trueman & d'Ozouville, 2010). Nonetheless, this general hot-cold nomenclature should be taken cautiously since the highlands may also show very wet during cool seasons and the low lands may also be very wet during cold ones.

Moreover, ENSO alters Pacific SSTs which in turn disrupt ITCZ conditions. During ENSO's warm phase, or El Niño high sea temperature in the eastern Pacific displace the ITCZ which results in more intense rainfall and hot seasons in the Galapagos (Snell & Rea, 1999). On the other hand, During ENSO's cold phase, or La Niña, the Islands experience abnormal colder seasons as well as droughts. In the last decades the most extreme El Niño events include those of 1975–1976, 1982–1983, 1986–1987, 1993–1994 and 1997–1998; additional recent high-rainfall events in 2002 and 2010 were also associate with these conditions. It is also important to note that the high precipitation observed in 2008 was not associated with El Niño conditions (Trueman & d'Ozouville, 2010). Apart from rainfall intensities, ENSO events can also modify seasonal characteristics. For example, El Niño conditions are understood to extend hot seasons by 1 or 2 months affecting as well cold seasons characteristics (Smith & Sardeshmukh, 2000).

Nonetheless, existing observations note that this connection between SSTs and rainfall patterns may not be very straightforward. For instance, by examining decadal records since the 1960s (Trueman & d'Ozouville, 2010) show that there is just a positive correlation between year-to-year SSTs and rainfalls during the hot season, throughout the Islands. On the other hand, cold season rainfall seems to have been particularly constant during the last decades with no linear and rather weak correlation with SST across the Islands. Moreover, in the decades of this new century observations suggest that yearly rainfall has diminished, and hot seasons have become hotter whereas cold seasons have become colder. Yet, Galapagos monthly mean SSTs have shown no pronounced trend and has remained relatively constant (Wolff, 2010). So recent intensifications of seasonal patterns cannot be merely explained by SSTs. Such patterns, in turn may be explained by the observed strong increase in cooling by upwelling within the Humboldt Current System which increasingly has influenced the Islands. As such, it is understood that the Humboldt Current System may also be one of the main drivers of the Galapagos interannual and decadal variability.

Agriculture in the Galapagos Islands

The land area in Galapagos 96.7% (799,771 ha) is area protected by the Galapagos National Park (PNG) and 3.3% (26,245 ha) corresponds to areas of human settlements (urban and rural), of which 19,010 ha (CGREG, 2014) are lands destined for agricultural activities, mostly occupied by natural or cultivated pastures (59%). The productive lands are organized in Agricultural Production Units (APUs), with around 755 UPAs in the four populated islands, distributed 357 in Santa Cruz, 260 in San Cristóbal, 127 in Isabela and 11 in Floreana, where the responsible of the land are 568 men and 187 women (CGREG, 2014).

Historically, along the human colonization processes, agriculture has been marked by intense periods of occupation and abandonment determined by changes in economic development and conservation in the Islands. Initially, enthusiastic colonists carried over-use of agricultural areas, later with the introduction of relative large-scale tourism and aging populations within farming households resulted in the abandonment of local agriculture (DNPG, 2014). The historical evidence of this intermittency shows that the agriculture of the islands has been seen more as an extractive activity and less linked to conservation or sustainable practices.

In terms of biodiversity, more than 640 vascular plants have been introduced by people to the Galapagos, about 90% of them deliberately since the discovery of the islands. This introduction of plant species translates into a rate of arrival and successful establishment of introduced species of 1.3 per year (CGREG, 2016); 90% of these species are considered useful plants (from an agricultural point of view). These include some fruit trees, vegetables and other crops, timber, medicinal and ornamental plants. Allauca, Valverde, and Tapia (2018) identified 147 crops planted in the 178 farms surveyed. Among the most frequent crops were reported: orange (77%), plantain (70%), lemon (59%), banana (55%), tangerine (52%), avocado (47%), cassava (47%), papaya (46%) and pineapple (41%). The number of crops planted ranges from 1 to 44, with an average of 11 crops, per farm. 46.6% of farmers grow between 1 and 10 species, 41% between 11 and 20, 9.5% between 21 and 25, and only 0.6% with 26, 29, 32, 39 and 44 crops.

Important for the biophysical conditions for agriculture, and in addition to the climatic conditions described below, there is a "shadow effect" observed in islands with altitudes above 500 m.a.s.l. This effect is characterized by the presence of greater humidity and precipitation on the slopes facing south, east and southeast, because of the winds that whip from the Pacific are loaded with moisture; leaving other sides with drier and more arid conditions most of the year. In terms of soils, few studies have been carried out for productive purposes, Zehetner (2016) sampled soil from 130 agricultural sites (fields and pastures) taken at a depth of 20 cm, on the four inhabited islands. Results the diversity in relation to soil properties and limitations. Younger soils (Isabela, Floreana and parts of Santa Cruz) are much less weathered and generally have higher pH values and higher contents of cationic macronutrients and micronutrients. On the other hand, older soils (especially in San Cristobal) have lower pH values and nutrient levels have decreased due to leaching.

The data also reveals that some important soil characteristics, such as pH, are strongly influenced by elevation, as the weather becomes wetter and more water is available for leaching of basic cations and the pH decreases with a higher elevation.

In terms of agricultural systems, the Institute of Agricultural Research of Ecuador (*Instituto de Investigaciones Agropecuarias*—INIAP) in describes the socio-economic organization of those families carrying out agricultural activities. INIAP collected information in 208 farms (APU) in the populated islands, through 21 variables related to socio-economic aspects, land use, livestock productivity, use of agricultural technology, pasture management, water availability and use of reservoirs, credit, migration, technical assistance, training of producers, among others. As a result of this study, three groups or clusters of households prevalent in the islands were identified: (a) Households with intermediate surface area dedicated to agriculture and livestock, and with low technology in agriculture, and less technical assistance and training, (b) Households with a smaller area dedicated to agriculture and livestock, with medium technology in agriculture, and greater technical assistance and training and (c) Households with larger area dedicated to agriculture and livestock, with low technology in agriculture, and intermediate technical assistance and training.

Methods: Climate Models for the Galapagos

In this climate trend analysis, we compiled data on mean monthly temperature and mean monthly precipitation changes on Galapagos based on downscaled climate projection. The projections were obtained from Ministerio del Ambiente (MAE, 2016) and they correspond to a public effort to synchronize, at the country level, climate data as part of climate change adaptation and mitigation programs under the United Nations Framework Convention on Climate Change (UNFCCC). These projections in turn are based on four downscaled Global Circulation Models—GCMs from phase 5 of the Coupled Model Intercomparison Project (CMIP5), which were selected of a set of 15 models that better represent climate in the tropical South America region. The four GCMs selected are CSIRO-Mk3-6-0 ($1.86° \times 1.87°$), GISS-E2-R ($2.5° \times 2.0°$), IPSL-CM5A-MR ($2.50° \times 1.26°$), MIROC-ESM ($2.81° \times 2.78°$). The twenty-first century simulations were performed, using the reference period of 1981–2005.

The future climate data are grouped in statistical and dynamical downscaling datasets. In the statistical downscaling approach, the authors (MAE, 2016) used the Bias-Correction method to predict monthly climate variables, considering the four Representative Concentration Pathways scenarios (RCP 2.6, RCP 4.5, RCP 6.0 and RCP 8.5; van Vuuren, Edmonds, Kainuma, et al., 2011). The bias-Correction approach aims to reduce the differences between the statistics of simulated model outputs and local observations, calculating an adjustment factor to correct coarse-scale GCM monthly biases (Walsh, 2011).

In the dynamical downscaling approach, they used the Weather Research and Forecasting (WRF) climate model, version 3.6.1. This model was used to estimate daily and monthly climate variables at regional-local scales, considering two RCP scenarios: RCP 4.5 and RCP 8.5. GCMs data was used to provide initial information to perform WRF simulations in four domains. The last two domains were run with a 10-km horizontal resolution covering the mainland Ecuador and Galapagos.

Finally, a multi-model ensemble was performed to better represent the most likely future conditions, characterizing the feasible ranges and uncertainty of future climate variables. For this purpose, the Ecuadorian Environment Agency applied the Reliability Ensemble Averaging—REA method which takes into account two "re-liability criteria": the performance of the model in reproducing present-day climate and the convergence of the simulated changes across models (Giorgi & Mearns, 2001; Tebaldi & Knutti, 2007).

Future Climate Change in the Galapagos

Projected Future Changes on Galapagos

The Galapagos climate projections include two periods: the historical period (1981–2005) and the future period (2011–2070). For this analysis, the future period was further divided in a multidecadal average for each variable into three times intervals: 2020–2040, 2041–2060, 2061–2070. The statistical and dynamical downscaling projections using multiple models and the ensemble model were analyzed in each period.

Range of Climate Change Projections

In general, the comparison of the results of the four global circulation models gives an initial indication of the range of changes in rainfall and temperature that may be expected in Galapagos relative to the period 2020–2070, given different scenarios of greenhouse gas emissions in the future (Fig. 1). Our results show a consistent trend in the future increase in precipitation and temperature in the islands. The annual mean temperature is projected to increase between 0.5 and 1.35 °C for the near-term future (period 2020–2040); 0.6 and 2.5 °C for the mid-term future (2041–2060); and 0.6 and 3.2 °C for the long-term future (2061–2070). Changes in annual precipitation typically show a future increase of approximately 25% when compared to the baseline period. The most extreme scenarios suggest that the precipitation could increase by 75% for the near-term future, 180% for the mid-term future, and 250% for the period 2061–2070, considering the CSIRO-Mk3-6-0 model.

The REA multi-model ensemble response indicates an increase in temperature in the four RCP scenarios for the period 2020–2070, ranging from 0.8–1.3 °C (RCP

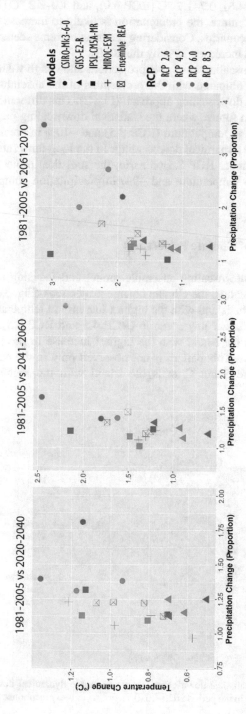

Fig. 1 Average absolute temperature and proportion precipitation changes on Galapagos Archipelago in the period 2020–2070 according to four different greenhouse gas concentration scenarios (RCPs) using four global circulation models and a multi-model ensemble. Reference period is 1981–2005

2.6), 1.1–1.8 °C (RCP 4.5), 0.7–1.7 °C (RCP 6.0), and 1.0–2.2 °C (RCP 8.5). For the mid and long-term future, the precipitation is likely to increase to 50% under RCP 4.5 and RCP 6.0 scenarios. Considering the most extreme scenario (RCP 8.5), the precipitation would increase ~75% by the 2070s.

The dynamical downscaling ensemble projections show high values in temperature and precipitation compared with the projections of the ensemble REA model derived from statistical downscaling approach (Fig. 2). This difference is more significant in the mid-term future, where the statistical downscaling projections show an increase of ~0.5 °C in temperature (RCP 8.5) and ~10% in precipitation (RCP 4.5) compared with the dynamical downscaling. In the long-term future, the difference is more evident under RCP 8.5 scenario, the statistical projections show an increase of ~0.8 °C in temperature and ~9% in precipitation compared with the dynamical approach.

Spatial Distribution of Climate Projections

Using the dynamical downscaling ensemble model and focusing on each of the inhabited Galapagos Islands, the climate change projections (Fig. 3) show that for all periods, Isabela is the island with the highest increase in temperature compared with the other islands under any scenario (RCP 4.5 and RCP 8.5). On the other hand, San Cristobal is the island with the highest increase in precipitation under RCP 8.5 for all periods. This pattern is not observed only in the near-term future under RCP 4.5, where Santa Cruz is the island with the highest increase in

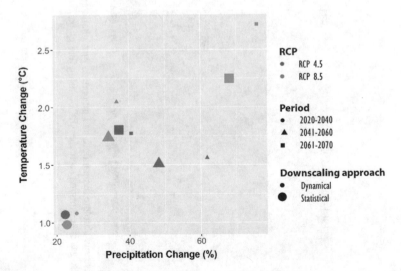

Fig. 2 Ensemble REA statistical downscaling versus ensemble dynamical downscaling projections in Galapagos Islands in the period 2020–2070 according to two greenhouse gas concentration scenarios (RCP 4.5, RCP 8.5)

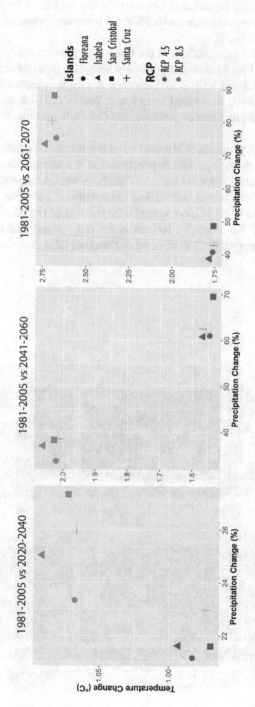

Fig. 3 Average absolute temperature and percentage precipitation changes on each inhabited Galapagos Islands in the period 2020–2070 according to two greenhouse gas concentration scenarios (RCP 4.5, RCP 8.5) using a multi-model ensemble. Reference period is 1981–2005

precipitation. An atypical behavior is observed in the mid-term future, where the precipitation projections are higher under RCP 4.5 scenario compared with RCP 8.5 extreme scenario.

The projected annual precipitation in the three time periods analyzed here using the ensemble model (Fig. 4) shows a gradually increase, from west to east, in the two scenarios (RCP 4.5 and RCP 8.5). However, the highest increase in precipitation (55–70%) is observed in the mid-term future under RCP 4.5, and considering the RCP 8.5, the highest increase in precipitation (75–90%) is observed in the long-term future.

The climate change projections showed similar patterns of change in temperature for the three periods. As opposed to precipitation, the annual mean temperature increases from the east to the west region (Fig. 5). A significant spatial change is observed in the mid-term future, where San Cristobal and Floreana show a lower increase in temperature (~1.4 °C) compared with the rest of the islands under RCP 4.5. Considering RCP 8.5 scenario, Isabela is the only island that show a higher increase in temperature (~2.2 °C) compared to the other islands.

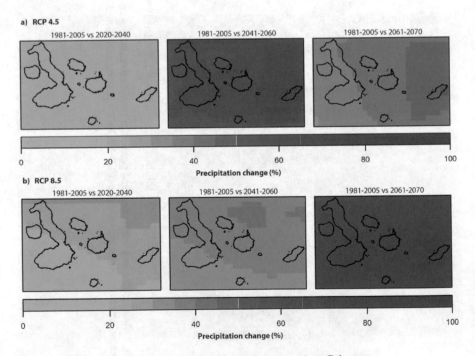

Fig. 4 Spatial distribution of precipitation change (percentage) on Galapagos

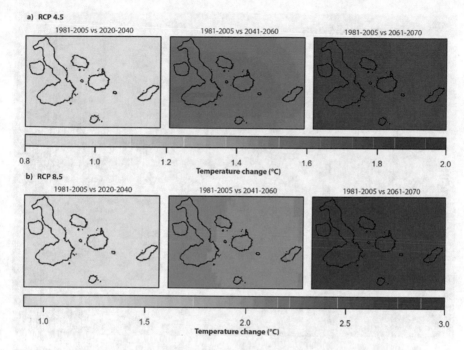

Fig. 5 Spatial distribution of temperature change (°C) on Galapagos

Extreme Precipitation Events

Regarding the precipitation, the frequency and intensity of rainfall events (e.g. El Niño Southern Oscillation—ENSO) are the ones that matter the most, particularly given that the water for agriculture is supplied by rainfall rather than by the irrigation system on the islands. In addition, these events can modify the natural Galapagos landscape, enhancing dispersal of some alien invasive species with high tolerance to environmental conditions (Hellmann, Byers, Bierwagen, & Dukes, 2008). It could reduce the spatial distribution of both native and endemic species, disrupting biological processes.

Based on long-term series of meteorological data of the Ecuadorian Meteorological Service, INAMHI (Instituto Nacional de Meteorología e Hidrología) station "San Cristobal" with data coverage for the period 1981–2018, the average annual precipitation in Galapagos is 507.53 mm/year, where heavy rainfall has been correlated with El Niño occurrences (Fig. 6a). The 1997–1998 ENSO event was a period where the total precipitation was 2541.6 mm in the period July 1997– June 1998.

Future projections of total annual precipitation under the lower emissions scenario (RCP 4.5) and the high emissions scenario (RCP 8.5) are shown in the Fig. 6b. Based on these projections, we expect that heavy precipitation events like the ones observed in the 1997–1998 El Niño, will occur more frequently in the mid-term

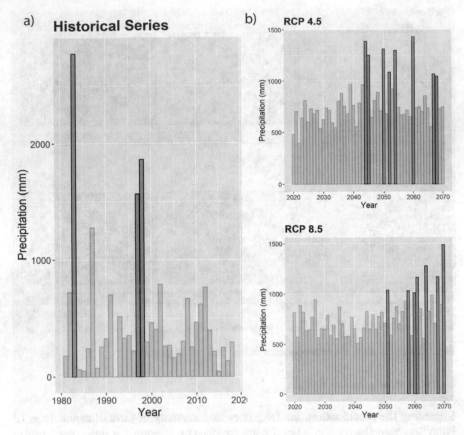

Fig. 6 (**a**) Total annual precipitation in San Cristobal-Galapagos (bars in dark grey represent the very strong El Niño events). Period 1981–2018. (**b**) Projected annual precipitation for the period 2020–2070 relative to the baseline period (1981–2005) on Galapagos, considering two GHG concentration scenarios (RCP 4.5 and RCP 8.5) and using a multi-model ensemble (bars in dark grey represent extreme precipitation events)

future under RCP 4.5. This pattern changes under RCP 8.5, where we expect that heavy precipitation events will occur more frequently in the long-term future.

It is important to consider the cascade of uncertainties derived from downscaling GCM projections, which is a significant constrain for arriving at appropriate adaptation actions (Vermeulen, Challinor, Thornton, et al., 2013; Harris et al., 2014). It is therefore absolutely necessary to conduct a more profound uncertainty analyses with more advanced techniques such as bootstrap analyses, etc.

Conclusion and Discussion: Linkages Between Climate Change and Agriculture in the Galapagos

Future climate scenarios for Galapagos show a consistent trend in the future increase in precipitation and temperature in the islands in the mid- and long-terms. The annual mean temperature is projected to increase between 0.5 and 1.35 °C for the near-term future (period 2020–2040); 0.6 and 2.5 °C for the mid-term future (2041–2060); and 0.6 and 3.2 °C for the long-term future (2061–2070), given different scenarios of greenhouse gas emissions in the future. In terms of the spatial distribution of the impact of climate change, there is a consist trend across models, with Isabela with higher increase in temperature and San Cristobal with higher increase in precipitation for all periods. In terms of intense El Niño events, models indicate that under RCP 4.5 that the frequency will increase in mid-term future scenarios, while for RCP 8.5, these events will be more frequent for the long-term future. Despite that different models, scenarios and downscaling approaches are consistent, there is high degree of uncertainty related to the magnitude and intensity of change, specially for agriculture purposes.

Oceanic islands have a very explicit link between agriculture, tourism and environment, including climate. In Galapagos, local agriculture production, either for subsistence or commercial purposes, is decreasing, while food imports is increasingly (Sampedro et al., 2018). In general, several are the factors identified repeatedly in different islands that limit local production, with strong impacts in human health. For example rural labor decrease due to other source of employment such is the case of tourism or public service, which offer more attractive incomes and less physical effort for young people, turning rural in urban populations; besides, aging farmers (Connell, 2015). Subdivision of land through inheritance that limit the agricultural area for production; the asymmetry in the flow of knowledge and information; the low levels of technology; and the insufficient private and public investment in agricultural production (Saint Ville, Hickey, & Phillip, 2015). Production incentive is decreasing due to the absence of barriers to market entry, a weak marketing infrastructure, unregulated importation, and inadequate food prices; limiting their ability to compete in domestic markets flooded with imported food, incrementing the consumption of processed food, causing health problems on population (Campbell, 2020; Lowitt et al., 2020). As an expected consequence, in Galapagos for example, food security has worsened in recent decades with strong consequences in human health (Thompson, Nicholas, Watson, Terán, & Bentley, 2019; Watson, Thompson, Bentley, & Hopping, 2019).

We see impacts of climate change as a synergistic force to other pressure over agricultural and food supply systems in oceanic islands. Climate change impacts can be divided in two categories, the first related to the physical environment factors and the second to the social, political, and institutional practices (Shea, 2003). In the specific case of agricultural lands in Galapagos, some relevant variables to considered are for example the CO_2 rise related to crop yield; the increase (or decrease) of solar radiation and temperature; the change on frequency and intensity of precipitation (inundation or drought), as a consequence of the latter, poor nutrient availability, invasion of exotic species, and pest interactions; increased contamination of groundwater and estuaries by saltwater incursion; increment in frequency and intensity in natural disasters or ENSO; increase frequency and intensity of cyclones and storm; among others (Barnett, 2011; Hay & Hay, 2003).

These physical impacts to agriculture, along with negative effects of climate change in other compartments of the social and economic system, affect the availability of resources for production such as water and labor; the availability and development of local infrastructure; and most of all the incentive of society to dedicate to agricultural activities. These are the climate change impacts on the social, political and institutional practices, which explicitly in the agricultural system translate on the limited local production capacities due negative effects on crop yield; damage on local infrastructure such as silos or granaries for storing food or seeds, road networks, and farm equipment; influence on agricultural labor dynamics and rural to urban migration; impacts on production in key sectors such as tourism and others, that affect employment and incomes, which in turn could suppress demand for locally grown foods sold in local markets; increase on the costs of inputs relative to the value of production, among others (Campbell, 2020; Pemberton, 2005).

When climate change effects are added to such a fragile agricultural system, negative impacts on agricultural production and on the other components of food security such as access to food and use of food are expected (Tai, Martin, & Heald, 2014). Yet, the way climate change affects the agricultural sector and the general food security in Pacific islands will depend on the adaptability of local agricultural and social systems. Also influenced by globalization, national policies and the socio-cultural changes (Nunn, Aalbersberg, Lata, & Gwilliam, 2014; Saint Ville, Hickey, Locher, & Phillip, 2020). In this regard, adapting and reducing impacts from climate change represent an official priority for all pacific, however, are constrained by the lack of commitment of their own resources to meet these challenges (McNamara & Westoby, 2011). And furthermore, most of the adaptation efforts are based on a top-down planning approach with limited effects, as the identification of impacts and the design of adaptation measures has proven to be more effective using a bottom-up approach (Garnaut, 2008).

Agriculture climate change adaptation initiatives are implemented by regional and government institutions, sometimes supported by external donors, including the developing community land-use plans; trialing new agroforestry and soil stabilization methods; soil management; water management; the distribution of crops varieties with suitable tolerant traits to climate change such as drought, salinity and water-logging, included: taro, sweet potato yam, banana, breadfruit, cassava, potato and vanilla; incentive people to eat more nutritious local foods, while reducing the amount of processed imported food, promoting specially crop included in the program of drought and salt tolerance, pest and disease resistant varieties that are adaptable to changing climate; and undertaking innovative climate adaptation education programs (Wairiu, Lal, & Iese, 2012).

On the other hand, several local adaptation actions are also taking place, for example, in a recent study Kurashima, Fortini, and Ticktin (2019) assess the conservation and restoration potential of indigenous agricultural systems as tools to improve community and landscape resilience in the face of climate change, using a polycultural approach to reduce the likelihood of total crop losses during extreme events, which enhances the spatial shifts of agricultural systems under climate change effects, to determine resilient areas for protection and restoration; aiming to increase food self-sufficiency.

This chapter explore preliminary results of climate change scenarios in the Galapagos, its potential effects on agriculture in the islands and a description of possible adaptation measures. While many the uncertainty in climate is large due the mismatch between the spatial resolution of the models and the set of farms within Galapagos, we can conclude that climate change will be an additional factor within the drivers of food insecurity and food stress in the islands. Fortunately, there is already many proven adaptation strategies that can be used.

References

Allauca, J., Valverde, M., & Tapia, C. (2018). *Conocimiento, Manejo y Uso de la agrobiodiversidad en la Isla San Cristóbal.* INIAP. Boletín Técnico 173. INDIGO 480 Publicaciones. Puerto Baquerizo Moreno, Galápagos-Ecuador. 76p.

Allen, M. R., Barros, V. R., Broome, J., Cramer, W., Christ, R., Church, J. A., ... Dubash, N. K. (2014). IPCC fifth assessment synthesis report-climate change 2014 synthesis report.

Barnett, J. (2011). Dangerous climate change in the Pacific Islands: Food production and food security. *Regional Environmental Change, 11*(1), 229–237.

Barnett, J., Evans, L. S., Gross, C., Kiem, A. S., Kingsford, R. T., Palutikof, J. P., ... Smithers, S. G. (2015). From barriers to limits to climate change adaptation path dependency and the speed of change. *Ecology and Society, 20*(3), 5.

Campbell, J. R. (2020). Development, global change and food security in Pacific Island countries. In *Food security in small island states*. Springer.

CGREG. (2014). *Censo de Unidades de Producción Agropecuaria de Galápagos. Consejo de Gobierno del Régimen Especial de Galápagos.* Puerto Baquerizo Moreno, Galápagos-Ecuador. 138p.

CGREG. (2016). *Plan de Desarrollo Sustentable y Ordenamiento Territorial del Régimen Especial de Galápagos-Plan Galápagos, 2016. Consejo de Gobierno del Régimen Especial de Galápagos.* Puerto Baquerizo Moreno, Galápagos-Ecuador.

Chavez, F. P., & Brusca, R. C. (1991). The Galapagos Islands and their relation to oceanographic processes in the tropical Pacific. In *Galapagos marine invertebrates* (pp. 9–33). Springer.

Connell, J. (2015). Vulnerable Islands: Climate change, tectonic change, and changing livelihoods in the Western Pacific. *The Contemporary Pacific, 27*(1), 1–36.

Connell, J., Lowitt, K., Saint Ville, A., & Hickey, G. M. (2019). Food security and sovereignty in small island developing states: Contemporary. In *Food security in small island states*. Springer.

dos Santos Coelho, C. A., & Ambrizzi, T.. 1999. *Climatological studies of the influences of El Niño Southern Oscillation events in the precipitation pattern over South America during austral summer.* Paper read at Sixth International Conference on Southern Hemisphere Meteorology and Oceanography.

DPNG. 2014. *Plan de Manejo de las Áreas Protegidas de Galápagos para el Buen Vivir. Dirección del Parque Nacional Galápagos.* Puerto Ayora, Galápagos-Ecuador.

Garnaut, R. (2008). *The Garnaut climate change review.* Cambridge: Cambridge University Press.

Giorgi, F., & Mearns, L. (2001). Calculation of average, uncertainty range, and reliability of regional climate changes from AOGCM simulations via the "reliability ensemble averaging" (REA) method. *American Meteorological Society, 15*, 1141–1158.

Hamann, O. (1979). On climatic conditions, vegetation types, and leaf size in the Galápagos Islands. *Biotropica, 11*(2), 101–122.

Hamann, O. (1985). The El Nino influence on the Galápagos vegetation.

Harris, R. M. B., Grose, M. R., Lee, G., Bindoff, N. L., Porfirio, L. L., & Fox-Hughes, P. (2014). Climate projections for ecologists. *Wiley Interdisciplinary Reviews: Climate Change, 5*(5), 621–637. https://doi.org/10.1002/wcc.291

Hartten, L. M., & Gage, K. S. (2000). ENSO's impact on the annual cycle: The view from Galápagos. *Geophysical Research Letters, 27*(3), 385–388.

Hay, J. E., & Hay, J. E. (2003). *Climate variability and change and sea-level rise in the Pacific Islands region: A resource book for policy and decision makers, educators and other stakeholders*. Tokyo: SPREP.

Hellmann, J., Byers, J., Bierwagen, B., & Dukes, J. (2008). Five potential consequences of climate change for invasive species. *Conservation Biology, 22*(3), 534–543. Retrieved from www.jstor.org/stable/20183419

Houvenaghel, G. T. (1974). Equatorial undercurrent and climate in the Galapagos Islands. *Nature, 250*(5467), 565.

IDEAM, PNUD, MADS, DNP, & CANCILLERIA (2015). Escenarios de Cambio Climático para Precipitación y Temperatura en Colombia.Instituto de Hidrología, Meteorología y Estudios Ambientales (IDEAM), Programa de las naciones Unidas para el Desarrollo (PNUD), Ministerio de Ambiente y Desarrollo Sostenible (MADS), Departamento Nacional de Planeación (DNP) y Cancillería. (2015). Bogota, Colombia.

Koutavas, A., & Lynch-Stieglitz, J. (2004). Variability of the marine ITCZ over the eastern Pacific during the past 30,000 years. In *The Hadley circulation: Present, past and future* (pp. 347–369). Springer.

Kurashima, N., Fortini, L., & Ticktin, T. (2019). The potential of indigenous agricultural food production under climate change in Hawai'i. *Nature Sustainability, 2*(3), 191.

Lowitt, K., Hickey, G. M., Saint Ville, A., Raeburn, K., Thompson-Colón, T., Laszlo, S., & Phillip, L. E. (2020). Knowledge, markets and finance: Factors affecting the innovation potential of smallholder farmers in the Caribbean community. In *Food security in small island states*. Springer.

MAE. (2016). *Proyecciones Climáticas de precipitación y temperature para el Ecuador, bajo distintos escenarios de cambio climático*. Quito: Ministerio del Ambiente del Ecuador.

McNamara, K. E., & Westoby, R. (2011). Local knowledge and climate change adaptation on Erub Island, Torres Strait. *Local Environment, 16*(9), 887–901.

Nunn, P. D., Aalbersberg, W., Lata, S., & Gwilliam, M. (2014). Beyond the core: Community governance for climate-change adaptation in peripheral parts of Pacific Island countries. *Regional Environmental Change, 14*(1), 221–235.

Pemberton, C. (2005). *Agricultural development and employment in the Caribbean: Challenges and future prospects*. Port of Spain, Trinidad: International Labour Organization.

PIRCA, 2016. Expert consensus on downscaled climate projections for the main hawaiian islands. Pacific Islands Regional Climate Assessment (PIRCA), Available on: https://www.eastwestcenter.org/system/tdf/private/2016pirca-hiprojections-lowres.pdf?file=1&type=node&id=35908, Last Accessed 1/3/2020

Rosegrant, M. W., & Cline, S. A. (2003). Global food security: Challenges and policies. *Science, 302*(5652), 1917–1919.

Sachs, J. P., & Ladd, S. N. (2010). Climate and oceanography of the Galapagos in the 21st century: Expected changes and research needs. *Galapagos Research, 67*, 50–54.

Saint Ville, A., Hickey, G. M., Locher, U., & Phillip, L. E. (2020). The role of social capital in influencing knowledge flows and innovation in St. Lucia. In *Food security in small island states*. Springer.

Saint Ville, A. S., Hickey, G. M., & Phillip, L. E. (2015). Addressing food and nutrition insecurity in the Caribbean through domestic smallholder farming system innovation. *Regional
Environmental Change, 15*(7), 1325–1339.

Sampedro, C., Pizzitutti, F., Quiroga, D., Walsh, S. J., & Mena, C. F. (2018). Food supply system
dynamics in the Galapagos Islands: Agriculture, livestock and imports. *Renewable Agriculture
and Food Systems*, pp. 1–15.

Santos, J. L. (2006). The impact of El Niño-southern oscillation events on South America.
Advances in Geosciences, 6, 221–225.

Shea, E. (2003). Living with a climate in transition: Pacific communities plan for today and tomor-
row. *Asia-Pacific Issues*, no. 66, p. 1.

Smith, C. A., & Sardeshmukh, P. D. (2000). The effect of ENSO on the intraseasonal variance
of surface temperatures in winter. *International Journal of Climatology, 20*(13), 1543–1557.

Snell, H., & Rea, S. (1999). The 1997–98 El Niño in Galápagos: Can 34 years of data estimate 120
years of pattern?, Not. *Galápagos, 60*, 111–120.

Tai, A. P., Martin, M. V., & Heald, C. L. (2014). Threat to future global food security from climate
change and ozone air pollution. *Nature Climate Change, 4*(9), 817.

Tebaldi, C., & Knutti, R. (2007). The use of the multi-model ensemble in probabilistic climate
projections. *Philosophical Transactions of the Royal Society A, 365*, 2053–2075. https://doi.
org/10.1098/rsta.2007.207

Thompson, A. L., Nicholas, K. M., Watson, E., Terán, E., & Bentley, M. E. (2019). Water, food,
and the dual burden of disease in Galápagos, Ecuador. *American Journal of Human Biology,
32*(1), e23344.

Trueman, M., & d'Ozouville, N. (2010). Characterizing the Galapagos terrestrial climate in the
face of global climate change. *Galapagos Research, 67*, 26–37.

USGCRP (2017) Climate Science Special Report: Fourth National Climate Assessment, Volume
I [Wuebbles, D.J., D.W. Fahey, K.A. Hibbard, D.J. Dokken, B.C. Stewart, and T.K. Maycock
(eds.)]. U.S. Global Change Research Program, Washington, DC, USA, 470 pp., https://doi.
org/10.7930/J0J964J6.

van Vuuren, D. P., Edmonds, J. A., Kainuma, M., Riahi, K., Thomson, A., Hibbard, K., ... Rose,
S. K. (2011). The representative concentration pathways: An overview. *Climatic Change,
109*(1), 5–31.

Vermeulen, S. J., Challinor, A. J., Thornton, P. K., Campbell, B. M., Eriyagama, N., Vervoort,
J. M., ... Smith, D. R. (2013). Addressing uncertainty in adaptation planning for agriculture. *Proceedings of the National Academy of Science of the United States of America, 10*,
8357–8362.

Wairiu, M., Lal, M., & Iese, V. (2012). Climate change implications for crop production in Pacific
Islands region. In *Food production–approaches, challenges and tasks* (pp. 67–86). Rijeka:
InTech.

Walsh, J. (2011). *Statistical downscaling.* NOAA Climate Services Meeting. Retrieved from http://
www.iarc.uaf.edu/sites/default/files/workshops/2011/noaa_climate_change_needs/Walsh-
StatisticalDownscaling.pdf

Watson, E., Thompson, A., Bentley, M., & Hopping, E.. (2019). *Social determinants of household
diet quality in Galapagos, Ecuador.* Paper read at American Journal of Human Biology.

Wolff, M. (2010). Galapagos does not show recent warming but increased seasonality. *Galapagos
Research, 67*, 38–44.

Zehetner, F. (2016). *Informe: Recomendaciones para la protección y el manejo sostenible del
suelo en las islas habitadas de Galápagos.* SENESCYT-Inst. de Investigación del Suelo, Univ.
de Recursos y Ciencias Naturales (BOKU), Viena, Austria-Parque Nacional Galápagos, Puerto
Ayora, Galápagos, Ecuador.

Galapagos is a Garden

Francisco Laso

Our Hands (and Minds) Shape Evolution's Eden

The famed Galapagos Islands, the first UNESCO Natural World Heritage Site, are undergoing a profound change in both ecological and socio-economic aspects (Watkins & Cruz, 2007). Since their discovery in 1535, the reason why humans have valued these islands has changed through time. The islands changed from a place for ships to rest and re-stock on food with tortoise meat, to a source of valued materials for trade such as tincture from lichen and fertilizer from guano, to a place where undesirable humans were cast away from the rest of society (Basset, 2009; Donoso, 2012; González, Montes, Rodríguez, & Tapia, 2008). In the 1800s, agriculture was already one of the main economic engines in Galapagos, and a necessity for the survival of human settlements (Chiriboga & Maignan, 2006a). However, the visits of naturalists like Charles Darwin and writers like William Beebe redirected people's attention to the archipelago's unique flora and fauna (Grenier, 2007; Quiroga, 2017). Galapagos' biota was special not only for its uniqueness, but because it thrived in isolation, and its evolutionary processes were readily visible in this *natural laboratory*. The establishment of the Galapagos National Park (GNP) in 1959 and its recognition as a UNESCO World Heritage Site in 1979 solidified tourism and conservation research as the central pillars for the values and economic development of the region (Romanova, Yakushenkov, & Lebedeva, 2013). Meanwhile, local populations were seen as a negative force that needed to be curbed if the natural wonders that made these islands famous were to survive (Groot de, 1983; Quiroga, 2013). Today, conservation is a crucial component of the discourse that per-

F. Laso (✉)
Department of Geography, University of North Carolina at Chapel Hill,
Chapel Hill, NC, USA
e-mail: laso@live.unc.edu

© Springer Nature Switzerland AG 2020
S. J. Walsh et al. (eds.), *Land Cover and Land Use Change on Islands*, Social
and Ecological Interactions in the Galapagos Islands,
https://doi.org/10.1007/978-3-030-43973-6_6

suades more than 250,000 visitors annually from around the world to spend thousands of dollars to experience the (so-called) *pristine* ecosystems (Benitez-Capistros, Hugé, Dahdouh-Guebas, & Koedam, 2016; Hennessy & Mccleary, 2011; Quiroga, 2009). However, the extensive and long-standing human presence, as well as the uncontrolled development of tourism infrastructure, is making pristine ecosystems an increasingly rare and often artificial experience (Epler, 2007; Epler Wood, Milstein, & Ahamed-Broadhurst, 2019; Watson, Trueman, Tufet, Henderson, & Atkinson, 2009). In fact, many conservationists consider densely inhabited islands like Santa Cruz and San Cristobal to have already been lost (Reck, 2017). Sixty years after the foundation of the GNP, the discourse that values the absence of local populations (but encourages the presence of tourists) to conserve pristine ecosystems has become dominant as both local populations and annual visitors have soared past the limits that scientists recommended for the region's sustainable development (Celata & Sanna, 2012; Hennessy & Mccleary, 2011). The archipelago's accelerated integration with the continent and international community forces scientists and conservation practitioners to re-examine prevailing discourses to collectively guide the future evolution of the archipelago towards ecologically and socioeconomically sustainable directions.

Our understanding of present-day socioecological conditions and dynamics of the Galapagos demands that we abandon the marketed myth of the pristine Galapagos as conservation's moral compass in favor of a more realistic, nuanced, and inclusive perspective (Hennessy & Mccleary, 2011; Valdivia et al., 2013; Walsh & Mena, 2013). Advances in ecological science and political ecology question the human/nature duality inherent in the pristine narrative and acknowledge empirical and theoretical evidence that suggests ecosystems are complex systems in constant flux (Folke, 2006; Lawesson, 1988; Scoones, 1999). Insights from integrated coupled-human-natural systems suggest that community involvement for active land management is sometimes necessary to maintain protected ecosystems, and this seems to be the case for the highlands of the Galapagos (Gardener, Atkinson, Rueda, & Hobbs, 2010; Liu et al., 2007). Therefore, recognizing our perception of what is natural as a profoundly cultural construction helps advance discussions that seek to include a broader segment of the population into the vision for the entire region.

Humans construct the meaning of all concepts, even those as fundamental and seemingly common-sensical as *nature*, in relation to some concepts and opposition to others (Haraway, 1992). However, meanings and narratives are not stagnant and we can use narrative analysis in order to answer the question: *how have conservation science's dominant narratives of the Galapagos changed through time and what are the implications of these changes?* Narrative analysis is a useful method of qualitative inquiry to understanding how people create meaning in their lives, focusing on both the content (what is being said) and the form (how it is being said) of texts or interviews (Jackson, Drummond, & Camara, 2007). First, I will review published literature and use a standard tool in semiotics to break down the narratives surrounding the widely-held imaginary of Galapagos as a *natural laboratory*; and second, I will contrast traditional static imaginaries of the pristine Galapagos with more dynamic narratives offered by present-day views of Galapagos as a *coupled human-natural system*. I will argue that a more inclusive narrative for conservation in the Galapagos would encourage more active participation in conservation work from local farmers and the wider community, potentially turning a relationship that has historically been rife with conflict[1] into one where the conservationists and the local food movement can both benefit.

Breaking Down Dualisms in Traditional and Modern Narratives of Galapagos

Previous authors have written about the evolution of conservation discourses in the Galapagos (Benitez-Capistros et al., 2016; Hennessy & Mccleary, 2011; Lu, Valdivia, & Wolford, 2013; Quiroga, 2009). In this chapter, I build upon previous narrative analyses using an adaptation of Greimas' square of oppositions or 'semiotic square' (Bonfiglioli, 2008; Greimas, 1983) as a tool to explore the discursive boundaries between different elements of the imaginary of Galapagos. The semiotic square is not frequently used to study human-environment interactions in the Galapagos, but it is well suited to articulate sets of complex relationships because this tool emphasizes how symbols acquire their meaning as part of a dynamic signification system (Greimas, 1970). Furthermore, the elements identified by the semiotic square represent not just abstract concepts, but real-life relationships that are deeply spatial and can be mapped. Therefore, the visual format is an efficient way to register conceptual blockages between elements. In other words, this tool helps us become aware of biases of particular narratives by identifying not only the elements that constitute them but also those that are excluded from them.

Before embarking on an exploration of biases found within conventional narratives of the Galapagos, I must acknowledge my positionality to avoid the illusion of an omniscient viewpoint. During my doctoral studies fieldwork, I spent a lot of time with farmers and there were several instances when differences in our positionality became painfully evident. Farmers are always on the move, so to avoid taking up too much of their time, I would offer my help in whatever activity they needed to accomplish while I was there, from clearing land from brush, planting grass, making holes in rocky volcanic soils, or milking cows. The developing blisters on my fingers would sometimes highlight my physical inexperience in these tasks, a reminder that my body was more adapted to reading and writing about agriculture

---

[1] Since the foundation of the GNP, there has been a lot of tension and even violent confrontations between conservation practitioners and Galapagos inhabitants (Celata & Sanna, 2012). Some of the reasons for conflict in rural areas include prohibition of resource extraction from protected areas (Valdivia et al., 2014), eradication of goats that served as a source of food for locals (Bocci, 2017), tortoises trampling fences and crops (Benitez-Capistros et al., 2018), finches raiding seeds and fruits (Gómez & Ramírez, 2017), and a general lack of representation of the agricultural sector in crafting policies and allocating funds (Chiriboga & Maignan, 2006b).

than actually practicing it. And yet, despite the relative embodied ignorance in my research subject, the documents I produce during my Ph.D. are aimed at changing people's perspectives about this activity. The content of my writings would be, in part, informed by the knowledge that farmers share with me during my time with them. As I would quickly find out, farmers had been advocating for contributing more actively to conservation in the Galapagos since the '90s. I was humbled by one such farmer and career politician who thanked me for paying attention to their perspective because researchers like me have mostly ignored their message for decades. I found it unfair that the words and actions of the very practitioners of agriculture in Galapagos can be validated by a non-*galapageño* simply because they are chronicled in a foreign language and printed in paper bearing the logo of a US academic institution.

As an ecologist-turned geographer, I encountered many occasions that led me to revise the narratives and assumptions that my formal training handed down. For example, I was surprised to realize that some farmers don't see guayaba trees (*Psidium guajava*) as "invasive," and I was downright shocked that most farmers referenced the endemic Galapagos finch whenever we talked about "agricultural pests." So while my background is in conservation science, I engage with coupled human-natural systems from a post-normal science, systems-based perspective (Colwell, 2001; Lister, 1998; Pickett, Cadenasso, & Grove, 2005). This view considers the complexities of systems theory and embraces the planet's biocomplexity, the dynamic web of interactions of living systems at all levels, by recognizing that there is a multiplicity of legitimate perspectives on any issue. In other words, I believe biodiversity is worth conserving, but I also understand humans and their multiple conceptions of the world are an integral part of the environment where the non-anthropogenic biodiversity resides (Sanderson et al., 2002; Sterling, Gómez, & Porzecanski, 2010). Under post-normal science, the network of peers that can offer useful perspectives on how to solve problems is extended beyond traditional scientific experts, especially to those who have most at stake in addressing these problems. Under systems-based thinking, I try to consider the physical, biological, and socioeconomic dimensions at local and global scales that influence land system dynamics (Turner, Lambin, & Reenberg, 2007). System thinking also differs from traditional conservation because coupled human-natural systems are not portrayed as being either *degraded* or *pristine*, but rather have multiple stable states, where some states may be more desirable than others for human life, and where human activities can push system states beyond irreversible thresholds (Sterling et al., 2010).

In this chapter, I first examined the traditional narrative of Galapagos as a *natural laboratory*, where certain human activities threaten a pristine Galapagos. Second, I compared and contrasted this traditional narrative with more modern narratives of Galapagos as a *coupled human-natural system*, using the relationship between the conservation and agricultural sectors as a case study to exemplify potential conflicts and opportunities of this perspective shift. For each narrative, first, I defined key concepts and their use in the context of the Galapagos (A, B), as well as their respective mutually constituting opposites $(\bar{A}, \bar{B})$. Second, I identified elements that break the contradiction between each component and its respective antithesis

$(A \& \bar{A}, \ B \& \bar{B})$ as well as something that breaks the contradiction between the two concepts $(A \& B)$ and between their respective opposites $(\bar{A} \& \bar{B})$. Finally, I identified an element that disrupts all four concepts $(A \& B \& \bar{A} \& \bar{B})$. I grounded the abstract spaces of this thought exercise using existing features from the Galapagos to make the gradients between binaries more palpable.

Galapagos is a Natural Laboratory

The narrative of Galapagos as a *natural laboratory* is deeply entrenched in the region's history and continues to play a crucial role in its socio-ecological development. Over half a century after the foundation of the GNP, socio-ecological conditions and dynamics in the Archipelago have dramatically changed and taken predictable paths. The revenue from a booming tourism industry has led to higher levels of immigration, population growth, consumption patterns, and economic inequality (Kerr, Cardenas, & Hendy, 2004; Taylor, Dyer, Stewart, Yunez-Naude, & Ardila, 2003). Economic incentives from the tourism industry have drained people and resources from other activities where humans actively transform their environment, such as agriculture (Valdivia et al., 2013). The extractive, tourism, and conservation industries often come into conflict about which activities are compatible with the conservation of *pristine* Galapagos, but the symbiotic relationship between the tourism and conservation industries often results in more power and more lenient regulations being granted to the tourism sector (Quiroga, 2013; Reck, 2017). And yet, all these processes and conflicts are invisibilized by the contradictory imaginary portrayed in nature documentaries and travel brochures, where human activities (A, Fig. 1) threaten Galapagos' unique biota, all the while remaining a *pristine* location in which tourists can become immersed (B, Fig. 1) (Hennessy & Mccleary, 2011; Kannar-Lichtenberger, 2018).

Galapagos is a Natural Location for Humans to do Science

Authors across multiple disciplines have questioned the categories that arise from this oft-dissected western dualism between humans (A, Fig. 1) and nature ($\bar{A}$, Fig. 1) (Haila, 2000; Rebecca Lave et al., 2014; Noel Castree, 2009). The romantic attraction to primitivism means that the gray areas between *humans* and *nature* is often occupied by the construction of the *noble savage* whose lives are 'more attune with nature' (Cronon, 1996). However, the limited human history of the Galapagos means that this conceptual niche was empty in the imaginary of the enchanted Islands; the first people who occupied the Galapagos were European and British pirates and whalers who valued the islands as a source of food and fuel. Far from being attuned with their environment, it was their activities that decimated tortoise

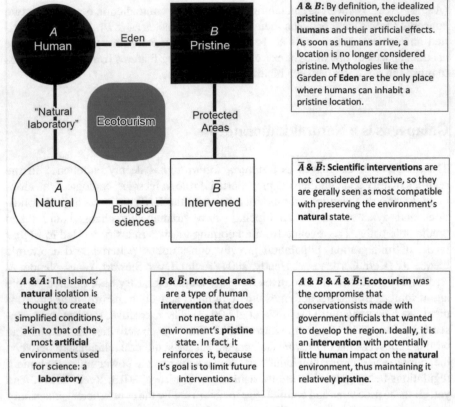

Fig. 1 Semiotic square analyzing the relationship between *human* (*A*) and *pristine* (*B*), as well as the discursive boundaries between these constructions, both with their respective opposites ($\bar{A}$ & $\bar{B}$) and with each other. The text boxes explain the relationships (dashed lines) and opposites (solid lines) of these constructions, as well as a term that disrupts all four corners of the semiotic square

populations and even drove some species to extinction (MacFarland, Villa, & Toro, 1974). Along with questioning the conceptual divide between humans and nature, authors have also encouraged us to question the assumption that there is a world of barely perceptible human disturbance (Denevan, 1992). Despite the ecologically transformative impact of humans since the archipelago's discovery in 1535, its anthropogenic change has been downplayed by scientists and visitors to the islands since the nineteenth century, who portrayed the archipelago as nearly untouched by man due to its isolation, and therefore, as a natural location for scientists to study evolutionary processes (Hennessy & Mccleary, 2011; Quiroga, 2009). In other words, in the absence of longstanding native populations, scientists were able to claim the islands as their own *natural* location to do their investigations. After all, the Galapagos Islands were the stage for one of the most widely circulated textbook accounts of the origins of modern science, when Darwin visited the islands and became inspired by its unique creatures to formulate the theory of evolution.

Charles Darwin's life and ideas are closely tied to the imaginary of the Galapagos. The legend that Darwin observed morphological differences among finches from different islands and that he deduced that they were all descendants from a common ancestor is one of the most popular secular myths in modern science, but historically, it is also wildly inaccurate. Darwin paid little attention to the finches during his 1935 visit on the HSM Beagle, and he didn't even mention them in his seminal publication *On the Origin of Species by Means of Natural Selection* (Darwin, 1859; Sulloway, 1982). Nevertheless, this narrative became so popular that the finches themselves became commonly known as 'Darwin's Finches.' It wasn't until years after Darwin's visit to the enchanted islands and after he consulted with ornithologists and herpetologists that Darwin was able to ascertain that the variability of mockingbirds and tortoises that he saw challenged the stability of species concepts held by creationists (Darwin, 1859; Sulloway, 2009). Historical inaccuracies notwithstanding, Darwin's legacy, in turn, was fundamental to the global construct of Galapagos as a Natural Laboratory, which became popularized as a classic stomping ground for biologists around the world (Mariscal, 1965; Quiroga, 2009). Out of reverence for this tradition, foreign evolutionary biologists were the first to argue for the need to conserve the archipelago's ecological and evolutionary processes by creating a wildlife preserve and a research station on the Galapagos (Bowman, 1960). After international pressure from scientists with a vision to conserve Galapagos as pristine, the Charles Darwin Research Station (CDRS) was founded in Belgium and the Ecuadorian government established the GNP, non-coincidentally, one hundred years after the publication of *Origin of Species* (Quiroga, 2009; Reck, 2017). Indeed, the foundation of CDRS and GNP to conserve Galapagos' flora and fauna was a direct homage to the birthplace of evolutionary thought, as a way to preserve a site of scientific history and pristine wilderness (Hennessy & Mccleary, 2011).

Conservation Organizations Attempt to Preserve 'Pristine' Galapagos from Degradation

Modern-day maps of the Galapagos have two distinct categories separated by a hard boundary: The GNP, constituting 97% of the territory dedicated towards conservation of native ecosystems, and Human Use Areas, or the remaining 3% where humans settlements have been established since the 1800s. In legislation and maps, the park boundary separates that which is meant to be kept pristine (B, Fig. 1) and that which has already been irreversibly Intervened ($\overline{B}$, Fig. 1) by human actions. In 1998 the Special law of Galapagos came into effect, promoting comprehensive environmental protection and economic development laws for the entire province (Congreso Nacional, 1998). This law established the Galapagos Marine Reserve, limited the ability of immigrants to secure jobs, promoted cooperation between public and non-profit sectors, and gave the GNP higher institutional power (Heylings & Cruz, 1998). Crafting this law was an elite-driven process. For example, no representatives of the agricultural sector were present during stakeholder meetings (Heylings & Cruz, 1998). Human activities shaped the landscapes of the Galapagos for centuries, so it was no surprise that the discourse of conservation was resisted by Galapagos' residents whose traditional extractive activities (logging of native trees, fishing sea turtles and sharks, using tortoises as a source of meat) were being outlawed within the GNP (Bowman, 1960; MacFarland et al., 1974; Quiroga, 2009). By 2015, the Galapagos Management Plan for Good Living established more graduated zoning laws of the Galapagos that divided protected areas into several categories to allow for increasing human intervention, depending on their perceived level of degradation (DPNG, 2014; Reck, 2017). At one end of the spectrum, areas that are perceived as *pristine or nearly pristine*, where no known human impact has taken place only allow non-extractive activities like scientific observation and monitoring. At the opposite end of the spectrum, *Transition* areas include agricultural zones outside of the GNP and aim to collaborate with other institutions and private landowners to implement sustainable models of development that strengthen their economic activities while engaging in restoration and conservation work (DPNG, 2014). Thus, these different zones and their permitted uses further blur the distinction between *pristine* and *intervened* territories.

The boundary demarcating a protected area lies between intervened and pristine realms both physically and conceptually, because these regions are not static and because maintaining pristine ecosystems ironically requires a lot of human intervention. Boundaries are permeable to humans, native and invasive flora and fauna, chemicals used on either side of the fence, and other forces (Valdivia, Wolford, & Lu, 2014). Because of this, boundaries must be negotiated, made, and maintained through arduous manual labor, chemical applications, and sometimes, disputes between locals and park officials (Benitez-Capistros, Camperio, Hugé, Dahdouh-Guebas, & Koedam, 2018; Gardener, Atkinson, & Renteria, 2010; Hennessy & Mccleary, 2011; Heylings & Cruz, 1998). One of the most severe transgressors of park boundaries are invasive plants, which threaten both protected and productive areas. Guézou et al. (2010) found that introduced plant species now outnumber native species, but that most taxa have not been introduced long enough to become a problem yet. With time, however, more introduced plants are likely to become naturalized, and a part of those will become invasive (Guézou et al., 2010). These results paint a sobering image of the prospect of excluding introduced plants within park boundaries, which is why conservation practitioners extort the urgency of early monitoring and control actions both in GNP and private lands to keep them cost-effective (C. Buddenhagen & Renteria, 2009; Guézou et al., 2010; Rentería & Buddenhagen, 2006). The biology of introduced plant species and the context in which they grow must meet strict criteria to attempt plant control or eradication programs (C. E. Buddenhagen & Tye, 2015). For example, eradication efforts are more likely to succeed in the initial stages of invasion when information exists about their distribution, when they grow in easily accessible areas, when plants are easily detectable before reproduction, and when they have not yet developed a persistent seed bank or spread over large areas (Gardener, Atkinson, & Renteria, 2010; Panetta, 2009). Quite the opposite of the absence of human interventions to maintain protected areas as pristine, the control of invasive plants requires continuous, long-term, and unwavering commitment from funding sources, conservation practitioners, and the public to succeed (Buddenhagen & Tye, 2015).

Galapagos is an Eden... for Tourism Operators?

The fairy-tale character of referring to Galapagos as pristine becomes evident by pointing out that, by definition, the idealized pristine (B, Fig. 1) environment excludes humans (A, Fig. 1) and their artificial effects. A location is no longer considered *pristine* as soon as humans are in the picture. Both elements can only coexist in a mythological location, like the garden of Eden inhabited by Adam and Eve. Meanwhile, according to the narrative of Galapagos as a Natural laboratory, the only type of intervention ($\bar{B}$, Fig. 1) that does not negate a location's natural status ($\bar{A}$, Fig. 1) are those from the environmental sciences. These activities are perceived as non-extractive and therefore, consistent with protecting a designated UNESCO world heritage site (Bowman, 1960; IUCN, 1979).

This narrative sees inhabited areas and local populations as the source of problems for the protected areas. Populated areas are where humans come from to chop wood from the GNP illegally, where humans come from every day to fish within the Marine Reserve, where introductions of invasive plants and animals occur, and where people generate waste and pollution (Lu et al., 2013). Legislative boundaries were created to impose regulations over the province, its inhabitants, and its immigrants (Bowman, 1960; Congreso Nacional, 1998). Meanwhile, conservation organizations promoted tourism ($A \& B \& \bar{A} \& \bar{B}$, Fig. 1) as an alternative to extractive economic activities, a turning point that led Galapagos to become the world-renowned tourist destination that it is today (Mariscal, 1965; Reck, 2017). Today, the tourism industry is arguably more powerful than the GNP itself. It has successfully managed to relax regulations on virtually any aspect of their operations: fleet and boat size, number of passengers and the length of their stay, permits to build hotels, the amount of revenue that must stay within the community, etc. (Hoyman & McCall, 2013; Watkins & Cruz, 2007). What began as a recommendation from the conservation sector is now seen as one of the primary threats to the integrity of the Galapagos.

The discourse of *ecotourism* in Galapagos follows a neoclassical economic tradition in that it falsely presents itself as a way for the economy to continue growing while not degrading the environment (Grenier, 2007; Taylor, Hardner, & Stewart, 2009). Tourism has grown exponentially since 1995 at a staggering average rate of 9% per year (Pizzitutti, Mena, & Walsh, 2014). Despite the swelling revenue flow that tourism represents for the region and the country as a whole, it has become apparent that the accelerating continentalization of the archipelago has resulted in uncontrolled population growth, widening economic inequality, increased civil unrest, higher rates of accidental species introductions, strained public services and infrastructure, and more frequent conflicts with the conservation sector (Epler, 2007; Hoyman & McCall, 2013; Watkins & Cruz, 2007). The pressure exerted by both tourists and locals on the archipelago's modern fossil-fuel fuel economy, water consumption, food production, and waste management network is directly related to the degradation of the natural wonders that made Galapagos world-famous (Cecchin, 2017; Kingston, Runciman, & McDougall, 2003). Furthermore, social inequalities and integration into global markets have caused the degradation of the social and environmental fabric of the Galapagos, as was seen in the case of the now collapsed sea cucumber industry and the riots sparked by its regulation (Bremner & Perez, 2002; Celata & Sanna, 2012; Wolford, Lu, & Valdivia, 2013). Tourism will inevitably have an impact on the environment and the local community, but the high growth rate from this loosely-regulated and highly-profitable economic sector makes changes particularly hard to manage (Cater, 1993; Proaño & Epler, 2007). There is so little planning that hotels remain at 50% occupancy rate, and yet more hotels are continuously being built, much to the anger of locals who have even violently protested against foreign-owned projects (Epler, Watkins, & Cárdenas, 2008; Shaw, 2015). Under specific standards for defining ecotourism, Galapagos' development model falls short, as protected areas and the demands from the tourism industry have displaced and overwhelmed traditional practices (farming, fishing) (Cater, 1993; Wallace & Pierce, 1996). In short, the regions' economic and social development promoted by the narrative of *Galapagos as a natural laboratory* has failed to encourage standards that can be considered to be consistent with its status as a UNESCO world heritage site.

Galapagos is a Coupled Human-Natural System

In more recent history, there has been a shift away from traditional static, environmental science-centric, and linear perspectives of pristine ecosystems conveyed in the *natural laboratory* narrative. Authors now have a more holistic framework to describe observed patterns from the Galapagos—that of a *coupled human-natural system* (Cecchin, 2017; González et al., 2008; Miller, Carter, Walsh, & Peake, 2014; Walsh & Mena, 2013, 2016). These frameworks, often known as systems thinking or complex systems science, focus on feedback structures between humans and their environment to understand the dynamics of a coupled human-natural system. From this perspective, Galapagos is a self-organizing socio-ecosystem with no inherent *correct* or *natural* state, but one that is a result of its history (Kay, 2000). Complexity studies use insights from both natural and social sciences to Identify feedback loops among its ecological and social elements (Holling, Gunderson, & Ludwig, 2002; Liu et al., 2007). Identifying feedback loops within a system highlights how its elements are connected, even if they are spatially or temporally separated. Identifying feedback loops can help us gauge a socio-ecosystem's resilience to disturbances and anticipate surprising behaviors and changes in stable states that the system can suddenly undergo (Turner et al., 2003). At the scale of socio-ecosystems, land-use change science has become particularly useful to untangle feedback loops from a spatial perspective (Walsh, Malanson, Messina, Brown, & Mena, 2011; Walsh et al., 2009). Therefore, we can understand the spatial relationships between the dominant elements of the narrative of Galapagos as a *coupled human-natural system* by focusing on the two most pervasive land uses, agriculture (*A*, Fig. 2) and conservation (*B*, Fig. 2).

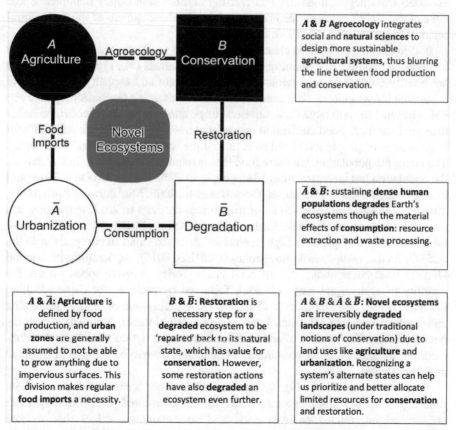

Fig. 2 Semiotic square analyzing the relationship between *agriculture* (A) and *conservation* (B), as well as the boundaries between these constructions, both with their respective opposites ($\bar{A}$ & $\bar{B}$) and with each other. The text boxes explain the relationships (dashed lines) and opposites (solid lines) between these constructions, as well as a term that disrupts all four corners of the semiotic square (A & B & $\bar{A}$ & $\bar{B}$)

Galapagos is No Place to Grow Food. Or is It?

The majority of human-use areas are covered by agricultural land, which is located in the humid highlands of all four inhabited islands. The humid highlands are also the most biologically productive zones and house the majority of floral diversity in the archipelago (Trueman & D'Ozouville, 2010). Agriculture (A, Fig. 2) was one of the main economic engines for the region before the founding of the GNP, but conservation (B, Fig. 2) practitioners viewed agricultural activities as antagonistic towards conservation and thus recommended against the development of this activity (Mariscal, 1965). Most conservation practitioners and public policy documents have traditionally framed agriculture within the context of habitat fragmentation and the introduction of invasive species (Bowman, 1960; Congreso Nacional, 1998; Mariscal, 1965). Perhaps, for this reason, the agricultural sector has usually been excluded from negotiations towards crafting conservation policy documents, and none of the GNP's entrance fee is allocated for the development of the agricultural sector (Congreso Nacional, 1998; Reck, 2017).

If agriculture (A, Fig. 2) is characterized by food production in rural areas, then its opposite and mutually defining element is urban areas ($\overline{A}$, Fig. 2). Urban areas are usually covered in impervious surfaces (concrete and asphalt), making them unsuitable for agriculture. Furthermore, there is a reinforcing feedback loop between both elements as well: large concentrations of people in urban areas consume rather than produce food. Food cultivation and animal husbandry would not exist without large masses of people who need to consume the food produced (and vice-versa). The larger the population, the more food that is required to sustain it. In Galapagos, the population has increased from 1346 people in 1950 to over 30,000 in 2018, and mostly concentrated in urban areas (González et al., 2008; Ministerio de Agricultura, 2018). Galapagos had over 300 ha of impervious surfaces in 2017, expanding at a rate of 3.3% per year (Benítez, Mena, & Zurita-Arthos, 2018). Though the rate of urbanization in Galapagos is high, the urban areas are small in comparison to the >25,000 ha that comprises its rural zones (CGREG, 2015); Agricultural Production Units (UPAs) cover about 75% of rural lands. Today, invasive species cover the majority of agricultural lands (28.5%), followed by pastures for cattle ranching (22.3%), while food crops cover about the same percentage of land as remaining native ecosystems (18.5% each, Laso, Benítez, Rivas-torres, Sampedro, & Arcenazario, 2020). The current trends of urbanization and agricultural land abandonment means most products consumed in Galapagos have to be shipped in from the mainland (Epler, 2007).

Food imports (A & $\overline{A}$, Fig. 2) are the link between agricultural areas, where food is produced, and urban areas, where food is consumed. Unfortunately, there are several interrelated environmental and socioeconomic crises that weaken domestic food security and have led to a dependence on regular food imports not only from the agricultural areas in the humid highlands, but from continental Ecuador, over 1000 km away. The average age of farmers is nearly 60 years old, as most younger people have abandoned rural areas in favor of tourism-related employment in urban areas (CGREG, 2015; Lu et al., 2013). The rural-urban migration is harder to affront because laws meant to make it harder for non-residents to secure jobs in Galapagos makes it prohibitively expensive for most farmers to hire someone from the mainland to help them tend their land (Guzman & Poma, 2015). Furthermore, climatic instability means that water availability has become a persistent issue, so purchasing freshwater and investing in irrigation infrastructure has become an expensive necessity to avoid crop failure. Meanwhile, the dependence of goods from the continent means more species are entering the islands every day, and with them are species that become agricultural pests or invasive species (Brewington, Rosero, Bigue, & Cervantes, 2013). To make matters worse for farmers, GNP policies prohibit farmers from using strong chemicals to combat agrarian pests or as fertilizer (CGREG, 2014). The law also prohibits owning industrial-grade machinery, so farmers must rely more heavily on manual labor to till their soil and keep their land clear from invasive plants. Regulations meant to curb the ownership of combustion-engine vehicles have also made it harder for farmers to have a reliable and affordable means of transportation to bring their products into town to sell in the local market (Toledo, 2014). And once products are in town, they must compete with all products that are shipped from the mainland. Products from the mainland are favored by restaurants that cater to tourists because mainland farmers tend to use industrial methods, so products are larger, more uniform, are available year round, and can be sold at prices that local products can hardly compete (Salvador Ayala, 2015).

The agricultural sector will not be able to contribute a more substantial portion to the region's food security if it does not overcome these and other environmental and socio-economic challenges. Estimates show that about 6600 MTs of food are produced locally every year, compared to >19,000 MTs of food products that are imported from the mainland (Ministerio de Agricultura, 2018). The agricultural food supply sourced from the continent was approximately 75% in 2017, and with no policy changes, this number will increase to 95% in just 20 years (Sampedro, Pizzitutti, Quiroga, Walsh, & Mena, 2018). This dependency of shipments from continental Ecuador creates a vulnerable food system because shipping vessels have regularly crashed or gone out of service, leaving the island in a state of emergency (Viteri Mejia, 2014). The local government recognizes the necessity to tighten the feedback loop between agricultural food supply and consumption by strengthening domestic food production and commercialization through technical training of the agricultural sector and creating markets where only certified locally produced food can be sold. More recently, they are attempting to reduce the import of food products from the mainland by introducing tariffs on imported products that can be sourced locally like yogurt or tomatoes (CGREG, 2015; Ministerio de Agricultura, 2018). One way that the Ministry of Agriculture and a select few individual farmers have tried to reconcile environmental regulations and agriculture is through organic farming techniques, but these require much manual labor, which is one of the limitations for farming in present-day rural areas. An even more drastic way to reduce the distance that food must be transported is by producing food where most people live. Urban farms blur the lines between agriculture and urbanization altogether. Leading a rural lifestyle and pursuing food production within the limits of urban regions is not a new concept (Wehrwein, 1942). Techniques developed in the last century (hydroponics, aeroponics, vertical farming), allow people to produce vast amounts of food over impervious surfaces and with significantly reduced water requirements (Despommier, 2013). A few people in Galapagos are already pursuing hydroponics, and both government agencies and NGOs are considering the economic and environmental viability of these techniques (Chiriboga & Maignan, 2006b; Toledo, 2014). Whatever the strategy, strengthening local food security is crucial for the conservation of Galapagos. Lawmakers should weigh the risk of locally produced crops becoming invasive against the risk of high volumes of imports from the mainland serving as vectors for new invasive species. Biosecurity agents are only able to inspect a tiny sample (1–2%) of the shipment of goods that arrive at the islands (Brewington et al., 2013), so continued accidental introduction of organisms through shipments of products is a near certainty.

Humans are the Crisis & Humans are the Solution

Conservation biology (*B*, Fig. 2) is a crisis-oriented field that values biodiversity and tries to prevent its loss, as well as mitigate the effects of declining diversity. It is also an interdisciplinary field with a deeply colonial history, and in its more traditional forms, it is still guided by myths similar to those described in Galapagos as a Natural Laboratory narrative. For example, it is a field that makes a value judgment between *natural* and *anthropogenic* biodiversity, preferring the former (Ginsberg, 1987). The value of biodiversity is not a myth; the utilitarian and intrinsic value of biodiversity is unquantifiable, as any human valuation system falls short of encompassing all ways in which maintaining the integrity of the web of life is essential to human life and the rest of the biosphere (Edwards & Abivardi, 1998; Gowdy, 1997). One of the core ideas guiding the field of conservation and the institution of natural reserves can be traced back to Malthus' 1798 book *Essay on the Principle of Population*. Malthus postulated that human populations tend to grow geometrically until they have exhausted their resources, bringing the system to collapse because resources' arithmetic regeneration cannot keep up with population growth (Malthus, 1798). Given the unrelenting expansion of resource extraction since the industrial era, it was not long before the threat of timber shortages, soil erosion, and changing rainfall patterns motivated government intervention to prevent resource depletion and ecosystem collapse. The practice of declaring *protected areas*, vast expanses of natural areas as absolute state property under scientific management, can be traced back to British colonies in India during the mid-1800s (Barton, 2001). Access to protected forests became restricted to most human activities like logging operations and paper mills, which allowed macro and microscopic fauna to thrive, so imperial foresters likened these areas to a 'Great Laboratory.' Not coincidentally, this label resonates with the reputation cultivated by scientists of the Galapagos as a *natural laboratory*.

The opposite and the mutually defining element of conservation is *degradation* ($\bar{B}$, Fig. 2) because environmental protection cannot exist without the idea that the environment can become degraded in the first place. Environmental degradation most often refers to artificial change or disturbance by humans, which causes a decrease in the *natural conditions* of an environment (Johnson et al., 1997). The main threat to Galapagos ecosystems is the threat of invasive plants. About 5.5% of the archipelago's territory has been 'completely degraded' either by direct human activities or by the substantial invasion of invasive plant species like *Psidium guajava*, *Rubus niveus*, *Cinchona pubescens*, and *Syzygium jambos* (Watson et al., 2009). *Degradation* is a context-dependent and value-laden term that depends on perspective: the conservation biologists can see a farm degrading the value of a surrounding forest while farmers can see an encroaching forest degrading the value of their farm. The world-famous Darwin's finches or the emblematic giant tortoises represent a striking example of how different people's valuation of conservation subjects and degradation agents. For scientists, finches and tortoises are the royalty of Galapagos as part of evolutionary biology's folklore. For farmers, they can be an agricultural pest who can decimate their crops (Benitez-Capistros et al., 2016; Chiriboga & Maignan, 2006b). However, this perspective is little known, despite how common and damaging crop-raiding can be for a farmer. It should also be noted that a decrease in *natural conditions* doesn't necessarily imply a challenge to endemic fauna. Giant tortoises of Santa Cruz have readily adopted invasive plants and fruits into their diet, actively searching energy-rich foods like *P. guajava*, the invasive guava fruit (Blake, Guezou, Deem, Yackulic, & Cabrera, 2015; Yackulic, Blake, & Bastille-Rousseau, 2017). Tortoises, in turn, have spread the seeds of invasive plants as far as climate allows these plants to grow (Ellis-Soto et al., 2017). Furthermore, *P. guajava* trees are considered a valuable asset for many farmers who depend on them to provide shade and fodder for their cattle during times of drought (Chiriboga, Fonseca, & Maignan, 2006; Peñafiel, 2017). Despite these benefits for humans and tortoises, *P. guajava* cannot escape its status as *invasive* because, much like the myth of *pristine* environments, underlying any definition of degradation is the nature and culture dualism (Haila, 2000). Human-use areas are, by definition, inherently degraded, no matter what humans do (or don't do) in the 3% of the territory that is not a designated protected area (Valdivia et al., 2014).

Paradoxically, ecosystems often require human intervention to remain the same or be returned to their *natural* state. Restoration (B & $\bar{B}$) is understood as the process of repairing damage caused by humans to the diversity and dynamics of indigenous systems (L. Jackson, Lopoukhine, & Hillyard, 1995). Some of the most ambitious restoration campaigns in the world have taken place in the Galapagos, including the removal of feral mammals from multiple islands, manual and chemical removal of introduced plants, manual planting of native plants, or repatriation of captive-bred tortoises to remote islands where they can once again fulfill their ecological roles (Atkinson et al., 2010; Brewington, 2013; Carrion, Donlan, Campbell, Lavoie, & Cruz, 2011; Guézou et al., 2010; Rowley, 2013). The idea that humans can 'repair' ecosystems arose when biologists still assumed ecosystems had a stable or pristine state that preceded human influence (Cronon, 1996; Denevan, 1992; L. Jackson et al., 1995). Today, scientists understand this is not a straightforward process because the evolution of coupled human-natural systems is deeply complex; not only can the same environment have multiple alternate states, but feedback mechanisms, thresholds, phenomena that occur at different scales, and non-linear behaviors means that human interventions often have unexpected consequences (Liu et al., 2007; Nogués-Bravo, Simberloff, Rahbek, & Sanders, 2016; Walsh et al., 2011). Furthermore, new technologies can challenge current classification systems of what belongs in a category and what does not (Hennessy, 2015). For example, Hennessy (2015) describes how genetic tests on tortoises from Wolf volcano in northern Isabela Island, once thought to be in their natural environment, revealed the tortoises were genetically hybrid and thus were suddenly and paradoxically considered *alien* Galapagos tortoises. Phylogenetics can give humans a glimpse into prehistory, and therefore informs our construction of the ideal *pristine* environment, but we must be wary about the economic, social, and environmental costs of pursuing this construction. At its worst, conservation and restoration actions can unintentionally further degrade ecosystems or create new problems (Bocarejo & Ojeda, 2016; Chauvenet, Durant, Hilborn, & Pettorelli, 2011; Dodd & Sharpley, 2016; R. Lave, Doyle, & Robertson, 2010; Mizrahi, 2012). Galapagos is not the exception: during rodent eradication campaigns in the island of Pinzón, 28 endemic Galapagos hawks (*Buteo galapagoensis*) likely died because scientists did not anticipate that rat poison residues would persist in lava lizards for at least 850 days (Clout, Martin, Russell, & West, 2019; Rueda, Campbell, Fisher, Cunninghame, & Ponder, 2016). Therefore, restoration lies between conservation and degradation, not only because it is a way that humans 'repair' the environment, but because sometimes restoration *breaks* the environment even more.

Conservation in the Anthropocene Requires a Paradigm Shift

Human *consumption* of resources ($\bar{A}$ & $\bar{B}$, Fig. 2) is the process that drives environmental degradation. This is the concept behind ecological footprints, which measure human appropriation of the biosphere's supply of ecosystem goods and services (Borucke et al., 2013). The exact relationship between human populations, consumption, and environmental degradation has been debated and revised since Malthus' original publication in 1798, and different schools of thought have contributed to our understanding of the main drivers of this relationship. At a superficial level, Malthus' view of finite resources is particularly acute in the case of island ecosystems (Malthus, 1798). It is not a coincidence that competition for limited resources and the observable boom and busts patterns inspired both economic theorists and early naturalists like Darwin himself. Classical economic logic to create a natural reserve and regulate population growth in Galapagos is firmly rooted in Malthus (Grenier, 2007; Lu et al., 2013). However, the relationship between human population and environmental degradation is not one-dimensional, but rather depends on consumer patterns of different occupations and individual choices

(K. Davis, 1963). There is a feedback loop between population density and the intensity of resource exploitation (Boserup, 1965). Furthermore, some authors denounced the integration to global markets as a primary cause for environmental degradation and social inequality, as they can make people dependent on outside products and lead to overexploitation of resources (Nietschmann, 1972). However, neoclassical economists do not see market integration as a cause for environmental degradation, but instead blame market distortions (regulations, taxes, subsidies) that don't allow the free market to innovate and find technological replacements. Additionally, political ecologists focus on how structural issues affect the distribution of resources and see poverty as the proximate cause of environmental degradation (Jolly, 1994). Rather than consider these as competing explanations for ecological degradation, their insights are tools we can use to analyze the complex relationships between humans and their environment. Whatever lens we use, it is clear that growth-oriented 'conscious consumerism' and greenwashed ecologically-friendly tourism will not reconcile environmental integrity with the high consumption rates of Ecuador's fastest-growing province. For ecotourism ventures to indeed be sustainable, they must reject the continuous increase in size and scale of their operations, reduce their reliance on external inputs, and focus on improving experiences rather than trying to draw more tourists (Conway & Timms, 2010). Marginal sites within the islands are most vulnerable to the economic, social, and environmental pressures from the increase in tourism infrastructure and connectivity with the continent (Timms & Conway, 2015). However, marginal areas are particularly well-positioned to counter the hard growth model from urban areas as well, as they have not yet developed in a particular direction (Conway & Timms, 2010; Timms & Conway, 2015). Given that rural areas are currently at the margins of where most infrastructure exists, they are also particularly well-positioned to lead with alternative models for development.

Agricultural zones are well-positioned to push for an alternative and more sustainable form of development for Galapagos, but only if the region's development follows agroecological principles. *Agroecology* (A & B, Fig. 2) blurs the lines between agriculture and conservation because it integrates concepts and methods from social and natural sciences to better understand the complexities that arise from particular sociocultural contexts and design more sustainable agricultural systems (Pettersson, 2015). For example, ecologists focus on energy and matter pathways as they flow among a community of species and their physical environment to define an *ecosystem* (Begon, Townsend, & Harper, 2006). Similarly, an *agroecosystem* is the basic unit of analysis in agroecology; as the name suggests, it is a type of ecosystem with food production as its core, but it extends beyond the boundaries of the farm to encompass the area that is affected by the material and energetic flows of the farm. Agroecology studies agroecosystems holistically, including all human and environmental components, focusing on the structures, dynamics, and functions of their interrelationships (Altieri & Nicholls, 2005). Agroecological principles are particularly useful in the context of the Galapagos, where laws prohibit industrial-scale agriculture and emphasize the integrity of surrounding ecosystems. For example, agroecology emphasizes the crucial role of smallholder farms for food security as an alternative to industrial food production models (Altieri, 2009; Perfecto & Vandermeer, 2010) as well as their potential for environmental conservation (Chappell & LaValle, 2009; Gliessman, 1992). An agroecological perspective is useful for breaking down the binaries of conservation and agricultural interests (Perfecto & Vandermeer, 2008). For example, despite fundamental distinctions between *pristine* vs. *degraded* habitats, research has shown that endangered species can inhabit agricultural areas and that these may be biologically rich and worthy of preservation, and not only because biodiversity may serve human needs (Vandermeer & Perfecto, 1997). Furthermore, the amount of time and resources that the agricultural sector invests in combating invasive plants on their land dwarfs the investments by the conservation sector. One study found that while public institutions invested less than $500,000 per year to rid the island of Santa Cruz from invasive plants, farmers of the same island collectively invested more than two million dollars per year to clean their lands (Toledo, 2014). Far from being a source of problems, as the *natural laboratory* narrative suggests, conservation organizations could hardly ask for better allies than farmers. In more recent administrations, public institutions included agroecological language in public policy documents and acknowledged the crucial role of agriculture, not only for the region's food security, but for its conservation as well (Guzman & Poma, 2015). Decision-makers can design effective strategies for simultaneously controlling invasive species and pursuing sustainable livelihoods by recognizing the key functions that farmers serve as part of the multispecies assemblage of both protected and productive areas. This recognition of humans as an integral part of the landscape is still a relatively new trend in conservation, much like the new terminology of *novel ecosystems*.

Proponents of the *novel ecosystem* concept (A & B & $\bar{A}$ & $\bar{B}$, Fig. 2) define it as a 'new' assemblage of species that arises spontaneously and irreversibly from the degradation and invasion of wild systems or from the abandonment of intensively managed systems like agricultural or urbanized lands (Hobbs et al., 2006; Mascaro et al., 2013). This theoretical construct has been harshly criticized for its lack of explicit thresholds for defining irreversible ecological events or for what constitutes a *historical* vs. *novel* ecosystem (Murcia et al., 2014). Traditional conservation practitioners often characterize this concept as defeatist for abandoning the struggle to go back to historical (pre-human) species assemblages if they are too resource or time-intensive. However, proponents argue that recognizing that there are degrees of ecological degradation (novel > hybrid > historical) allows us to consider a greater diversity of management options for conservation, and that the irreversibility of degraded ecosystems is often determined by the limited resources that can be devoted to restoring degraded ecosystems (Hobbs, Higgs, & Harris, 2014). If critics of the concept of novel ecosystems argue that it is a managerial vision that detracts from 'true' conservation work, it is not because there is no way to identify *novel* land cover categories in management policy plans, but because this category challenges the *wild* vs. *domesticated* duality (Cole & Yung, 2010; Cronon, 1996; Haila, 2000). Novel ecosystems indeed disrupt all corners of the coupled human-natural system semiotic square (Fig. 2). It is a concept that is meant to aid conservation decision-making, but it is also the result of degradation, whether that be from the abandonment of intensively managed lands, like urban and rural areas, or the invasion of protected areas (Hobbs et al., 2006). However, the conservation sector should not shy away from it. Despite its many limitations, the novel ecosystem concept represents an acknowledgment of the conscious and unconscious ways that humans have shaped the environment through history, and this blurring of what constitutes as natural vs. artificial allows us to talk about the role of humans in the natural landscape without the stigma that traditional conservation places on human interventions. This recognition also signals to the broader inclusion of underrepresented human-centric interests at the conservation negotiating table. The tourism sector has always had a privileged position in shaping the narrative and vision for the Galapagos islands, and it might be time this changed. Most tourism ventures could hardly be defined as *ecotourism*, and as long as tourism agencies continue to profit from the *natural laboratory* narrative, people will keep turning a blind eye to their erosion of the islands' natural and social capital. Accepting land uses such as agriculture that drive novel ecosystems will no doubt be uncomfortable and inconvenient to reconcile with traditional narratives, but this is a similar concession to the one that had to be made to include the tourism sector in the imaginary of the Galapagos in the first place.

Integrating agriculture in the vision for Galapagos can no doubt bring about fruitful collaborations between agriculture and conservation, especially if done in conjunction with ecotourism. For example, decision-makers could support farmers who have turned to tourism not only as paying visitors of their farm but as a source of volunteers to tend their crops. This alternative form of tourism is already being applied in several farms of the Galapagos, in mainland Ecuador, and on tropical islands worldwide as a short-term fix for the labor market failures that have become common in rural areas (Terry, 2014; Wearing & McGehee, 2013). Organic farming usually requires more manual labor than industrial methods that rely on agrochemicals like pesticides, so strategies for achieving the policy guidelines for responsible food production would greatly benefit from an influx in agricultural volunteers (Mostafanezhad, Suryanata, Azizi, & Milne, 2015). Volunteer tourism is also compatible with environmental restoration in island environments (Chao, 2014), so there is a clear opportunity to channel the demand for volunteer farming opportunities in a world-renowned tropical location with the control of invasive plants. Volunteer tourism programs are effective at engaging not only the volunteers but the local populations that host them (Chao, 2014). Volunteer tourism may not represent a permanent solution to the structural issues that have led the agricultural sector to be in crisis, but it can serve as a stepping stone to reaching a more balanced distribution of resources between the different economic sectors so that supplying the local demand for responsibly-grown good becomes an economically attractive option for farmers. Conservation scientists must once again reach out beyond academic circles and find common ground with local populations in order to steer development in the Galapagos in a more sustainable direction. One such occasion when this occurred was when tourism agencies proposed to establish a hotel in Tortuga Bay, one of the most beautiful sites of Santa Cruz; conservation practitioners opposed this idea because it would prevent access to locals who did not have a car while irreversibly degrading Tortuga Bay (Bowman, 1960; Reck, 2017). As a triumph for conservation and a direct result of putting the interests of the local community front and center, the only land access to Tortuga Bay to this day is a pedestrian path. If conservation practitioners become aligned with the interests of marginalized groups at the boundary of protected and productive areas, similar potential alliances that seek to prioritize local populations while pushing against accelerated models of economic growth will no doubt abound.

Conclusion: Galapagos is a Garden with Human and Non-human Gardeners

> For gardeners of an early age, the concept of the frame was both practical and emotional. From Eden on, gardens have been places of refuge—worlds set apart, nature heightened or nature perfected, but first, nature *enclosed*. The world beyond the garden might be arid or barren, or dark, tangled, and inhabited with beasts. Alien bands with blue painted faces might lurk there, hostile of intent. But within the garden, life was ordered and fertile, soft, and safe. Our word *garden* derives, in fact, from the Old High German word *gart*, an enclosure or safe place, and the etymology of the word is preserved in *kindergarten*, a safe place for children. (Joe Eck, in *Elements of Garden Design*, 2005)

It is perhaps no coincidence that Eck's (2005) quote about garden design can be directly translated to the scale of the Enchanted Islands, where the national park serves as a *frame* to construct, manage, and progress toward the postcard-like vision of the islands: a safe place for endemic native fauna and flora, heightening its nature. Considering the borders of the GNP and the agricultural zone as frames also implies creativity rather than negation, a bringing in rather than a leaving out. Humans not only frame, but inhabit and (ideally) tend what thrives within this frame, and thus they too are part of the nature that should be heightened.

In this chapter, I used semiotic squares to expand upon the constructs that give meaning to two critical narratives of the Galapagos: the traditional *natural laboratory* narrative, and the more modern *coupled human-natural system* narrative. The narratives are not mutually exclusive, but the differences between them are significant because the modern narrative guides the conversation about what constitutes as conservation of the GNP towards new directions and considers previously overlooked or undervalued allies by the *natural laboratory* narrative, namely, the agricultural sector. Under traditional conservation narratives, the Galapagos is no place to grow food. Evolution was meant to be the only gardener in this earthly garden of Eden because scientists highlighted the ways agricultural activities threatened endemic ecosystems. The pristine character of evolutionary processes, which are readily visible in the *natural laboratory*, was meant to be studied by scientists and enjoyed by tourists alone. Scientists who first promoted ecotourism as a 'sustainable' alternative to other forms of development did not anticipate the environmental degradation, economic inequality, and social unrest that the tourism industry would bring to their evolutionary Eden in mere decades. The *natural laboratory* narrative placed tourism in a privileged seat at the conservation negotiating table while marginalizing other sectors, including agriculture. This bias has resulted in lax enforcement of regulations on tourism ventures and conservation practitioners turning a blind eye to the adverse effects of this industry. Now that the harmful effects of tourism have become apparent, and that farmers clearing their lands have proven to be crucial for the control of invasive plants, our narrative needs to be updated. For the sake of conserving this evolutionary Eden and providing security to its inhabitants, our imaginary of the Galapagos should include human gardeners as well.

The modern narrative of Galapagos as a *coupled human-natural system* recognizes that human agency is a driver of social and ecological patterns of all kinds, beyond the conservation to degradation spectrum. The ideas that guide traditional conservation favor outdated exclusionary tactics to prevent human populations from degrading the environment. However, economists, systems scientists, demographers, and political ecologists have since revised traditional assumptions and proposed that the relationship between human populations and environmental degradation is multi-dimensional, involving people's individual choices and consumption patterns, as well as different modes of production. Modern conservation scientists understand that the abandonment or prohibition of traditional land uses for the sake of conservation causes social discontent and is in itself unsustainable while further exacerbating people's reliance on the import-dependent development model. Meanwhile, embracing the *human* side from coupled human-natural systems in Galapagos allows decision-makers to move beyond exclusionary tactics for environmental protection and consider new solutions to socioecological crises. One particularly fruitful way of including humans into the conservation paradigm is by applying agroecological principles to support local food production and responsible agricultural practices simultaneously.

Supporting agroecological food production as part of conservation plans is one of the most direct ways decision-makers can include humans into the imaginary of the Galapagos. Agroecology harnesses the strengths of ecological and social sciences to understand the food production system from a functional perspective. Local food security and conservation are intricately intertwined under this perspective. Developing local capacity to provide food for residents and visitors makes the islands less dependent on imports from the mainland, thus reducing the likelihood that new organisms will enter the island, some of which are likely to become invasive species or agricultural pests. Real inclusion of the agrarian sector into the collective vision for this territory means giving farmers a seat at the negotiation table when crafting new legislation and investing a portion of the income from the tourism industry outside of the protected areas to address common problems. For example, decision-makers could take the lead from individual farms in the Galapagos that have already turned to volunteer tourism as a way to overcome the lack of affordable labor and help establish an alternative form of tourism; a portion of visitors could serve several more functions for local communities than they currently do under traditional tourism models: from controlling the spread of invasive plants on agricultural and national park lands to helping produce the food that is necessary to sustain themselves and others on the island.

Humans may be an integral and often irreversible agent of pattern, but other species partake in our place-making habits as well. From crop-raiding finches to and seed-dispersing giant tortoises, there are many non-human gardeners whose agency is also shaping the evolution of Galapagos landscapes. Humans are often caught off guard by other species' capacity to take advantage of the conditions we have created for our imaginary of what belongs on each side of the boundary surrounding a protected area. Invasive plants are the best example of how other species have made their home in places where we did not expect them. Ultimately, a weed is just a plant where the gardener does not want it. There is nothing inherently unnatural about a plant growing anywhere its seeds are dispersed by the wind, water or animals, including people. Once again, it is human agency (or the sudden halt of it) that promotes the invasion of territories by invasive plants. Novel ecosystems emerge from these interactions between humans and their environment. Regardless of whether humans remove said 'weeds' to plant crops or to reforest with native vegetation, both are *artificial* results that require upkeep. No matter which side of the GNP boundary we focus on, it is human hands and minds that shape the multispecies assemblage of both the productive and protected areas. Authors have urged conservationists and land managers to organize priorities not based on species origin, but rather on the function that they serve. Do species produce benefits or harm to biodiversity, human health, ecological services, and local economies? (M. Davis, 2011). Indeed, acknowledging that novel ecosystems can have value is a concession that conservationists must make to accommodate other economic activities that are needed for the region's socioeconomic diversification, but it is no different from the concession conservationists once made to include tourism as part of the vision of the Galapagos. A functional perspective allows us to see the current state of affairs based on relationships rather than idealized spaces and consider associations that were previously off-limits. Much like a gardener, we must pick which elements we want in our little plot of land and realize that it is our action or inaction that determines what will flourish in our Galapagos Garden.

References

Altieri, M. (2009). Agroecology, small farms, and food sovereignty. *Monthly Review-an Independent Socialist Magazine, 61*(3), 102–113.

Altieri, M., & Nicholls, C. (2005). *Agroecology and the search for a truly sustainable agriculture* (1st ed.). Mexico D.F., Mexico: United Nations Environment Programme.

Atkinson, R., Trueman, M., Guézou, A., Jaramillo, P., Paz, M., Sanchez, J., … Silva, M. (2010). Native gardens for Galapagos—Can community active help to prevent future plant invasions? *Galapagos Report 2009–2010* (pp. 159–163).

Barton, G. (2001). Empire forestry and the origins of environmentalism. *Journal of Historical Geography, 27*(4), 529–552. https://doi.org/10.1006/jhge.2001.0353

Basset, C. (2009). *Galapagos at the crossroads: Pirates, biologists, tourists and creationists battle for Darwin's cradle of evolution*. Washington, DC: National Geographic.

Begon, M., Townsend, C., & Harper, J. (2006). *Ecology: From individuals to ecosystems* (4th ed.). Malden, MA: Blackwell Publishing.

Benitez-Capistros, F., Camperio, G., Hugé, J., Dahdouh-Guebas, F., & Koedam, N. (2018). Emergent conservation conflicts in the Galapagos Islands: Human-giant tortoise interactions in the rural area of Santa Cruz Island. *PLoS ONE, 13*(9), 1–27. https://doi.org/10.1371/journal. pone.0202268

Benitez-Capistros, F., Hugé, J., Dahdouh-Guebas, F., & Koedam, N. (2016). Exploring conservation discourses in the Galapagos Islands: A case study of the Galapagos giant tortoises. *Ambio*, 1–19. https://doi.org/10.1007/s13280-016-0774-9

Benítez, F., Mena, C., & Zurita-Arthos, L. (2018). Urban land cover change in ecologically fragile environments: The case of the Galapagos Islands. *Land, 7*(1), 21. https://doi.org/10.3390/land7010021

Blake, S., Guezou, A., Deem, S. L., Yackulic, C. B., & Cabrera, F. (2015). The dominance of introduced plant species in the diets of migratory Galapagos tortoises increases with elevation on a human-occupied island. *Biotropica, 47*(2), 246–258. https://doi.org/10.1111/btp.12195

Bocarejo, D., & Ojeda, D. (2016). Violence and conservation: Beyond unintended consequences and unfortunate coincidences. *Geoforum, 69*, 176–183. https://doi.org/10.1016/j. geoforum.2015.11.001

Bocci, P. (2017). Tangles of care: Killing goats to save tortoises on the Galapagos Islands. *Cultural Anthropology, 32*(3), 424–449. https://doi.org/10.14506/ca32.3.08

Bonfiglioli, S. (2008). Aristotle's non-logical works and the square of oppositions in semiotics. *Logica Universalis, 2*(1), 107–126. https://doi.org/10.1007/s11787-007-0021-z

Borucke, M., Moore, D., Cranston, G., Gracey, K., Iha, K., Larson, J., … Galli, A. (2013). Accounting for demand and supply of the biosphere's regenerative capacity: The National Footprint Accounts' underlying methodology and framework. *Ecological Indicators, 24*, 518–533. https://doi.org/10.1016/j.ecolind.2012.08.005

Boserup, E. (1965). The conditions of agricultural growth. *Population Studies, 20*. https://doi. org/10.2307/2172620

Bowman, R. I. (1960). *A biological reconnaissance of the Galapagos Islands during 1957* (pp. 1–56). United Nations Educational, Scientific and Cultural Organization.

Bremner, J., & Perez, J. (2002). A case study of human migration and the sea cucumber crisis in the Galapagos Islands. *Ambio: A Journal of the Human Environment, 31*(4), 306–310. https:// doi.org/10.1579/0044-7447-31.4.306

Brewington, L. (2013). Mapping invasion and eradication of feral goats in the Alcedo region of Isabela Island, Galapagos. *International Journal of Remote Sensing, 34*(7), 2286–2300. https:// doi.org/10.1080/01431161.2012.743695

Brewington, L., Rosero, O., Bigue, M., & Cervantes, K. (2013). *The quarantine chain— Establishing an effective biosecurity system to prevent the introduction of invasive species into the Galapagos Islands*. WildAid.

Buddenhagen, C., & Renteria, J. (2009). The control of a highly invasive tree Cinchona pubescens in Galapagos. *Weed Technology, 18*, 1194–1202. https://doi.org/10.1614/0890-037X(2004)01 8[1194:TCOAHI]2.0.CO;2

Buddenhagen, C. E., & Tye, A. (2015). Lessons from successful plant eradications in Galapagos: Commitment is crucial. *Biological Invasions, 17*(10), 2893–2912. https://doi.org/10.1007/ s10530-015-0919-y

Carrion, V., Donlan, C. J., Campbell, K. J., Lavoie, C., & Cruz, F. (2011). Archipelago-wide island restoration in the Galápagos Islands: Reducing costs of invasive mammal eradication programs and reinvasion risk. *PLoS One, 6*(5), e18835. https://doi.org/10.1371/journal.pone.0018835

Cater, E. (1993). Ecotourism in the third world: Problems for sustainable tourism development. *Tourism Management, 14*(2), 85–90. https://doi.org/10.1016/0261-5177(93)90040-R

Cecchin, A. (2017). Material flow analysis for a sustainable resource management in island ecosystems: A case study in Santa Cruz Island (Galapagos). *Journal of Environmental Planning and Management, 60*(9), 1640–1659. https://doi.org/10.1080/09640568.2016.1246997

Celata, F., & Sanna, V. S. (2012). The post-political ecology of protected areas: Nature, social justice and political conflicts in the Galápagos Islands. *Local Environment, 17*(Dec.), 977–990. https://doi.org/10.1080/13549839.2012.688731

CGREG. (2015). Censo de Unidades de Producción Agropecuaria de Galápagos 2014. In *Consejo de Gobierno del Régimen Especial de Galápagos (CGREG)*. Retrieved from http://sinagap.agricultura.gob.ec/censo-unidades-produccion-agropecuaria-galapagos-2014

CGREG, C. de G. del R. E. de G. (2014). *Políticas Agropecuarias para Galápagos* (pp. 1–45). San Cristobal: Consejo de Gobierno del Régimen Especial de Galápagos.

Chao, R. F. (2014). Volunteer tourism as the approach to environmental management. A case study of Green Island in Taiwan. *Journal of Environmental Protection and Ecology, 15*(3), 1377–1384.

Chappell, M. J., & LaValle, L. A. (2009). Food security and biodiversity: Can we have both? An agroecological analysis. *Agriculture and Human Values, 28*(1), 3–26. https://doi.org/10.1007/s10460-009-9251-4

Chauvenet, A. L. M., Durant, S. M., Hilborn, R., & Pettorelli, N. (2011). Unintended consequences of conservation actions: Managing disease in complex ecosystems. *PLoS ONE, 6*(12). https://doi.org/10.1371/journal.pone.0028671

Chiriboga, R., Fonseca, B., & Maignan, S. (2006). Producto 1: Zonificación agroecológica de las zonas agropecuarias en relación con las especies invasoras. In *Proyecto ECU/00/G31 "Control de las especies invasoras en el Archipiélago de las Galápagos"*. Galapagos, Ecuador: United Nations Development Programme (UNDP).

Chiriboga, R., & Maignan, S. (2006a). Producto 2 (A): Historia de las relaciones y elementos de la reproduccion social agraria en Galapagos. In *Proyecto ECU/00/G31 "Control de las especies invasoras en el Archipiélago de las Galápagos"*. Puerto Ayora: United Nations Development Programme (UNDP).

Chiriboga, R., & Maignan, S. (2006b). Producto 3 (1/2) Mapeo de Actores Relacionados con el sector agropecuario y el fenomeno de las especies invasoras en Galapagos. *Especies Invasoras de Las Galapagos*. MAE, GEF, INGALA, UNDP.

Clout, M. N., Martin, A. R., Russell, J. C., & West, C. J. (2019). Island invasives: Scaling up to meet the challenge. *Occasional Paper SSC* (Vol. 62).

Cole, D. N., & Yung, L. (2010). *Beyond naturalness: Rethinking park and wilderness stewardship in an era of rapid change* (D. N. Cole & L. Yung, Eds.). Washington, DC: Island Press.

Colwell, R. R. (2001). Biocomplexity. ELS, 1–8. https://doi.org/10.1038/npg.els.0003437.

Congreso Nacional. (1998). *Ley de regimen especial para la conservacion y desarrollo sustentable de la provincia de Galapagos*. Quito, Ecuador: Registro Oficial No. 278.

Conway, D., & Timms, B. F. (2010). Re-branding alternative tourism in the Caribbean: The case for 'slow tourism'. *Tourism and Hospitality Research, 10*(4), 329–344. https://doi.org/10.1057/thr.2010.12

Cronon, W. (1996). The trouble with wilderness: Or, getting back to the wrong nature. *Environmental History, 1*(1), 7–28.

Darwin, C. (1859). *On the origin of species by means of natural selection*. https://doi.org/10.2307/2485224

Davis, K. (1963). The theory of change and response in modern demographic history. *Population Index, 29*(4), 345–366.

Davis, M. (2011). Don't judge species on their origins. *Nature, 474*(9), 153–154. https://doi.org/10.1038/474153a

Denevan, W. (1992). The pristine myth: The landscape of the Americas in 1492. *Annals of the Association of American Geographers, 82*(3), 369–385.

Despommier, D. (2013). Farming up the city: The rise of urban vertical farms. *Trends in Biotechnology, 31*(7), 388–389. https://doi.org/10.1016/j.tibtech.2013.03.008

Dirección del Parque Nacional Galapagos. (2014). *Plan de Manejo de las Áreas Protegidas de Galápagos para el Buen Vivir*. Puerto Ayora, Galapagos, Ecuador.

Dodd, R. J., & Sharpley, A. N. (2016). Conservation practice effectiveness and adoption: Unintended consequences and implications for sustainable phosphorus management. *Nutrient Cycling in Agroecosystems, 104*(3), 373–392. https://doi.org/10.1007/s10705-015-9748-8

Donoso, S. (2012). *Piratas en Galapagos (1680–1720)*. Quito, Ecuador: Editorial Ecuador.

Eck, J. (2005). *Elements of garden design*. New York, NY: North Point Press.

Edwards, P. J., & Abivardi, C. (1998). The value of biodiversity: Where ecology and economy blend. *Biological Conservation, 83*(3), 239–246. https://doi.org/10.1016/S0006-3207(97)00141-9

Ellis-Soto, D., Blake, S., Soultan, A., Guézou, A., Cabrera, F., & Lötters, S. (2017). Plant species dispersed by Galapagos tortoises surf the wave of habitat suitability under anthropogenic climate change. *PLoS ONE, 12*(7), 1–16. https://doi.org/10.1371/journal.pone.0181333

Epler, B. (2007). *Tourism, the economy, population growth, and conservation in Galapagos* (p. 75). Charles Darwin Foundation.

Epler, B., Watkins, G., & Cárdenas, S. (2008). Tourism and the Galápagos economy. *Galapagos Report 2006–2007* (pp. 42–47).

Epler Wood, M., Milstein, M., & Ahamed-Broadhurst, K. (2019). Destinations at risk: The invisible burden of tourism. The Travel Foundation.

Folke, C. (2006). Resilience: The emergence of a perspective for social–ecological systems analyses. *Global Environmental Change, 16*(3), 253–267. https://doi.org/10.1016/j.gloenvcha.2006.04.002

Gardener, M. R., Atkinson, R., & Renteria, J. L. (2010). Eradications and people: Lessons from the plant eradication program in Galapagos. *Restoration Ecology, 18*(1), 20–29. https://doi.org/10.1111/j.1526-100X.2009.00614.x

Gardener, M. R., Atkinson, R., Rueda, D., & Hobbs, R. J. (2010). Optimizing restoration of the degraded highlands of Galapagos: A conceptual framework. *Galapagos Report 2009–2010* (pp. 164–169).

Ginsberg, J. R. (1987). What is conservation biology? *Trends in Ecology & Evolution, 2*(9), 262–264. https://doi.org/10.1016/0169-5347(87)90031-0

Gliessman, S. R. (1992). Agroecology in the tropics: Achieving a balance between land use and preservation. *Environmental Management, 16*(6), 681–689. https://doi.org/10.1007/BF02645658

Gómez, E. G., & Ramírez, N. B. G. (2017). Agricultura en las islas Galapagos, contradicciones y retos. *La Jornada Del Campo, 112*, 1–3. Retrieved from https://www.jornada.com.mx/2017/01/21/cam-islas.html

González, J. A., Montes, C., Rodríguez, J., & Tapia, W. (2008). Rethinking the Galápagos Islands as a complex social-ecological system: Implications for conservation and management. *Ecology and Society, 13*(2), 13. https://doi.org/10.5751/ES-02557-130213

Gowdy, J. M. (1997). The value of biodiversity: Markets, society, and ecosystems. *Land Economics, 73*(1), 25–41.

Greimas, A. J. (1970). *On meaning: Selected writings in semiotic theory*. Minneapolis: University of Minnesota Press.

Greimas, A. J. (1983). *Structural semantics: An attempt at a method*. Lincoln: University of Nebraska Press.

Grenier, C. (2007). Conservación contra natura. Las Islas Galápagos. In *Travaux de l'Institut Français d'Études Andines* (Vol. 233). https://doi.org/10.4000/books.ifea.5519

Groot de, R. S. (1983). Tourism and conservation in the Galapagos Islands. *Biological Conservation, 26*(4), 291–300. https://doi.org/10.1016/0006-3207(83)90093-9

Guézou, A., Trueman, M., Buddenhagen, C. E., Chamorro, S., Guerrero, A. M., Pozo, P., & Atkinson, R. (2010). An extensive alien plant inventory from the inhabited areas of Galapagos. *PLoS ONE, 5*(4), 1–8. https://doi.org/10.1371/journal.pone.0010276

Guzman, J. C., & Poma, J. E. (2015). Bioagricultura: Una oportunidad para el buen vivir insular. In *Informe Galapagos 2013–2014* (pp. 25–29). Puerto Ayora, Galapagos: GNPD, GCREG, CDF, GC.

Haila, Y. (2000). Beyond the nature-culture dualism. *Biology and Philosophy, 15*(2), 155–175. https://doi.org/10.1023/A:1006625830102

Haraway, D. (1992). The promises of monsters. In L. Grossberg, C. Nelson, & P. Treichler (Eds.), *Cultural studies* (pp. 295–337). New York and London: Routledge.

Hennessy, E. (2015). The molecular turn in conservation: Genetics, pristine nature, and the redis-
    covery of an extinct species of Galápagos giant tortoise. *Annals of the Association of American
    Geographers, 105*(1), 87–104. https://doi.org/10.1080/00045608.2014.960042
Hennessy, E., & Mccleary, A. L. (2011). Nature's Eden? The production and effects of "pristine"
    nature in the Galápagos Islands. *Island Studies Journal, 6*(2), 131–156.
Heylings, P., & Cruz, F. (1998). Common property, conflict and participatory management in the
    Galapagos Islands. *Crossing Boundaries Conference*.
Hobbs, R. J., Arico, S., Aronson, J., Baron, J. S., Bridgewater, P., Cramer, V. A., … Zobel, M. (2006).
    Novel ecosystems: Theoretical and management aspects of the new ecological world order. *Global
    Ecology and Biogeography, 15*(1), 1–7. https://doi.org/10.1111/j.1466-822X.2006.00212.x
Hobbs, R. J., Higgs, E. S., & Harris, J. A. (2014). Novel ecosystems: Concept or inconvenient
    reality? A response to Murcia et al. *Trends in Ecology and Evolution, 29*(12), 645–646. https://
    doi.org/10.1016/j.tree.2014.09.006
Holling, C. S., Gunderson, L. H., & Ludwig, D. (2002). In search of a theory of adaptive change.
    In *Panarchy*. Island Press.
Hoyman, M. M., & McCall, J. R. (2013). Is there trouble in paradise? The perspectives of
    Galapagos community leaders on managing economic development and environmental conser-
    vation through ecotourism policies and the Special Law of 1998. *Journal of Ecotourism, 12*(1),
    33–48. https://doi.org/10.1080/14724049.2012.749882
IUCN. (1979). *UNESCO Advisory Body Technical Evaluation—Galapagos*. Retrieved from
    https://whc.unesco.org/en/list/1/documents
Izurieta, A., Tapia, W., Mosquera, G., & Chamorro, S. (2014). *Plan de Manejo de las Áreas
    Protegidas de Galápagos para el Buen VIvir*.
Jackson, L., Lopoukhine, N., & Hillyard, D. (1995). Ecological restoration: A definition and com-
    ments. *Restoration Ecology, 3*(2), 71–75. Retrieved from http://www3.interscience.wiley.com/
    journal/119255787/abstract
Jackson, R. L., Drummond, D. K., & Camara, S. (2007). What is qualitative research? *Qualitative
    Research Reports in Communication, 8*(1), 21–28. https://doi.org/10.1080/17459430701617879
Johnson, D. L., Ambrose, S. H., Bassett, T. J., Bowen, M. L., Crummey, D. E., Isaacson, J. S.,
    … Winter-Nelson, A. E. (1997). Meanings of environmental terms. *Journal of Environment
    Quality, 26*(3), 581. https://doi.org/10.2134/jeq1997.00472425002600030002x
Jolly, C. L. (1994). Four theories of population change the environment. *Population and
    Environment: A Journal of Interdisciplinary Studies, 16*(1), 61–90. https://doi.org/10.1007/
    BF02208003
Kannar-Lichtenberger, L. (2018). Beyond the verbiage: Consumerism through tourism and its
    manifestations in small islands and remote places. *Environment and Ecology Research, 6*(5),
    471–478. https://doi.org/10.13189/eer.2018.060507
Kay, J. (2000). Ecosystems as Self-organizing Holarchic Open Systems : Narratives and the
    Second Law of Thermodynamics. In *Handbook of Ecosystem Theories and Management* (pp.
    135–160). CRC Press.
Kerr, S., Cardenas, S., & Hendy, J. (2004). Migration and the environment in the Galapagos: An
    analysis of economic and policy incentives driving migration, potential impacts from migra-
    tion control, and potential policies to reduce migration pressure. *Motu Working Paper Series
    No. 03-17*.
Kingston, P. F., Runciman, D., & McDougall, J. (2003). Oil contamination of sedimentary shores
    of the Galápagos Islands following the wreck of the Jessica. *Marine Pollution Bulletin, 47*(7–
    8), 303–312. https://doi.org/10.1016/S0025-326X(03)00159-0
Laso, F. J., Benítez, F. L., Rivas-torres, G., Sampedro, C., & Arce-nazario, J. (2020). Land cover
    classification of complex agroecosystems in the non-protected highlands of the Galapagos
    Islands. *Remote Sensing, 12*(65), 1–39. https://doi.org/10.3390/rs12010065
Lave, R., Doyle, M., & Robertson, M. (2010). Privatizing stream restoration in the US. *Social
    Studies of Science, 40*(5), 677–703. https://doi.org/10.1177/0306312710379671

Lave, R., Wilson, M. W., Barron, E. S., Biermann, C., Carey, M. A., Duvall, C. S., … Van Dyke, C. (2014). Intervention: Critical physical geography. *Canadian Geographer, 58*(1), 1–10. https://doi.org/10.1111/cag.12061

Lawesson, J. E. (1988). Stand level dieback and regeneration of forests in the Galapagos Islands. *Vegetatio, 77,* 87–93.

Lister, N. (1998). A systems approach to biodiversity conservation planning. *Environmental Monitoring and Assessment, 49,* 123–155. Retrieved from http://link.springer.com/article/10.1023/A:1005861618009

Liu, J., Dietz, T., Carpenter, S. R., Alberti, M., Folke, C., Moran, E., … Taylor, W. W. (2007). Complexity of coupled human and natural systems. *Science, 317*(5844), 1513–1516. https://doi.org/10.1126/science.1144004

Lu, F., Valdivia, G., & Wolford, W. (2013). Social dimensions of 'Nature at Risk' in the Galápagos Islands, Ecuador. *Conservation and Society, 11*(1), 83. https://doi.org/10.4103/0972-4923.110945

MacFarland, C. G., Villa, J., & Toro, B. (1974). The Galapagos giant tortoises (Geochelone elephantopus) Part I: Status of the surviving populations. *Biological Conservation, 6*(2), 198–212. https://doi.org/10.1016/0006-3207(74)90068-8

Malthus, T. R. (1798). An essay on the principle of population, as it affects the future improvement of society. *Contemporary Sociology, 2.* https://doi.org/10.2307/2064821

Mariscal, R. N. (1969). Charles Darwin and conservation in the Galápagos islands. *Biological Conservation, 2*(1), 44–46. https://doi.org/10.1016/0006-3207(69)90114-1.

Mascaro, J., Harris, J. A., Lach, L., Thompson, A., Perring, M. P., Richardson, D. M., & Ellis, E. C. (2013). Origins of the novel ecosystems concept. In *Novel ecosystems: Intervening in the new ecological world order.* https://doi.org/10.1002/9781118354186.ch5

Miller, M. L., Carter, R. W. B., Walsh, S., & Peake, S. (2014). A conceptual framework for studying global change, tourism, and the sustainability of iconic national parks. *George Wright Forum, 31*(3), 256–269.

Ministerio de Agricultura. (2018). *Plan Agropecuario Ecosostenible Para Galapagos 2018–2022 (Enero 2018).* Dirección Provincial Agropecuaria de Galápagos.

Mizrahi, M. (2012). *Potential for unintended consequences in an Ecuadorian hook exchange program.* University of Washington.

Mostafanezhad, M., Suryanata, K., Azizi, S., & Milne, N. (2015). "Will Weed for Food": The political economy of organic farm volunteering in Hawai'i. *Geoforum, 65,* 125–133. https://doi.org/10.1016/j.geoforum.2015.07.025

Murcia, C., Aronson, J., Kattan, G. H., Moreno-Mateos, D., Dixon, K., & Simberloff, D. (2014). A critique of the "novel ecosystem" concept. *Trends in Ecology and Evolution, 29*(10), 548–553. https://doi.org/10.1016/j.tree.2014.07.006

Nietschmann, B. (1972). Hunting and fishing focus among the Miskito Indians, eastern Nicaragua. *Human Ecology, 1*(1), 41–67. https://doi.org/10.1007/BF01791280

Noel Castree. (2009). Nature and society. In *The Sage handbook of geographical knowledge* (Vol. 30, pp. 73–79). https://doi.org/10.1163/156854289X00453

Nogués-Bravo, D., Simberloff, D., Rahbek, C., & Sanders, N. J. (2016). Rewilding is the new Pandora's box in conservation. *Current Biology, 26*(3), R87–R91. https://doi.org/10.1016/j.cub.2015.12.044

Panetta, F. D. (2009). Weed eradication—An economic perspective. *Invasive Plant Science and Management, 2*(4), 360–368. https://doi.org/10.1614/ipsm-09-003.1

Peñafiel, J. (2017). *Posibilidad de adopcion y de permanencia en el tiempo de sistemas silvopastoriles en la isla Santa Cruz, Galapagos.* Universidad Central del Ecuador. Retrieved from http://www.dspace.uce.edu.ec/handle/25000/15102

Perfecto, I., & Vandermeer, J. (2008). Biodiversity conservation in tropical agroecosystems: A new conservation paradigm. *Annals of the New York Academy of Sciences, 1134,* 173–200. https://doi.org/10.1196/annals.1439.011

Perfecto, I., & Vandermeer, J. (2010). The agroecological matrix as alternative to the land-sparing/
    agriculture intensification model. *Proceedings of the National Academy of Sciences of the
    United States of America, 107*(13), 5786–5791. https://doi.org/10.1073/pnas.0905455107
Pettersson, J. (2015). Introduction: Agroecology as a transdisciplinary, participatory, and action-
    oriented approach. In *Agroecology: A transdisciplinary, participatory and action-oriented
    approach* (pp. 1–21). https://doi.org/10.1002/pssb.201300062
Pickett, S. T. A., Cadenasso, M. L., & Grove, J. M. (2005). Biocomplexity in coupled natural–
    human systems: A multidimensional framework. *Ecosystems, 8*(3), 225–232. https://doi.
    org/10.1007/s10021-004-0098-7
Pizzitutti, F., Mena, C., & Walsh, S. (2014). Modelling tourism in the Galapagos Islands: An agent-
    based model approach. *Journal of Artificial Societies and Social Simulations, 17*(2014), 1–25.
Proaño, M. E., & Epler, B. (2007). Tourism in Galapagos: A strong growth trend. *Galapagos
    Report 2006–2007* (pp. 31–35). Retrieved from https://www.galapagos.org/wp-content/
    uploads/2012/04/socio4-tourism-in-galapagos.pdf
Quiroga, D. (2009). Crafting nature: The Galapagos and the making and unmaking of a "natural
    laboratory". *Journal of Political Ecology, 16*, 123–140.
Quiroga, D. (2013). Changing views of the Galapagos. In S. Walsh & C. Mena (Eds.), *Science and
    conservation in the Galapagos Islands* (p. 243). https://doi.org/10.1007/978-1-4614-5794-7
Quiroga, D. (2017). Darwin, emergent process, and the conservation of Galapagos ecosystems.
    In *Darwin, Darwinism and conservation in the Galapagos Islands* (pp. 135–150). https://doi.
    org/10.1007/978-3-319-34052-4
Reck, G. (2017). The Charles Darwin Foundation: Some critical remarks about its history and
    trends. In *Darwin, Darwinism and conservation in the Galapagos Islands* (pp. 109–133).
    https://doi.org/10.1007/978-3-319-34052-4
Rentería, J., & Buddenhagen, C. (2006). Invasive plants in the Scalesia pedunculata forest at Los
    Gemelos, Santa Cruz, *Galápagos. Galapagos Research, 64*, 31–35. https://doi.org/10.1016/j.
    agee.2006.01.012.
Romanova, A. P., Yakushenkov, S. N., & Lebedeva, I. V. (2013). Media coverage of cultural heri-
    tage and consumerism in modern society. *World Applied Sciences Journal, 24*(1), 103–112.
    https://doi.org/10.5829/idosi.wasj.2013.24.01.13178
Rowley, A. (2013). Restoration of Pinta Island through the repatriation of giant tortoises. *Testudo,
    7*(4), 50–66.
Rueda, D., Campbell, K. J., Fisher, P., Cunninghame, F., & Ponder, J. B. (2016). Biologically sig-
    nificant residual persistence of brodifacoum in reptiles following invasive rodent eradication,
    Galapagos Islands, Ecuador. *Conservation Evidence, 13*, 38.
Salvador Ayala, G. (2015). *Análisis del sistema de producción y abastecimiento de alimentos en
    Galápagos*. Facultad Latinoamericana de Ciencias Sociales (FLACSO).
Sampedro, C., Pizzitutti, F., Quiroga, D., Walsh, S. J., & Mena, C. F. (2018). Food supply system
    dynamics in the Galapagos Islands: Agriculture, livestock and imports. *Renewable Agriculture
    and Food Systems*, 1–15. https://doi.org/10.1017/S1742170518000534
Sanderson, E., Jaiteh, M., Levy, M., Redford, K., Wannebo, A., & Woolmer, G. (2002).
    The human footprint and the last of the wild. *BioScience, 52*(10), 891. https://doi.
    org/10.1641/0006-3568(2002)052[0891:THFATL]2.0.CO;2
Scoones, I. (1999). New ecology and the social sciences: What prospects for a fruitful engage-
    ment? *Annual Review of Anthropology, 28*(1), 479–507. https://doi.org/10.1146/annurev.
    anthro.28.1.479
Shaw, J. (2015). Galápagos rebellion against foreign investment in hotels, golf courses, luxury tour-
    ism. Retrieved September 5, 2019, from The Ecologist website: https://theecologist.org/2015/
    jun/25/galapagos-rebellion-against-foreign-investment-hotels-golf-courses-luxury-tourism
Sterling, E. J., Gómez, A., & Porzecanski, A. L. (2010). A systemic view of biodiversity and
    its conservation: Processes, interrelationships, and human culture. *BioEssays: News and
    Reviews in Molecular, Cellular and Developmental Biology, 32*(12), 1090–1098. https://doi.
    org/10.1002/bies.201000049

Sulloway, F. J. (1982). Darwin and his finches: The evolution of a legend. *Journal of the History of Biology, 15*(1), 1–53. https://doi.org/10.1007/BF00132004

Sulloway, F. J. (2009). Tantalizing tortoises and the Darwin-Galapagos legend. *Journal of the History of Biology, 42*(1), 3–31. https://doi.org/10.1007/s10739-008-9173-9

Taylor, J. E., Dyer, G. A., Stewart, M., Yunez-Naude, A., & Ardila, S. (2003). The economics of ecotourism: A Galápagos Islands economy-wide perspective. *Economic Development and Cultural Change, 51*(4), 977–997. https://doi.org/10.1086/377065

Taylor, J. E., Hardner, J., & Stewart, M. (2009). Ecotourism and economic growth in the Galapagos: An island economy-wide analysis. *Environment and Development Economics, 14*(2), 139–162. https://doi.org/10.1017/S1355770X08004646

Terry, W. (2014). Solving labor problems and building capacity in sustainable agriculture through volunteer tourism. *Annals of Tourism Research, 49*, 94–107. https://doi.org/10.1016/j.annals.2014.09.001

Timms, B. F., & Conway, D. (2015). Slow tourism at the Caribbean's geographical margins. *Tourism Geographies, 6688*, 37–41. https://doi.org/10.1080/14616688.2011.610112

Toledo, A. (2014). *Rentabilidad de la Producción Agrícola en Santa Cruz, Galápagos* [Technical Report]. Conservation International.

Trueman, M., & D'Ozouville, N. (2010). Characterizing the Galapagos terrestrial climate in the face of global climate change. *Galapagos Research, 67*, 26–37.

Turner, B. L., Kasperson, R. E., Matson, P. A., McCarthy, J. J., Corell, R. W., Christensen, L., … Schiller, A. (2003). A framework for vulnerability analysis in sustainability science. *Proceedings of the National Academy of Sciences of the United States of America, 100*(14), 8074–8079. https://doi.org/10.1073/pnas.1231335100

Turner, B. L., Lambin, E. F., & Reenberg, A. (2007). The emergence of land change science for global environmental change and sustainability. *Proceedings of the National Academy of Sciences of the United States of America, 104*(52), 20666–20671. https://doi.org/10.1073/pnas.0704119104

Valdivia, G., Wolford, W., & Lu, F. (2014). Border crossings: New geographies of protection and production in the Galápagos Islands. *Annals of the Association of American Geographers, 104*(3), 686–701. https://doi.org/10.1080/00045608.2014.892390

Valdivia, G., Wolford, W., Polo, P., Pena, K., Nelson, J., Perry, S., Lu, H., Hansanungrum, N., Walsh-Dilley, M. (2013). *Supporting Local Food Systems : New Geographies of Conservation and Production in Galápagos.* Cornell University Student Multidisciplinary Applied Research Team (SMART).

Vandermeer, J., & Perfecto, I. (1997). The agroecosystem: A need for the conservation biologist's lens. *Conservation Biology, 11*(3), 591–592. https://doi.org/10.1046/j.1523-1739.1997.07043.x

Viteri Mejia, C. (2014). Propuestas de Política Pública para la Restauración el Paisaje Agrícola en las Islas Galápagos; Documento de Discusión. *Conservation International.* Conservation International.

Wallace, G. N., & Pierce, S. M. (1996). An evaluation of ecotourism in Amazonas, Brazil. *Annals of Tourism Research, 23*(4), 843–873. https://doi.org/10.1016/0160-7383(96)00009-6

Walsh, S., Malanson, G., Messina, J., Brown, D., & Mena, C. (2011). Biocomplexity. In *The Sage handbook of biogeography* (pp. 469–488). London: SAGE Publications Ltd.

Walsh, S., & Mena, C. (2013). Perspectives for the study of the Galapagos Islands: Complex systems and human–environment interactions. In *Science and conservation in the Galapagos Islands* (p. 243). https://doi.org/10.1007/978-1-4614-5794-7

Walsh, S., & Mena, C. (2016). Interactions of social, terrestrial, and marine sub-systems in the Galapagos Islands, Ecuador. *Proceedings of the National Academy of Sciences, 2006*(13), 201604990. https://doi.org/10.1073/pnas.1604990113

Walsh, S., Miller, B., Breckheimer, I., Mccleary, A., Guzman-Ramirez, L., Caplow, S., & Jones-smith, J. (2009). Complexity theory and spatial simulation models to assess population-environment interactions in the Galapagos Islands. *Galapagos Science Symposium*, March 2016, 145–148.

Watkins, G., & Cruz, F. (2007). Galapagos at risk—A socioeconomic analysis. Retrieved from http://www.cometogalapagos.com/Galapagos_at_Risk.pdf

Watson, J., Trueman, M., Tufet, M., Henderson, S., & Atkinson, R. (2009). Mapping terrestrial anthropogenic degradation on the inhabited islands of the Galapagos Archipelago. *Oryx, 44*(1), 79. https://doi.org/10.1017/S0030605309990226

Wearing, S., & McGehee, N. G. (2013). Volunteer tourism: A review. *Tourism Management, 38*, 120–130. https://doi.org/10.1016/j.tourman.2013.03.002

Wehrwein, G. S. (1942). The rural-urban fringe. *Economic Geography, 18*(3), 217–228. https://doi.org/10.2307/141123

Wolford, W., Lu, F., & Valdivia, G. (2013). Environmental crisis and the production of alternatives: Conservation practice(s) in the Galapagos Islands. In *Science and conservation in the Galapagos Islands* (pp. 87–104). London: Springer.

Yackulic, C. B., Blake, S., & Bastille-Rousseau, G. (2017). Benefits of the destinations, not costs of the journeys, shape partial migration patterns. *Behavioral Ecology*. https://doi.org/10.1111/1365-2656.12679

Evaluating Land Cover Change on the Island of Santa Cruz, Galapagos Archipelago of Ecuador Through Cloud-Gap Filling and Multi-sensor Analysis

Yang Shao, Heng Wan, Alexander Rosenman, Francisco J. Laso, and Lisa M. Kennedy

Introduction

Tropical islands are vitally important ecosystems that support high biological diversity and productivity as well as contributing significantly to human societies (Walsh & Mena, 2013). The Galapagos Archipelago is well known for its natural beauty, history, diverse wildlife, and conservation efforts. Santa Cruz, San Cristobal, and Isabela Islands in the Galapagos Archipelago of Ecuador are among the most popular tourist attractions in the world. Since 1991, the annual number of visitors to the Archipelago increased by 9% per year (Epler, 2007), and an estimate of over 220,000 tourists visited Galapagos in 2015 (Izurieta, 2017). Partly due to growing tourism and related job opportunities, population growth rates in the previous two decades remained high at around 6.4% per year, mainly driven by migration from mainland Ecuador (Baine, Howard, Kerr, Edgar, & Toral, 2007; Epler, 2007). Human migration and tourism have already affected fragile and sensitive islands (Brewington, Frizzelle, Walsh, Mena, & Sampedro, 2014 and are increasingly impacting island ecosystems with the introduction of invasive plant species (Tye, 2001; Walsh et al., 2008). In addition, community expansion and infrastructure development further disturb the natural environment as a consequence of population migration of tourists and residents (Benítez, Mena, & Zurita-Arthos, 2018; Walsh et al., 2010).

Despite such dynamics, land cover map products currently available for the Galapagos Islands are limited in both spatial and temporal extents (Benítez et al., 2018; McCleary, 2013). The global land cover maps derived from the Moderate

Y. Shao (✉) · H. Wan · A. Rosenman · L. M. Kennedy
Department of Geography, Virginia Tech, Blacksburg, VA, USA
e-mail: yshao@vt.edu

F. J. Laso
Department of Geography, University of North Carolina at Chapel Hill, Chapel Hill, NC, USA

© Springer Nature Switzerland AG 2020
S. J. Walsh et al. (eds.), *Land Cover and Land Use Change on Islands*, Social and Ecological Interactions in the Galapagos Islands, https://doi.org/10.1007/978-3-030-43973-6_7

167

Resolution Imaging Spectroradiometer (MODIS) have a spatial resolution of 500-m, a scale inadequate for detailed characterization of tropical island landscapes. The effort of 30-m resolution global land cover mapping has major challenges for tropical areas (Yu, Liu, Zhao, Yu, & Gong, 2018). For example, we observed that recently available 30-m FROM-GLC (Finer Resolution Observation and Monitoring of Global Land Cover) map products (Gong et al., 2013) have a high classification error rate for the Galapagos Islands. Most local land cover mapping efforts focus only on a selected portion of the Galapagos Islands due to limited image availability due to cloud cover. For a selected intensive study area on Isabela Island, Walsh et al. (2008) classified high spatial resolution QuickBird data using Object-Based Image Analysis to characterize invasive plant species. McCleary (2013) evaluated land use/cover change in southern Isabela Island, Galapagos, between 2004 and 2010. A recent study by Rivas-Torres, Benítez, Rueda, Sevilla, and Mena (2018) generated a complete map coverage of native and invasive vegetation in protected areas of the Galapagos Archipelago using Landsat 8 OLI scenes from 2015 to 2016. Detailed vegetation types (e.g., evergreen forest, shrub, deciduous forest, tallgrass, invasive species) were characterized. However, agricultural lands and urban areas were masked-out from their study. At present, no ready-to-use dataset is available for decadal-scale land change analysis. It is not clear how agricultural and urban and built-up lands have changed over time, nor how various socio-economic and bio-physical variables affect Galapagos' land change trajectories.

One of the most significant challenges for land cover mapping of Galapagos Islands is persistent cloud cover. For example, within the entire Landsat archive, we only found one near-cloud-free (3.1% cloud/shadow) image for Santa Cruz Island, acquired on 10 March 2019 by Landsat-8 satellite. Cloud-free data for a single date are difficult to find for humid and tropical regions (Martinuzzi, Gould, & González, 2007; White, Shao, Kennedy, & Campbell, 2013). Martinuzzi et al. (2007) needed a total of 18 Landsat ETM+ images from 1999 to 2003 to generate a near-cloud-free image for Puerto Rico and its adjacent islands. The integration of multi-senor satellite data (e.g., Landsat, SPOT and Sentinel-2) could increase the chance of obtaining near-cloud-free imagery. Recently available Sentinel-2 imagery features five-day temporal resolution and can be easily combined with Landsat-8 data to offer even higher temporal coverage. Over 20 years of archived SPOT satellite data, albeit commercial, can be explored to supplement the freely available Landsat data. Additionally, Landsat-7 ETM+ SLC-off satellite images (after 2003) are underused due to apparent data gaps (Chen, Zhu, Vogelmann, Gao, & Jin, 2011; Maxwell, Schmidt, & Storey, 2007). With careful design, the portions of imagery with good quality data points can be routinely integrated for land cover mapping and change detection (Wan, Shao, Campbell, & Deng, 2019).

Land cover mapping for cloud-prone region typically involves data pre-processing procedures of cloud detection, radiometric normalization, and cloud gap-filling. Radiometric normalization is a key step for generating consistent radiometric measurements across all scenes before image mosaicing or gap-filling (Chander, Xiong, Choi, & Angal, 2010). Generally, one image with the least cloud cover is used as the reference image and the remaining subject images are radiometrically normalized for cloud gap-filling. A number of techniques are available for both semi-automated and automated radiometric normalization and most of them involve pseudo-invariant target detection and linear regression between subject and reference images (Chander et al., 2010; Du, Cihlar, Beaubien, & Latifovic, 2001; Roy et al., 2008; Song, Woodcock, Seto, Lenney, & Macomber, 2001; Syariz et al., 2019; Yang & Lo, 2000). Both Ordinary Least Squares (OLS) and Major Axis (MA) regressions have been widely used in previously published studies (Goslee, 2011).

In tropical landscapes, vegetation formations, phenological patterns, and dry-moist conditions are often complex in both spatial and temporal domains (Helmer & Ruefenacht, 2007). Linear regression-based image normalization may not perform well when subject and reference images have large phenological and illumination variations (Janzen, Fredeen, & Wheate, 2006; Yang & Lo, 2000). Accordingly, more aggressive normalizations are preferred to force the subject image to be similar to the reference image. Several studies have applied neural network and regression tree algorithms to relative radiometric normalizations (Helmer & Ruefenacht, 2007; Sadeghi, Ebadi, & Ahmadi, 2013; Velloso, de Souza, & Simoes, 2002). In a recent land cover change study, Seo, Kim, Eo, Park, and Park (2017) provided a detailed comparison of normalization performance and suggested random forest regression performed better than traditional linear regression and other selected machine learning algorithms. To date, few studies (except Helmer & Ruefenacht, 2007) have examined whether these aggressive algorithms are applicable in complex tropical landscapes.

This study focused on decadal land cover mapping (2000, 2009, and 2019) for Santa Cruz Island in the Galapagos Archipelago. We were particularly interested in mapping key land cover types of urban and built-up, agriculture, and invasive plant species. The spatial distributions of these land cover classes and their change trends are currently unavailable, but very important for identifying threats to the island ecosystem. To overcome the challenge of cloud contamination, we combined images from multiple satellites including Landsat-8, Sentinel-2, Landsat-7, and SPOT 4 for decadal land cover mapping. The mapping year of 2009 was relatively challenging because we needed to conduct radiometric normalization and gap-filling using three cloudy Landsat-7 ETM+ SLC-off scenes as input. We were particularly interested in the comparison of random forest regression with a commonly used linear regression (Major Axis) with respect to radiometric normalization in a complex tropical island setting.

Methods

Study Area

The island of Santa Cruz, with a local population of roughly 15,700, encompasses an area of 986 km² located in a chain of islands collectively known as the Galapagos Islands. Moist highlands reaching 864 m above mean sea level, occupy the center of the island, with southern slopes dominated by agriculture. January through June is marked

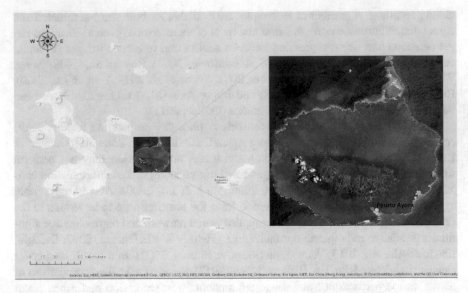

Fig. 1 Study area of Santa Cruz Island. Landsat-8 OLI image (10 March 2019) is presented in RGB true color combination

by the wet season with temperatures reaching up to 32 °C and heavy precipitation until April (Grant, 1985). July through December, with noticeably lower temperatures (down to 15 °C) is known as the dry season, though this is a misnomer; precipitation during this season is greater than the "wet season" but mainly comes from fog drip (Grant, 1985). The coastal zone and low elevations are arid, with increasing orographic precipitation upslope; spatial variation in precipitation has a strong influence on vegetation and soil patterns on Santa Cruz (Taboada, Rodríguez-Lado, Ferro-Vázquez, Stoops, & Cortizas, 2016). The dormant volcano in the island's center, Los Gemelos, has produced comparatively rich soils suitable for agriculture. Significant weathering has been observed on high altitude southern slopes (Taboada et al., 2016). Native vegetation would include evergreen and deciduous forests and shrublands, grasslands, and mangroves, with high proportions of endemic and rare species, but much of it has been replaced by agriculture and invasive vegetation (Rivas-Torres et al., 2018). Invasive species, such as *Cedrela odorata*, *Rubus niveus*, and others, behave aggressively in the agricultural zone and surrounding areas of Santa Cruz (Rivas-Torres et al., 2018) (Fig. 1).

Satellite Data and Cloud Masking

Previous remote sensing studies suggested that images from the warm-wet season (December to June) are more useful for land cover mapping of the Galapagos Islands, because vegetation boundaries are relatively easier to identify (Rivas-Torres et al., 2018). We thoroughly examined the USGS EarthExplorer Landsat archive and the

Table 1 Satellite images used for land cover mapping

Mapping year	Satellite	Image acquisition	Cloud and SLC-off (% of total land pixels)	Type
2019	Landsat-8 OLI	03/10/2019	3.1	Principle
	Sentinel-2	03/08/2019	11.2	Gap-filling
2009	Landsat-7 ETM+	04/07/2009	38.5	Gap-filling
		03/22/2009	36.5	Principle
		03/06/2009	44.3	Gap-filling
2000	SPOT-4	03/03/2000	0	Principle

Cloud/shadow pixels were identified by Fmask cloud masking algorithm. For 2009 and 2019 mapping years, one scene was used as the principle image and its cloud/gap areas need to be gap-filled with other supporting images

Copernicus Open Access Hub for Sentinel-2 to determine feasible mapping years. Cloud-free images for Santa Cruz Island were not available in both archives. Table 1 shows selected images from multiple satellite data sources. There is only one near-cloud-free Landsat-8 image (3.1% cloud/shadow, 10 March 2019), which can be supplemented with a Sentinel-2 image (11.2% cloud/shadow, 8 March 2019) to obtain a seamless 2019 land cover map. For the mapping year of 2009, all Landsat-5 images have complete cloud cover over the humid highlands and thus are practically unusable. The only feasible images are from Landsat-7 ETM+: three images from March to April 2009 need to be combined to generate a seamless image. Additionally, we obtained a SPOT 4 HRVIR (High-Resolution Visible and InfraRed) image from 3 March 2000 to extend our decadal land cover mapping backwards in time. This SPOT 4 image is cloud-free, which is rare for our study area. The SPOT 4 image has 20-m spatial resolution and four spectral bands covering part of the visible (green and red), near infrared, and thermal infrared spectral regions.

The cloud/shadow pixels for each image were identified as ancillary data to support the land cover mapping. We used the Fmask algorithm (Zhu, Wang, & Woodcock, 2015) to automatically identify the cloud/shadow pixels. The most recent Fmask 4.0 version includes stand-alone on both the Windows and Linux environments (Qiu, He, Zhu, Liao, & Quan, 2017). The package is easy to use and allows users to adjust cloud probability threshold to achieve optimal cloud/shadow masking. In our study, minor manual editing was needed to remove thin clouds and cloud shadows missed by Fmask detection.

Land Cover Mapping for 2000 and 2019

Our image classification scheme was similar to the one developed by McCleary (2013) for Isabela Island, Galapagos. Seven land cover classes were considered including: invasive plant species, urban and built-up, agriculture, forest/shrub, dry vegetation, barren, and water. On Santa Cruz Island, Cuban cedar (*Cedrala odorata*), a commercially-valued introduced tree, has become one of the most troublesome of many invasive species. Dry vegetation includes both dry grassland and shrubs, mainly distributed in the dry lowland areas (McCleary, 2013; Rivas-Torres et al., 2018).

Training polygons for image classification were manually selected using visual interpretation of Landsat-8, Sentinel-2, and SPOT-4 imagery, supported by limited field GPS points, and available map references. We compared bi-temporal 2000 and 2019 images and focused on common/no-change areas for training data selection, thus a common training dataset could be used for both 2000 and 2019 mapping. Specific number of training data points varied from around 350 (barren class) to 5200 (forest/shrub class). The classifier training and image classification were implemented in freely available RStudio software package. The entire workflow was streamlined using R scripts and several key R libraries. Specifically, R rgdal (Bivand, Keitt, Rowlingson, & Pebesma, 2014) and raster (Hijmans et al., 2015) packages were used for importing training polygons and remote sensing images. The R caret library (Kuhn, 2008) was then used to automate parameter tuning, cross-validation, and model selection. Among numerous available classification algorithms, we selected Random Forest (RF) classification algorithm due to its overall good performance reported in remote sensing and machine learning communities (Rodriguez-Galiano, Ghimire, Rogan, Chica-Olmo, & Rigol-Sanchez, 2012; Wang, Shao, & Kennedy, 2014; Shao, Campbell, Taff, & Zheng, 2015).

We note that the total number of spectral bands and the spatial resolution used in the classification varied depending on input images. For our Sentinel-2 image, we included a total of ten spectral bands covering visible, vegetation red edge, NIR, and SWIR regions. Spectral bands with 10-m spatial resolution were resampled to 20-m resolution for spatial consistency. For Landsat-8 and SPOT-4 imagery, the input spectral bands were six and four, respectively. The accuracies of the classification were assessed by using google earth imagery and commercial satellite imagery (e.g., WorldView-2 and QuickBird) as references. For each land cover class of interest, we randomly selected 25 points for visual interpretation and accuracy assessment. The accuracy assessment was conducted for 2019 only, because high resolution reference data were not available for year 2000 mapping.

Radiometric Normalization and Land Cover Mapping for 2009

The 2009 Land cover mapping was more challenging because we needed to merge three cloudy Landsat 7 ETM+ SLC-off scenes to obtain a near cloud/gap free image mosaic (Fig. 2). The image acquired on 22 March 2009 has the least amount of cloud/SLC-off pixels (36.5% of total land pixels). The other two images from 6 March 2009 and 7 April 2009 have 38.5% and 44.3% cloud/SLC-off pixels, respectively.

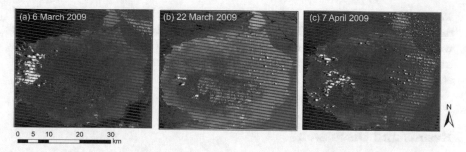

Fig. 2 Three Landsat-7 ETM+ SLC-off images were used for radiometric normalization, gap-filling, and 2009 land cover mapping

Before cloud/SLC-off gap-filling, radiometric normalization algorithms were used to minimize differences of digital numbers among three images. The 22 March 2009 image was treated as a reference image. The other two images were subject images that need to be radiometrically normalized. For each subject-reference image pair, we masked out all cloud/SLC-off pixels from both images and all the remaining valid data points were treated as pseudo-invariant target for radiometric normalization. It is reasonable to consider all pixels as no-change as all three images involved in the radiometric normalization had close acquisition dates.

We examined two radiometric normalization approaches. The first one was a commonly used Major Axis regression available through R Landsat package (Goslee, 2011). Major Axis regression is preferred over ordinary least squares because linear covariation of subject-reference images is of main interest. Major Axis regression was conducted band-by-band to estimate model coefficients and then predict new (radiometrically normalized) digital numbers for subject image. In the second radiometric normalization approach, we used more aggressive random forest regression to enhance the similarity of subject-reference image pairs. The general form is the same as those proposed by Helmer and Ruefenacht (2007):

$$y_{ref\_i} = f\left(x_{s1}, x_{s2}, x_{s3}, x_{s4}, x_{s5}, x_{s7}\right)$$

where y_{ref_i} is digital number from the ith band of reference image. x_{s1}–x_{s7} represent digital numbers from six spectral bands of subject image. The f function here denotes random forest regression, although any approximation function could be used to link subject-reference signals. Only 10% of no-change points were used to train the random forest regression function. The 10-fold cross validation was implemented in R caret package for estimating the prediction error. The trained random forest regression models were then applied to all pixels in the subject image, band-by-band, to generate radiometrically normalized values. Finally, the cloud/SLC-off pixels in the 22 March reference image were sequentially replaced by radiometrically normalized data values from the 6 March and 7 April subject images.

Following the radiometric normalization and gap-filling, the near-gap-free image mosaic was used as input for 2009 land cover mapping. The training data points were randomly selected from the no-change areas during 2000 to 2019, derived from previously classified 2000 and 2019 maps. A total of 5% of no-change pixels were randomly sampled and their class labels were used as targets for classification training. The trained classifier was applied to the near-gap-free input image to generate 2009 land cover map.

Results and Discussion

Land Cover Mapping for 2019

We compared 2019 land cover maps derived from Landsat-8 and Sentinel-2 images (Fig. 3). As we expected, spatial distributions of invasive species, agriculture, and built-up classes were similar because the acquisition dates for the two images were only two days apart and we used similar training data points for image classifications. Agricultural lands were primarily located in the upland area, known as an agricultural zone (Guézou et al., 2010). Rich volcanic soils and higher moisture levels make the upland zone more suitable for cultivation compared to areas at lower elevations (Schofield, 1989). Introduced invasive tree species (Cuban Cedar) were mixed with agricultural lands and appeared to thrive in the area neighboring the southwest corner of the agricultural zone. Built-up areas were mainly distributed in the town of Puerto Ayora, located on the southern shore of the Island.

A close comparison of two land cover maps suggested that Sentinel-derived map preserved more spatial details due to its higher spatial resolution (20 m). Small vegetation patches within agricultural fields were better retained in the Sentinel-derived map. However, because the Landsat-8 image has smaller cloud/shadow cover percentage, we used the Sentinel-derived land cover map to gap-fill the cloud/shadow areas in the Landsat-derived land cover map (see Results section "Land Cover Change in Santa Cruz Island").

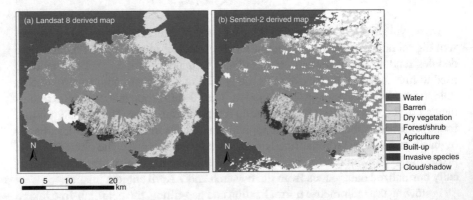

Fig. 3 Comparison of 2019 land cover maps derived from Landsat-8 (**a**) and Sentinel-2 (**b**) images

We conducted an accuracy assessment for the 2019 land cover map. The overall accuracy was 84% (Kappa = 0.81). The user's accuracies for invasive species, built-up, and agricultural lands were 96%, 98%, and 78%, respectively. Agriculture class was more difficult to map as they were often mixed with forest/shrub lands in the upland region. There were also relatively large confusions between forest/shrub and dry vegetation classes. In complex tropical islands, vegetation formations can change abruptly from dry/moist lowland to wet mountainous areas, thus making it challenging for remote sensing mapping (Helmer & Ruefenacht, 2007). Barren class had low user's accuracy (54%), but the total area of barren land was only 0.7% of land area, so it was not essential to further improve its accuracy (Table 2).

Radiometric Normalization of 2009 Landsat ETM+ Images

The 22 March 2009 Landsat image was gap-filled using radiomerically normalized data from the 6 March and 7 April subject images. Figure 4 compares results from two radiometric normalization approaches of Major Axis and random forest regressions. Only a portion of the study area is shown for a better visualization. Using a radiometrically normalized 6 March image for gap-filling, visual assessment of the gap-filled images suggested that both radiometric normalization methods performed well. There was almost no apparent color difference at the junction of original and gap-filling pixels (Fig. 4a, b). Narrower SLC-off gaps were still present after 6 March to 22 March gap-fill procedure.

Figure 4c, d shows the results of gap-filling procedure using the 7 April image as the subject image. The periodic color stripes are apparent in Fig. 4c (Major Axis regression), suggesting questionable results from this radiometric normalization method. Random forest-based radiometric normalization generated much smoother gap-filling results (Fig. 4d). The differences between reference and normalized subject images were further quantified using root-mean-square error (RMSE) and the coefficient of determination (R^2) (Table 3).

There are a number of factors that may affect the performance of radiometric normalization, including normalization algorithm, topographic relief, similarity between the source and reference images (Yang & Lo, 2000). For the image pair of 7 April and 22 March, the original band-by-band R^2 ranged from 0.55 to 0.88 (RMSE: 5.30–7.76). Major Axis regression only slightly reduced RMSE values (3.96–6.96) and R^2 remained to be the same due to the nature of linear transformation. As a comparison, Random forest-based normalization reduced RMSE values to the range of 2.59–4.80 and R^2 values improved to 0.80–0.94. These results are consistent with previously published studies (e.g., Seo et al., 2017). In complex tropical landscapes, more aggressive normalization algorithms such as random forest regression appeared to be more effective in reducing subject-reference scene differences.

Table 2 Accuracy assessment for 2019 land cover map

	Reference							%Correct	%Commission
	Invasive species	Built-up	Agriculture	Forest/shrub	Dry veg	Barren	water		
Invasive	48	0	0	2	0	0	0	96	4
Built-up	0	49	0	0	0	1	0	98	2
Agriculture	1	2	39	6	2	0	0	78	22
Forest/shrub	0	1	0	43	6	0	0	86	14
Dry veg	0	0	0	3	37	0	10	74	26
Barren	0	2	0	8	7	27	6	54	46
Water	0	0	0	0	0	0	50	100	0
%Correct	98	91	100	70	71	100	76	Overall = 84	
%Omission	2	9	0	30	29	0	24	Kappa = 0.81	

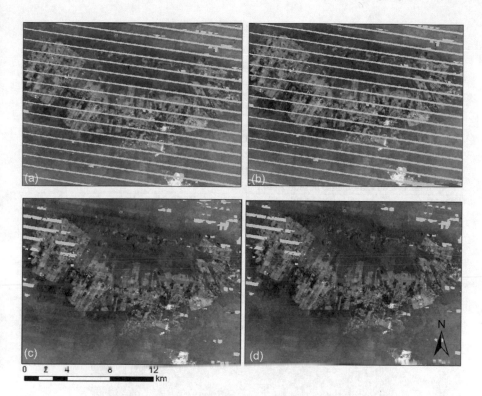

Fig. 4 Comparisons of two radiometric normalization approaches: (**a**) Major Axis regression with the 6 March image as the subject image; (**b**) random forest regression with the 6 March image as the subject image; (**c**) Major Axis regression with the 7 April image as the subject image; (**d**) random forest regression with the 7 April image as the subject image

Land Cover Mapping for 2009

We further evaluated radiometric normalization methods' impacts on 2009 land cover mapping. Figure 5a, b compares land cover mapping results using 2009 gap-filled images. Although the overall land cover distribution appears to be similar, there is a clear discrepancy for invasive class. The Major Axis radiometric normalization resulted in higher confusion between invasive species and forest/shrub classes (Fig. 5a). There were some apparent class discontinuities at the junction of original and gap-filling pixels. In the areas located to the north of agricultural zone, more forest/shrub lands were falsely classified as invasive class (Cuban Cedar). Using random forest-based radiometric normalization led to a more reasonable land cover map, indicated by rather contiguous spatial distribution of invasive species and fewer misclassification of forest/shrub lands in the upland region (Fig. 5b). Although it was not possible to conduct a thorough accuracy assessment due to lack of high resolution reference data, the visual assessment of classification maps, combined with previously described statistics (RMSE and R^2 in section "Radiometric

Table 3 Comparison of RMSE and R^2 for radiometric normalization methods using two Landsat image pairs

	B1	B2	B3	B4	B5	B7
6 March–22 March radiometric normalization						
RMSE_o	4.03	3.73	7.10	9.22	7.96	6.37
RMSE_ma	2.98	2.81	3.95	6.64	6.33	4.16
RMSE_rf	2.14	2.12	3.02	3.88	3.91	3.08
R^2_o	0.82	0.76	0.87	0.88	0.71	0.84
R^2_ma	0.82	0.76	0.87	0.88	0.71	0.84
R^2_rf	0.90	0.85	0.92	0.96	0.88	0.91
7 April–22 March radiometric normalization						
RMSE_o	5.56	5.30	6.20	7.45	7.76	5.99
RMSE_ma	4.40	3.96	5.61	6.62	6.96	5.21
RMSE_rf	2.67	2.59	3.71	4.69	4.80	3.64
R^2_o	0.61	0.55	0.73	0.88	0.64	0.73
R^2_ma	0.61	0.55	0.73	0.88	0.64	0.73
R^2_rf	0.85	0.80	0.88	0.94	0.81	0.86

Calculated against reference image of 22 March 2009, RMSE_o, RMSE_ma, and RMSE_rf represent RMSE values for the original subject image, Major Axis normalized image, and random forest normalized image. Corresponding R^2 values are denoted as R^2_o, R^2_ma, and R^2_rf

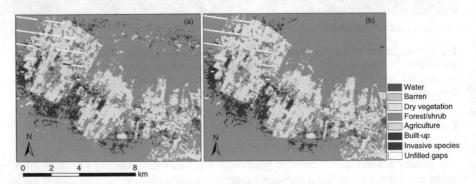

Fig. 5 Comparison of image classification results using two radiometric normalizations: (**a**) Major Axis radiometric normalization and gap-filling; (**b**) Random forest radiometric normalization and gap-filling

Normalization of 2009 Landsat ETM+ Images"), indicate random forest-based radiometric normalization performed better than traditional linear regression. It should be noted that there were still some small gaps in the 2009 land cover map even after merging three Landsat-7 ETM+ SLC-off images. These small gaps were subsequently filled with land cover labels from the 2000 map.

Land Cover Change in Santa Cruz Island

The total built-up areas in the Santa Cruz Island were 0.98-km², 1.64-km², and 2.81-km² for year 2000, 2009, and 2019, respectively (Table 4). The urban built-up areas almost tripled during this 20-year study period. Agricultural lands increased from 51.50-km² in 2000 to 53.64-km² in 2009 and then to 54.91-km² in 2019. Total area of invasive tree species also increased substantially from 5.79-km² to 19.16-km². Invasive species have been expanding from agricultural zones to native forest/shrub lands, especially in the areas neighboring the southwest corner of the agricultural zone (Fig. 6).

Table 4 Area distributions of three land cover types in the Santa Cruz Island (km²)

	2000	2009	2019
Built-up	0.98	1.64	2.81
Agricultural	51.50	53.64	54.91
Invasive species	5.79	10.22	19.16

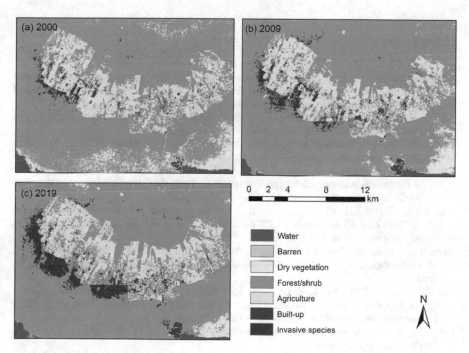

Fig. 6 Land cover maps for 2000, 2009, and 2019

Conclusion

In the cloud-prone tropical island of Santa Cruz, Landsat archived images alone were not sufficient for decadal land change analysis. To overcome cloud contamination and data availability issues, this study combined remote sensing images from multiple satellite sensors (Landsat-7, Landsat-8, Sentinel-2, and SPOT-4) to generate land cover maps for 2000, 2009, and 2019. Land cover change analysis showed that total areas of urban and built-up, invasive plant species, and agriculture classes all increased during the study period. These human-induced land cover changes need to be closely monitored, because they are increasingly impacting the island's ecosystems. For 2009 land cover mapping, two radiometric normalization methods were examined as a key data processing step for cloud/SLC-off gap filling. The random forest-based normalization achieved better performance compared to a commonly used Major Axis regression method. This was supported by calculated statistical measures (RMSE and R^2), visual evaluation of gap-filled image, and resultant land cover maps. The random forest-based normalization was more aggressive in reducing differences in the subject and reference image pairs and appeared to be more effective for complex tropical landscapes.

References

Baine, M., Howard, M., Kerr, S., Edgar, G., & Toral, V. (2007). Coastal and marine resource management in the Galapagos Islands and the Archipelago of San Andres: Issues, problems and opportunities. *Ocean & Coastal Management, 50*(3–4), 148–173.

Benítez, F., Mena, C., & Zurita-Arthos, L. (2018). Urban land cover change in ecologically fragile environments: The case of the Galapagos Islands. *Land, 7*(1), 21.

Bivand, R., Keitt, T., Rowlingson, B., & Pebesma, E. (2014). rgdal: Bindings for the geospatial data abstraction library. *R package version 0.8-16.*

Brewington, L., Frizzelle, B. G., Walsh, S. J., Mena, C. F., & Sampedro, C. (2014). Remote sensing of the marine environment: Challenges and opportunities in the Galapagos Islands of Ecuador. In *The Galapagos marine reserve* (pp. 109–136). Springer.

Chander, G., Xiong, X. J., Choi, T. J., & Angal, A. (2010). Monitoring on-orbit calibration stability of the Terra MODIS and Landsat 7 ETM+ sensors using pseudo-invariant test sites. *Remote Sensing of Environment, 114*, 925–939.

Chen, J., Zhu, X., Vogelmann, J. E., Gao, F., & Jin, S. (2011). A simple and effective method for filling gaps in Landsat ETM+ SLC-off images. *Remote Sensing of Environment, 115*(4), 1053–1064.

Du, Y., Cihlar, J., Beaubien, J., & Latifovic, R. (2001). Radiometric normalization, compositing, and quality control for satellite high resolution image mosaics over large areas. *IEEE Transactions on Geoscience and Remote Sensing, 39*(3), 623–634.

Epler, B., 2007. *Tourism, the economy, population growth, and conservation in Galapagos.* Charles Darwin Foundation.

Gong, P., Wang, J., Yu, L., Zhao, Y., Zhao, Y., Liang, L., … Li, C. (2013). Finer resolution observation and monitoring of global land cover: First mapping results with Landsat TM and ETM+ data. *International Journal of Remote Sensing, 34*(7), 2607–2654.

Goslee, S. C. (2011). Analyzing remote sensing data in R: The landsat package. *Journal of Statistical Software, 43*(4), 1–25.

Grant, P. R. (1985). Climatic fluctuations on the Galapagos Islands and their influence on Darwin's finches. *Ornithological Monographs, 36*, 471–483.

Guézou, A., Trueman, M., Buddenhagen, C. E., Chamorro, S., Guerrero, A. M., Pozo, P., & Atkinson, R. (2010). An extensive alien plant inventory from the inhabited areas of Galapagos. *PLoS One, 5*, e10276.

Helmer, E. H., & Ruefenacht, B. (2007). A comparison of radiometric normalization methods when filling cloud gaps in Landsat imagery. *Canadian Journal of Remote Sensing, 33*(4), 325–340.

Hijmans, R. J., van Etten, J., Cheng, J., Mattiuzzi, M., Sumner, M., Greenberg, J. A., ... Hijmans, M. R. J. (2015). Package 'raster'. *R package.*

Izurieta, J. C. (2017). Behavior and trends in tourism in Galapagos between 2007 and 2015. *Galapagos Report 2015–2016* (pp. 83–39). Puerto Ayora, Galapagos, Ecuador: GNPD, GCREG, CDF and GC.

Janzen, D. T., Fredeen, A. L., & Wheate, R. D. (2006). Radiometric correction techniques and accuracy assessment for Landsat TM data in remote forested regions. *Canadian Journal of Remote Sensing, 32*(5), 330–340.

Kuhn, M. (2008). Building predictive models in R using the caret package. *Journal of Statistical Software, 28*(5), 1–26.

Martinuzzi, S., Gould, W. A., & González, O. M. R. (2007). Creating cloud-free Landsat ETM+ data sets in tropical landscapes: Cloud and cloud-shadow removal. *Gen. Tech. Rep. IITF-32., 32*. US Department of Agriculture, Forest Service, International Institute of Tropical Forestry.

Maxwell, S. K., Schmidt, G. L., & Storey, J. C. (2007). A multi-scale segmentation approach to filling gaps in Landsat ETM+ SLC-off images. *International Journal of Remote Sensing, 28*(23), 5339–5356.

McCleary, A. L. (2013). Characterizing contemporary land use/cover change on Isabela Island, Galapagos. In *Science and Conservation in the Galapagos Islands* (pp. 155-172). Springer, New York, NY.

Qiu, S., He, B., Zhu, Z., Liao, Z., & Quan, X. (2017). Improving Fmask cloud and cloud shadow detection in mountainous area for Landsats 4–8 images. *Remote Sensing of Environment, 199*, 107–119.

Rivas-Torres, G. F., Benítez, F. L., Rueda, D., Sevilla, C., & Mena, C. F. (2018). A methodology for mapping native and invasive vegetation coverage in archipelagos: An example from the Galápagos Islands. *Progress in Physical Geography: Earth and Environment, 42*(1), 83–111.

Rodriguez-Galiano, V. F., Ghimire, B., Rogan, J., Chica-Olmo, M., & Rigol-Sanchez, J. P. (2012). An assessment of the effectiveness of a random forest classifier for land-cover classification. *ISPRS Journal of Photogrammetry and Remote Sensing, 67*, 93–104.

Roy, D. P., Ju, J., Lewis, P., Schaaf, C., Gao, F., Hansen, M., & Lindquist, E. (2008). Multi-temporal MODIS–Landsat data fusion for relative radiometric normalization, gap filling, and prediction of Landsat data. *Remote Sensing of Environment, 112*(6), 3112–3130.

Sadeghi, V., Ebadi, H., & Ahmadi, F. F. (2013). A new model for automatic normalization of multitemporal satellite images using Artificial Neural Network and mathematical methods. *Applied Mathematical Modelling, 37*(9), 6437–6445.

Schofield, E. K. (1989). Effects of introduced plants and animals on island vegetation: Examples from Galápagos Archipelago. *Conservation Biology, 3*, 227–239.

Seo, D., Kim, Y., Eo, Y., Park, W., & Park, H. (2017). Generation of radiometric, phenological normalized image based on random forest regression for change detection. *Remote Sensing, 9*(11), 1163.

Shao, Y., Campbell, J. B., Taff, G. N., & Zheng, B. (2015). An analysis of cropland mask choice and ancillary data for annual corn yield forecasting using MODIS data. *International Journal of Applied Earth Observation and Geoinformation, 38*, 78–87.

Song, C., Woodcock, C. E., Seto, K. C., Lenney, M. P., & Macomber, S. A. (2001). Classification and change detection using Landsat TM data: When and how to correct atmospheric effects? *Remote Sensing of Environment, 75*, 230–244.

Syariz, M. A., Lin, B. Y., Denaro, L. G., Jaelani, L. M., Van Nguyen, M., & Lin, C. H. (2019). Spectral-consistent relative radiometric normalization for multitemporal Landsat 8 imagery. *ISPRS Journal of Photogrammetry and Remote Sensing, 147*, 56–64.

Taboada, T., Rodríguez-Lado, L., Ferro-Vázquez, C., Stoops, G., & Cortizas, A. M. (2016). Chemical weathering in the volcanic soils of Isla Santa Cruz (Galápagos Islands, Ecuador). *Geoderma, 261*, 160–168.

Tye, A., (2001). Invasive plant problems and requirements for weed risk assessment in the Galapagos Islands. *Weed risk assessment*, pp. 153–175.

Velloso, M. L. F., de Souza, F. J., & Simoes, M. (2002, June). Improved radiometric normalization for land cover change detection: An automated relative correction with artificial neural network. In *IEEE International Geoscience and Remote Sensing Symposium* (Vol. 6, pp. 3435–3437). IEEE.

Walsh, S. J., McCleary, A. L., Heumann, B. W., Brewington, L., Raczkowski, E. J., & Mena, C. F. (2010). Community expansion and infrastructure development: Implications for human health and environmental quality in the Galápagos Islands of Ecuador. *Journal of Latin American Geography, 9*, 137–159.

Walsh, S. J., McCleary, A. L., Mena, C. F., Shao, Y., Tuttle, J. P., González, A., & Atkinson, R. (2008). QuickBird and Hyperion data analysis of an invasive plant species in the Galapagos Islands of Ecuador: Implications for control and land use management. *Remote Sensing of Environment, 112*(5), 1927–1941.

Walsh, S. J., & Mena, C. F. (2013). Perspectives for the study of the Galapagos Islands: Complex systems and human–environment interactions. In *Science and conservation in the Galapagos Islands* (pp. 49–67). Springer.

Wan, H., Shao, Y., Campbell, J. B., & Deng, X. W. (2019). Mapping annual urban change using time series Landsat and NLCD. *Photogrammetric Engineering and Remote Sensing, 85*(10), 715–724.

Wang, H., Shao, Y., & Kennedy, L. M. (2014). Temporal generalization of sub-pixel vegetation mapping with multiple machine learning and atmospheric correction algorithms. *International Journal of Remote Sensing, 35*(20), 7118–7135.

White, J., Shao, Y., Kennedy, L., & Campbell, J. (2013). Landscape dynamics on the island of La Gonave, Haiti, 1990–2010. *Land, 2*(3), 493–507.

Yang, X., & Lo, C. P. (2000). Relative radiometric normalization performance for change detection from multi-date satellite images. *Photogrammetric Engineering and Remote Sensing, 66*, 967–980.

Yu, L., Liu, X., Zhao, Y., Yu, C., & Gong, P. (2018). Difficult to map regions in 30 m global land cover mapping determined with a common validation dataset. *International Journal of Remote Sensing, 39*, 4077–4087.

Zhu, Z., Wang, S., & Woodcock, C. E. (2015). Improvement and expansion of the Fmask algorithm: Cloud, cloud shadow, and snow detection for Landsats 4–7, 8, and Sentinel 2 images. *Remote Sensing of Environment, 159*, 269–277.

Human and Natural Environments, Island of Santa Cruz, Galapagos: A Model-Based Approach to Link Land Cover/Land Use Changes to Direct and Indirect Socio-Economic Drivers of Change

Francesco Pizzitutti, Laura Brewington, and Stephen J. Walsh

Introduction

The impacts of growing human populations and exponentially increasing power of technology have radically altered the rate of change of Earth ecosystems since the appearance of *H sapiens* on the planet (Vitousek, Mooney, Lubchenco, & Melillo, 1997). In the twentieth and twenty-first centuries, globalization has accelerated and expanded the alteration of world's ecosystems (Lambin & Meyfroidt, 2011). Human modification of the environment is now impacting worldwide biodiversity and eco-systems health, contributing to the global climate change, and altering the living conditions of many human societies (Steffen et al., 2004). Furthermore, by directly hampering ecological sustainability and ecosystems services, human impacts have put the ability of natural environments to sustain societal needs at risk (Díaz et al., 2019). Global human impacts on natural ecosystems is often manifested as land cover/land use change (LCLUC) (Foley et al., 2005). Expansion of agricultural and urban areas, soil degradation, and coastal erosion are among the LCLUC effects of human transformations that can be assessed through satellite remote sensing (Lambin et al., 2001; Rivas-Torres, Benítez, Rueda, Sevilla, & Mena, 2018). Changes introduced by humans and manifested as LCLUC are particularly evident in island ecosystems (Zhao et al., 2004). Due to their vulnerability to climate change, habitat modification, ecosystem exploitation, alien species invasions, and disease introductions, island ecosystems are among the most vulnerable

F. Pizzitutti (✉)
University of North Carolina at Chapel Hill, Chapel Hill, NC, USA

L. Brewington
East-West Center, Honolulu, HI, USA

S. J. Walsh
Center for Galapagos Studies, University of North Carolina at Chapel Hill, Chapel Hill, NC, USA

© Springer Nature Switzerland AG 2020
S. J. Walsh et al. (eds.), *Land Cover and Land Use Change on Islands*, Social and Ecological Interactions in the Galapagos Islands, https://doi.org/10.1007/978-3-030-43973-6_8

environments to human changes and globalization (Graham, Gruner, Lim, & Gillespie, 2017). For this reason, island ecosystems provide an opportunity to study the impacts of globalization on ecosystem processes and to design strategies for enhancing the sustainability of human-natural systems (Kerr, 2005; Lim & Cooper, 2009).

Over the last century, tourism has become one of the main driving forces of environmental change in many islands around the world (Oreja Rodríguez, Parra-López, & Yanes-Estévez, 2008; Orams, 2002). Increased human presence in islands either directly due to tourism or indirectly through migration in response to economic tourism development, has led to profound impacts affecting island ecosystems worldwide in many ways, including increased ecosystem exploitation, agricultural intensification, deforestation, invasive species introductions and contamination (Macdonald, Anderson, & Dietrich, 1997). The changes introduced by tourism to islands can be detected and evaluated through remote sensing data (Rivas-Torres et al., 2018) and assembled in longitudinal studies of LCLUC. These temporal LCLUC trajectories can be useful in identifying the relevant drivers of change and trends in island environments (Walsh et al., 2008; Mccleary, 2013).

The Galapagos Islands environment is the perfect example of relatively pristine, and unique island ecosystem that has attracted an increasing number of tourists in recent decades, leading to dramatic changes in human and natural systems despite the fact that almost the entire territory of the archipelago is under a regime of integral natural protection (Lu, Valdivia, & Wolford, 2013; González, Montes, Rodríguez, & Tapia, 2008). The economic boom associated with tourism growth attracted sustained immigration to the islands from the Ecuadorian mainland composed of people seeking improved living and working conditions (Kerr, Cardenas, & Hendy, 2004). Although LCLU information is of striking importance for managing tourism and protected areas (Kennedy et al., 2009), LCLU data for the Galapagos Islands are often incomplete and fragmented. Furthermore, a complete analysis of past LCLUC trends is not available for extended areas of the archipelago. One of the first published studies of land cover in the Galapagos was a map produced by the Galapagos National Institute (INGALA; (Ministerio de Agricultura y Ganaderia, PRONAREG, ORSTOM, & INGALA, 1987). This map did not include some of the inhabited island of Galapagos and was more recently updated by a higher resolution map produced by The Nature conservancy (TNC) and the Ecuadorian government using remote sensing data collected in 2000 (The Nature Conservancy (TNC) & Centro de Levantamientos Integrados de Recursos Naturales por Sensores Remotos (CLIRSEN), 2006). The combined 1987 INGALA maps and 2000 TNC map were used to create the first study of LCLUC in Galapagos, which was limited to the rural areas of Santa Cruz (SC) and San Cristobal Islands (Villa & Segarra, 2011) and to study the anthropogenic degradation of the main islands (Watson, Trueman, Tufet, Henderson, & Atkinson, 2010). A series of studies (Mccleary, 2013; Walsh et al., 2008) analyzed LCLUC in the rural areas of Isabela Island as the local working force shifted from agriculture to tourism and impacted LCLUC in Isabela by increasing invasive species presence. More recently, Rivas-Torres et al. (2018) developed a methodology to identify and map the spread of aggressive invasive vegetation in island biomes, which showed that SC was the most infested island in the Galapagos archipelago.

System Dynamics (SD) is a modeling technique created by J.W. Forrester (Forrester, 1961) that combines system thinking with flow charts formalism to symbolize model parameters and rates of change. The casual connections between systems elements are described by equations that are translated in a typical formalism of "stock and flow" diagrams. SD and systems thinking have been used in several studies (Walker, Greiner, McDonald, & Lyne, 1998; McGrath, 2010; Georgantzas, 2003; Law et al., 2012; Hernández & León, 2007) to represent the dynamics of tourism destinations, including islands, with the goal of analyzing the interactions between natural and human environments and integrating different actions and relative impacts. Recently a series of studies were based on SD models to analyze the complex interactions between different dynamic aspects of the Galapagos, including tourism, population, invasive species introductions, tourism infrastructure (Pizzitutti et al., 2016), food supply system (Sampedro, Pizzitutti, Quiroga, Walsh, & Mena, 2018), and the connections between tourism development, the labor market, and demography (Espin, Mena, & Pizzitutti, 2019). In these studies, SD models were used to analyze how different scenarios of tourism development influenced the islands as an integrated system.

In this paper, we combined remote sensing analysis of past LCLUC trends with socio-economic data from SC Island to build an SD model that represented the main drivers of LCLUC and projected possible scenarios of LCLUC for the next 20 years. LCLUC over the last 30 years were analyzed and connected with the corresponding socio-economic drivers. The detected changes, together with relevant data on tourism, demography, migration, the economy, and the labor market in SC, were then used to inform the SD model. The model goal was to effectively represent the interactions between SC human communities and the relevant LCLUC. The SD modelling approach presented here provides valuable insights into the mechanisms through which tourism influences the economic and demographic development and exerts social and ecological pressures on the SC Island system observed through LCLUC. Moreover, the SD model for SC offers an understanding of possible future development and relative LCLUC impacts.

Methods

Galapagos and Santa Cruz Island

The Galapagos archipelago is located in the Pacific Ocean 1000 km west of the South American coast. Administratively the Galapagos Islands are part of the Republic of Ecuador and were declared the first UNESCO World Heritage Site in 1976 (UNESCO, 2019) due to the irreplaceable value of its unique ecosystems that inspired Charles Darwin to write his seminal work "On The Origin of Species" (Darwin, 1859). To protect the fragile natural environment of Galapagos, the Ecuadorean Government declared 97% of the archipelago to be a nature reserve and created the Galapagos National Park (GNP) in 1959. The remaining 3% of surface area was dedicated for human settlement and agricultural activities. The archipelago was further recognized by UNESCO as a Biosphere Reserve in 1984 and as a Ramsar site in 2011 (UNESCO, 2019). In 1998 the Ecuadorean government created the Galapagos Marine Reserve (GMR) to protect the 133,000 km^2 of marine ecosystem around the islands, establishing one of the largest marine reserves in the world.

With the advent of the Galapagos commercial tourism industry, in the 1970s, the annual tourist arrivals in the archipelago increased from a few hundred to 275,817 visitors in 2018 and, correspondingly, the population, according with the most recent National Population Census (INEC, National Institute of Statistics and Census, 2019), increased from 4037 inhabitants of 1974 to 25,214 inhabitants of 2010. The increasing of permanent and temporary human communities in Galapagos led to numerous impacts and challenges, including high rates of alien species introductions (Tye, Atkinson, & Carrion, 2008), native ecosystem losses due to plant invasions (Rivas-Torres et al., 2018), land use/cover changes due to urban and agricultural expansion (Watson et al., 2010), depletion of environmental resources (Bassett, 2009), and environmental contamination (Lougheed, Edgar, & Snell, 2002; Wikelski, Wong, Chevalier, Rattenborg, & Snell, 2002).

SC Island is 979 km^2 in area and is located in the center of the archipelago (Fig. 1). After the first permanent modern settlement in SC was established, in the middle of the nineteenth century (Lundh, 1995), the SC population grew slowly until the tourism boom when, from 1577 inhabitants in 1974, the population reached the 15,393 inhabitants of 2010 (INEC, National Institute of Statistics and Census, 2019). The projected population of SC for 2018 was 21,052 inhabitants. Like other

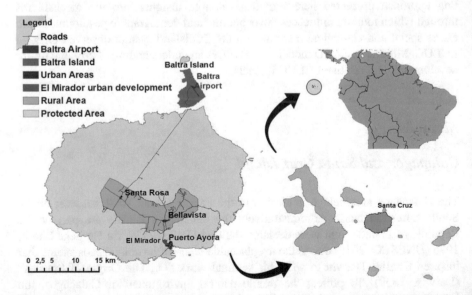

Fig. 1 The islands of Santa Cruz (left), the Galapagos archipelago (bottom right) and the location of the Galapagos archipelago in South America (top right)

inhabited Galapagos Islands, SC is strictly subdivided into protected areas (88%) and inhabited areas (12%). Human activities have direct and indirect impacts on SC's protected areas, including the displacement of native land cover by invasive vegetation (Rivas-Torres et al., 2018) and predation of native animal species by introduced animals like rats, dogs, cats, goats, and pigs (Kramer, 1983). Human impacts in the inhabited areas of SC are more evident and include different kinds of infrastructure, agricultural activities, and pastures. Administratively, SC is a canton that is subdivided into three parishes: the capital urban parish of Puerto Ayora (11,974 inhabitants in 2010) and the two rural parishes of Bellavista (2425 inhabitants) and Santa Rosa (994 inhabitants). SC's central geographic location in the Galapagos archipelago has made this island the main transportation hub. The main international airport of Galapagos is located just north of SC on Baltra Island and receives the 75% of visitors. Nearly all tourism cruises of Galapagos begin in SC and it is estimated that over the last 30 years SC received an average of 76% of all yearly touristic arrivals in Galapagos (Gobierno Autonomo Descentralizado Municipal de Santa Cruz, 2012). In 2018, the estimated number of tourists at the same time in SC (i.e. the average number of tourists in SC on a given day of the year) was 3070, corresponding to 208,435 annual tourist arrivals.

SC Land Cover Mapping

Decadal land cover mapping (1990–2019) for SC Island was implemented using multi-sensor satellite data analysis of Landsat archive, SPOT, and Sentinel-2 imagery (Shao, Wan, Rosenman, Laso, & Kennedy, 2020). We explored the entire USGS EarthExplorer Landsat archive and the Copernicus Open Access Hub for available Landsat 5, 7, 8 and Sentinel-2 satellite images. Cloud-free satellite images were difficult to find for SC. We focused on 1990, 2000, 2009, and 2019 imagery for a decadal land cover mapping strategy. One cloud-free SPOT 4 HRVIR image from 3 March 2000 was available for the study area. For other years, we needed to combine multiple images to gap-fill cloud/image gap areas prior to mapping. For example, the 2009 data required three Landsat ETM+ images to generate a cloud-free mosaic. For 2019, a near cloud-free (3.1% cloud cover) Landsat-8 OLI image was supplemented by a Sentinel-2 image. We used the Fmask algorithm (Zhu, Wang, & Woodcock, 2015) to automatically identify cloud/shadow pixels. Radiometric normalization was used to develop seamless image mosaics as inputs for land cover mapping. All Landsat, SPOT, and Sentinel-2 images were collected during the warm-wet season (December to June), during which vegetation boundaries are better defined for land cover mapping (Rivas-Torres et al., 2018).

We considered seven land cover classes: invasive plant species, built-up, agriculture, forest/shrub, dry vegetation, barren, and water. This classification scheme was similar to McCleary et al.'s (Mccleary, 2013) land cover mapping scheme developed for Isabela Island, Galapagos. The invasive plant species class mainly included Cuban cedar (*Cedrela cedar*). Training data points for each land cover class were obtained through visual interpretation of satellite images and high resolution Google Earth images. We used a Random Forest classifier for image classification (Shao et al., 2020). Random Forest is on of the most competitive classification algorithms with respect to performance and ease of use (Cooner, Shao, & Campbell, 2016). Using commercial high-resolution imagery (WorldView-2) and Google Earth imagery as a reference, we conducted accuracy assessments for the 2019 land cover map. The overall accuracy was 84% (Kappa = 0.81). Similar accuracy levels are expected for other mapping years because the same image classification procedures were used and training data were selected to be as consistent as possible.

The SD Model for LCLUC on SC Island

The SC island SD model can be conceptually separated into two modules: the first is the socio-economic module (Fig. 2, right side of the diagram) that describes the LCLU drivers of change including tourist arrivals, demography, the labor market, immigration, and emigration. The second module (Fig. 2, left side of the diagram) represents the LCLU dynamics. The model considers four classes of land cover: native land cover that includes both naturally bare zones and zones covered with native vegetation, built-up land cover (human infrastructure, buildings, roads, etc.), agricultural land cover (cultivated areas, pastures), and invasive vegetation land cover. As showed in Fig. 2, the model includes a simplified representation of SC population dynamics that is based on several hypotheses. First, tourism and the Ecuadorean mainland unemployment are the only exogenous variables to the model. No other exogenous variables, such climate change, changes to governmental policies, or investments to strengthen environmental conservation or foster the development of specific economic sectors are considered. Excluding climate change among the exogenous drivers of change was not based on the limited importance of this variable to determine the future trajectories of LCLUC in SC, but rather because past climate change impacts on Galapagos ecosystems, tourism, and demography have been small and are not quantifiable, making them difficult to project into the next 20 years. A second hypothesis was that the feedback loop between tourism and population was not counterbalanced by an opposing loop connected to the loss of

Fig. 2 LCLUC SD model of SC Island level zero diagram. Exogenous variables are in orange. Land cover variables are in rectangular boxes

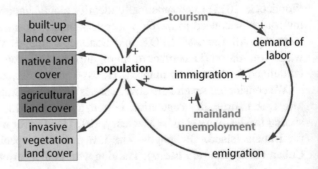

the island's natural ecosystem touristic appeal. This second hypothesis implies that the increase in tourism arrivals in SC will always translate into increases in labor demand, so if the level of mainland Ecuadorian unemployment is low enough, immigration to the island will result in demographic growth. It is clear that over the long term, the tourism increase will reduce the island's natural environment touristic appeal and consequently the number of visiting tourists. The counter-balancing effects of natural capital loss were represented in detail by Espin et al. (2019) but were not considered in the model presented here because their influence over the next 20 years are difficult to evaluate and potentially irrelevant.

The Socio-Economics Module

As shown in Fig. 2, the principal, direct driver of LCLUC change considered in the model is the total population or "floating population" of SC. The floating population is made up of the resident population plus the number of tourists at the same time in Santa Cruz. Increases in the total SC population will result in a corresponding increase in human impacts on the environment and in a consequent increase in LCLUC. Size changes of SC populations are connected directly with the network of tourism development, mainland unemployment, emigration to the mainland, and natural demographic dynamics. In Fig. 3 we showed some the factors that are directly connected with resident population changes in the SD model, including immigration, emigration, births, and deaths. Both immigration and emigration to

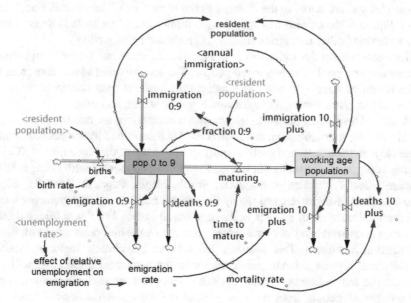

Fig. 3 Stock-flow diagram corresponding to SC resident population dynamics. In standard SD graphical format, the "shadow variables" that pertain to other views or sub-modules of the model are shown between <...> brackets

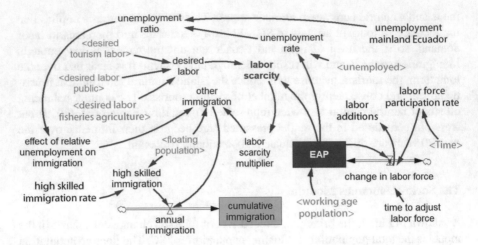

Fig. 4 Stock-flow diagram corresponding to SC immigration, labor scarcity, and unemployment dynamics

SC were determined by the ratio between unemployment in SC and on the Ecuadorean mainland. The resident population is divided in two segments: children ("Pop 0 to 9") and the working age population.

As shown in Fig. 4, the working age population variable multiplied by the labor participation force variable equals the "EAP" (Economically Active Population) whose changes are equal to the changes in the labor force ("labor addictions" variable). Figure 4 shows also that the "annual immigration" variable is derived from two variables: "other immigration" and "high skilled immigration".

The labor market dynamics shown in Fig. 5 contain five stocks: four employment sectors and one pool of unemployed people. The employment sector data from the national census were aggregated to represent the most relevant sectors for SC labor dynamics: tourism, fishing and agriculture, public sector, and others.

A desired level of employment in each sector determines the number of employed people. For the tourism sector, the desired level is determined by the desired number of employees per tourist and the total number of tourists at the same time in SC. The fishing and agriculture sector (fa in Fig. 5) has been connected with changes in the tourism sector (González et al., 2008; Walsh, Engie, Page, & Frizzelle, 2019), whose desired level are determined by the desired number of employees per tourist and the total number of tourists at the same time in SC. For the public and other sectors, the desired level of employment considers a desired number of employees per floating inhabitant. This assumes that a larger population, including visiting tourists, will require a larger public sector to manage local affairs and will also increase the demand for goods and services from other sectors. Each sector will hire the number of people from the unemployed stock to reach a desired level. The unemployed pool changes following the employees' turnovers and firings from all

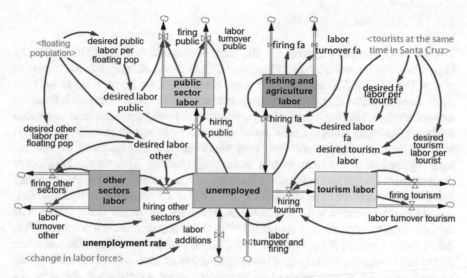

Fig. 5 Stock-flow diagram corresponding to SC labor market dynamics. The acronym "fa" indicates fisheries and agriculture

sectors, and labor additions correspond to changes in the labor force. As the pool of unemployed people decreases, it will be more difficult for each sector to hire the desired levels of labor, which over time will increase immigration pressure to increase the population and therefore the labor force through the unemployment ratio variable.

The population and employment sectors interact through the migration forces to and from SC based on two effects (Fig. 4). The first effect comes from the internal dynamics of supply and demand in the SC labor market that are represented by the labor availability ratio. When there is less labor availability due to an increase in demand or decrease in supply, the job market will be more attractive, increasing the immigration rate to SC. The second effect comes from the external dynamics between the SC and Ecuadorian labor markets represented by the unemployment ratio. Lower unemployment in SC will increase the incentives of mainland Ecuadorians to immigrate only if the labor conditions in the islands are perceived as better. If Ecuador improves its labor conditions, fewer people will be motivated to immigrate to SC. The same logic applies to the emigration rate. Low unemployment ratios in SC will motivate people to stay on the Island, only if the labor conditions are perceived as better than on the mainland. An additional factor determining immigration to SC is highly skilled immigrants, as some jobs, especially in tourism, require highly qualified professionals that cannot be found locally (Kerr, Cardenas, & Hendy, 2004). The highly skilled migration value is not affected by the labor dynamics described above because over the analyzed time horizon, a certain percentage of the jobs have always needed to be filled by highly qualified immigrants.

The LCLUC Module

The LCLUC module connects the SC human dimension to LCLUC under the hypothesis that the main direct and indirect LCLU driver of change in SC is human population (Fig. 6). The built-up land cover clearly follows the increase in SC residents since an expanding floating population requires more and variable infrastructure for tourism, housing, roads, transportation, etc. The agricultural land cover also follows population change trends because increases in tourist arrivals, correlated with resident population increases, may encourage the displacement of workers from agriculture to tourism with a consequent decline agricultural land cover (Sampedro et al., 2018). On the other hand, an increase in the island population may favor the development of production of local food and stimulate agricultural activities (Gobierno Autonomo Descentralizado Municipal de Santa Cruz, 2012). The area of invasive vegetation is connected with the Galapagos human presence (Tye, Atkinson, & Carrion, 2008; Pizzitutti et al., 2016) with probabilities of introduction that increase with the increasing of human population and spread speeds that depend on environmental conditions and invasive plant species adaptation to the archipelago environment conditions.

Model Data and Fit to Historical Data

The data used to inform the model were obtained from several sources. The population, births, deaths, labor for SC and unemployment in Ecuador data were from the national censuses of the Ecuadorean National institute of Statistics and Census (INEC, National Institute of Statistics and Census, 2019). LCLU data were derived from the classification of remote sensing images presented in this paper.

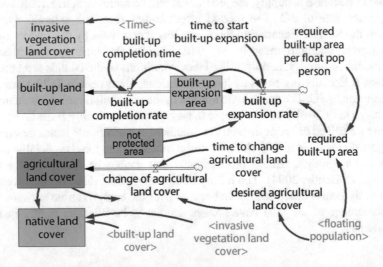

Fig. 6 Stock-flow diagram corresponding to SC LCLUC dynamics

Data on tourism arrivals in SC were not directly available because toursits visiting Galapagos are counted only when they board flights from the Ecuadorean mainland. For this reason, no reliable numeric or temporal statistics on tourist presence by island are available. To extimate the number of tourists at the same time in SC from tourism arrival statistics in Galapagos collected by the Ecuadorean Ministry of Tourism (Parque Nacional Galápagos & Observatorio de Turismo de Galápagos, 2018), we took advantage of the fact that the Population National Censuses of Ecuador were carried out following a "de facto" method: all residents present on the day of the census are counted, regardless of whether they are permanent or temporary. The non-permanent residents in Galapagos are composed mainly of tourists and people traveling to the islands for work, along with few Galapagos residents traveling from one island to another. Supposing that the proportion of non-permanent residents to tourists at a given time were approximately the same in SC and in the whole archipelago, T_{SC}, the number of tourists at the same time in SC can be expressed as:

$$T_{SC} = \frac{T_G \, nR_{SC}}{nR_G} \tag{1}$$

where T_G is the number of toursits at the same time in Galapagos and nR_{SC} and nR_G are the numbers of non-residents counted by the census in SC and Galapagos, respectively. Because the number of tourists at the same in Galapagos can be calculated by combining annual tourism arrivals and statistics on their stays (Pizzitutti et al., 2016), and the numbers of non-residents in SC and Galapagos are known from the National Censuses, it is possible to use Eq. (1) to extimate the number of toursits at the same time in SC. The resulting number of tourists at the same time in SC in the last 30 years was equal to 75.5% of the total Galapagos tourism presence.

Many model parameters were not known, including those describing the relationships between unemployment and labor scarcity with immigration and emigration, those describing the connections between desired levels of employment and the floating population and tourists at the same time and the parameters connecting population with LCLUCs. To determine the right values for those parameters we calibrated the model using historical data from a 30-year period, from 1990 to 2019. The calibration was carried out by varying the values of the unknown parameters cited above and comparing the model outcomes with historical data relative to the calibration period. The selected calibrated values corresponded to those that best reproduced the historical data.

Scenarios of Tourism Growth

As noted, the model only considered two main exogenous variables: tourism arrivals in SC and unemployment on the Ecuadorean mainland. Here we supposed that mainland unemployment will remain unchanged in the next 20 years and only tourism will modulate the changes in SC population dynamics over that time. To project tourism development outcomes over the next 20 years, we considered three possible scenarios of change for annual tourists visiting SC. The computation of tourist arrivals in SC for the three selected scenarios were based on projections of the number of annual arrivals across the whole archipelago assuming that SC will continue to draw the same proportion of total Galapagos tourists reported above (75.5%). The first scenario was a Moderate Growth scenario based on the assumption that annual Galapagos tourists will change by the same number of persons each year in the future as the average annual absolute change observed over the last 20 years. The average annual absolute change observed over the last 20 years was 10,551 tourists per year, an increase of 6584 for foreign tourists and 3967 for domestic tourists each year. The second scenario was a High Growth scenario based on the assumption that the number of tourists in Galapagos will grow by the same average annual growth rate during each year in the future as it did over the last 20 years. The average annual growth rate for foreign tourists over this time period was 6.8% and the average annual growth rate for domestic tourists was 10.7%. The third and last scenario was a scenario in which a Zero Growth policy was imposed on tourist arrivals so that the number of arriving tourists remained constant and equal to the 2018 value. Table 1 shows the projected number of annual arrivals in Galapagos and the corresponding arrivals in SC for each tourism growth scenario.

Past and Projected LCLUC Trends

Past Trends of LCLUC in SC

We subdivided the LULC classification for SC into three main areas: the urban area composed of Puerto Ayora's administrative zone, the rural area composed of the Santa Rosa and Bellavista parishes, and the remaining protected areas. As showed

Table 1 Projected annual tourist arrivals in Galapagos and SC under the three tourism development scenarios

	2018	2028	2038
Annual number of tourists visiting Galapagos (tourists)			
Moderate growth	182,037	381,327	486,837
High growth		611,713	1,398,609
No growth		182,037	182,037
Annual number of tourists visiting SC (tourists)			
Moderate growth	137,438	287,902	367,562
High growth		461,834	1,055,950
No growth		137,438	137,438

Table 2 SC LCLUC from 1990 to 2019

	1990	2000	2009	2019
Urban area				
Invasive vegetation	0	0	0	0
Built-up	0.49	0.79	1.21	1.24
Agricultural	0	0	0	0
Native	1.35	0.95	0.63	0.59
Rural area				
Invasive vegetation	0	3.67	5.62	4.01
Built-up	0.01	0.08	0.19	0.91
Agricultural	45.88	50.96	53.45	54.32
Native	67.66	58.84	54.29	54.27
Protected area				
Invasive vegetation	0	2.11	4.68	15.14
Built-up	0.03	0.06	0.13	0.37
Agricultural	0.31	0.18	0.12	0.32
Native	860.54	856.27	855.31	844.64
Total				
Invasive vegetation	0	5.78	10.3	19.15
Built-up	0.53	0.93	1.53	2.52
Agricultural	46.19	51.14	53.57	54.64
Native	929.55	916.06	910.23	899.3

Total area (km²) occupied by four land cover types across three administrative zones: urban, rural, and protected areas

in Table 2 and Fig. 7, no invasive vegetation or agricultural land cover were detected in the urban area. The main LULCC in this area over the last 30 years was the progressive broadening of built-up land cover at the expense of native land cover. This increase of built-up land cover reflected the increase in tourism flows and related expansion of coastal SC communities, as almost all tourism activities in SC are concentrated in the urban coastal area of Puerto Ayora. We note that built-up land cover not only increased in the SC urban area, but also in the rural and protected areas, denoting a growing demand for available land for residential infrastructure. The lack of available space in Puerto Ayora has driven SC residents to seek more affordable accommodations between Puerto Ayora and the village of Bellavista located only 5 km away (Fig. 7). This explain the increase in built-up land cover in the rural area around Bellavista that cannot be correlated with economic development in that zone because the main economic activities, with the exception of agriculture, are concentrated in Puerto Ayora (Gobierno Autonomo Descentralizado Municipal de Santa Cruz, 2012). The slight increase of built-up land cover within the protected area of SC corresponded to the establishment of the residential area of "el Mirador" (see Fig. 1) and other residential conglomerations between Puerto Ayora and Bellavista in zones that were formerly part of the GNP and that were added to the municipality of Puerto Ayora administrative area as result of a land swap agreement between the GNP and the municipality of Puerto Ayora (Gobierno

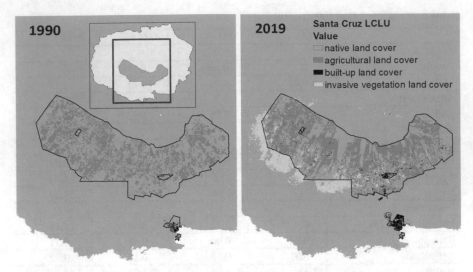

Fig. 7 SC maps of LCLU in 1990 and 2019. Only the island areas where LCLU were observed are showed

Autonomo Descentralizado Municipal de Santa Cruz, 2012). This new residential area of el Mirador is extended 0.71 km² and only 52% is currently occupied by newly built structures, but due to the high pressure for residential infrastructure in SC, it is likely that the remaining available space will be occupied quickly in the coming years.

LCLUC in the SC rural area showed an opposite trend over the last 30 years to what has been previously observed in Galapagos rural areas where highly attractive working opportunities in the coastal towns due to rapid tourism development have pulled a substantial part of the agricultural working force to tourism (González et al., 2008; Walsh & Mena, 2013; Mccleary, 2013). This agricultural working force shift has led to a progressive abandonment of agricultural areas that has facilitated the diffusion of introduced vegetation (Kerr, Cardenas, & Hendy, 2004). Our LCLUC data showed that in the SC rural areas, agricultural land cover increased by 18% over the last 30 years, a change that corresponds to an expansion from the 40% to the 47% of the total rural area. This increase in the extension of farmed lands in SC was due to an intense flow of farmers that immigrated, mostly illegally, from poor, rural areas of the Ecuadorean mainland (Gobierno Autonomo Descentralizado Municipal de Santa Cruz, 2012). These immigrants have restored abandoned farms in the area around denoted as "El Cascajo" along the eastern border of the SC rural area, and around the parish of Santa Rosa taking advantage, probably, of the active and growing population of Puerto Ayora that created a market for vegetables and fruits. Active farming in rural SC was also confirmed by the relatively small extension of invasive vegetation distribution in the areas where agriculture activities increased. According to the 1987 Galapagos map published by the National Institute of Galapagos INGALA (Ministerio de Agricultura y Ganaderia et al., 1987), around

1990 the invasive vegetation cover in the rural area of SC was negligible. After 1990, invasive vegetation expanded slowly up to cover the 3.5% of the total rural area surface with remarkable expansions around the densely populated parish of Bellavista. A more remarkable and alarming extension of invasive vegetation was observed in the protected area, around the south western border of the agricultural area, where the invasive vegetation land cover extension increased from 0 km^2 in 1990 to 15.14 km^2 in 2019. This increase was mostly due to human disturbance of native ecosystems that facilitated the spread of invasive vegetation and their continued transmission from rural areas to the adjacent GNP protected areas (Rivas-Torres et al., 2018).

SD Model Simulation Outputs

Population and Labor Dynamics

As shown in Figs. 4 and 5, labor and population dynamics were strictly interconnected in the SD model for SC Island. On one hand, labor was directly determined by the increase in tourist and resident numbers, while on the other hand, employment rates and labor scarcity influenced net migration to SC that contributed to population change. Migration had distinct effects on the SC labor-population systems. The first effect of migration, coming from the internal dynamics of supply and demand in the SC labor market, was represented by the labor scarcity ratio (EAP/desired labor). When there was less labor available due to an increase in desired labor, the job market was more attractive and increased the immigration rate to SC. The high growth scenario showed a decreasing labor scarcity ratio, which means that demand exceeded supply, increasing immigration. The moderate growth scenario showed a constant ratio and in the no growth scenario the labor scarcity ratio increased, reflecting a continuous excess of supply over demand. The second effect of immigration stemmed from the external dynamics between SC and Ecuador's labor markets represented by the unemployment ratio (unemployment in SC/unemployment in the Ecuadorean mainland). Higher labor demand increased the stock of employed people in the three scenarios. More employment incentivized Ecuadorians from the mainland to immigrate, but only if the labor conditions in SC were perceived as better than in Ecuador. The relative unemployment was reflected in the unemployment ratio. In the high tourism growth scenario, the unemployment ratio decreased considerably fostering immigration and decreasing emigration. In the moderate growth scenario, the unemployment ratio was almost constant and in the no growth scenario, the unemployment ratio doubled by 2038. This trend reduced immigration and increased emigration, reducing the labor force in SC and decreasing unemployment.

The interplay between the labor market, migration, and population was evident when population changes are considered. Table 3 shows model outputs for the variables related to population dynamics for the three tourism growth scenarios.

Table 3 SD model outputs: SC labor market dynamics

	2018	2028	2038
Unemployment ratio			
Moderate growth	0.27	0.25	0.27
High growth		0.02	0.01
No growth		0.41	0.56
Labor scarcity ratio (EAP/desired labor)			
Moderate growth	1.2	1.2	1.2
High growth		1	1
No growth		1.3	1.4

The unemployment ratio is defined as the ratio between unemployment in SC and unemployment in Ecuador. The labor scarcity ratio is the ratio between the number of people in the Economically Active Population (EAP) and the desired labor force

The resident population increased in all three scenarios, even under the no growth scenario. Immigration was the main driver of population growth in the high and moderate scenarios of tourism arrivals because the net migration flow (immigration–emigration) increased exponentially under the high growth scenario and linearly under the moderate growth scenario. In the third no growth scenario, the resident population increase was driven by natural demographic dynamics. By 2038, the population is expected to be 4.7 times greater than the 13,938 inhabitants in the 2010 National Census under the accelerated growth scenario, 2.3 times under the moderate tourism growth scenario, and 1.6 times under the no growth scenario. Tourists who are present at the same time that make up the non-resident component of SC's total floating population will increase over the next 20 years not only under the moderate and high tourism growth scenarios, but also in the no growth scenario because in recent decades the lengths of tourist stays in Galapagos have increased (Pizzitutti et al., 2016) (Table 4).

SD Island LCLUC

As shown in Fig. 6, the SD model for SC Island connects the fluctuating population with LCLUC, linking socio-economic changes with corresponding LCLU impacts. The model connects built-up and agricultural land cover with direct changes in the human population, while invasive vegetation land cover is represented as a linear projection of past trends. Native land cover is calculated as the difference between the remaining land cover types and the total Island surface. As shown in Table 5, the three scenarios demonstrated an increase in built-up land cover generated by infrastructure needs of the growing SC floating population. The moderate and no growth scenarios had respective increases of 1.2 and 0.68 km^2, while the high growth scenarios predicted an increase of 4.18 km^2. These built-up land cover increases are greater than the available extension of 0.34 km^2 in the new residential area "El Mirador". Therefore, the simulation outputs seems to suggest that new areas will be

Table 4 SD model outputs: SC population dynamics

	2018	2028	2038
Resident (people)			
Moderate growth	17,981	24,655	33,385
High growth		30,762	66,299
No growth		24,230	30,464
Net migration (immigration–emigration) (people)			
Moderate growth	225	351	438
High growth		1783	4171
No growth		192	184
Tourists at the same time (tourists)			
Moderate growth	3419	4832	6217
High growth		7656	15,485
No growth		3634	3829

Table 5 SD model outputs: SC Island LCLUC

	2019	2028	2038
Invasive vegetation (km²)			
Moderate Growth	19.5	24.0	30.5
High Growth		24.0	30.5
No Growth		24.0	30.5
Built-up (km²)			
Moderate Growth	2.52	2.7	3.7
High Growth		3.1	6.7
No Growth		2.5	3.2
Agricultural (km²)			
Moderate Growth	54.6	59.1	64.4
High Growth		62.7	84.9
No Growth		58.1	61.5
Native (km²)			
Moderate Growth	899.5	889.9	877.2
High Growth		881.1	853.6
No Growth		888.9	880.6

converted from protected to urban zoning and continue the trend of expansion of coastal communities toward the GNP protected areas.

Given that the three scenarios considered in this study predicted an expansion in SC human communities and that agricultural production has followed the growing population, the model predicted an increase in agricultural LCLU. Under the high, moderate, and no growth scenarios over the next 30 year, the SD model projected agricultural increases of 30.3, 9.8, and 6.9 km², respectively. Actively farmed areas will increase from 48% of the total rural area of SC in 2018, to 75% under the high growth scenario and 57% and 54% for the moderate and no growth scenarios, respectively. Considering that more extensive active farming helps control invasive vegetation that disseminate from the rural areas into protected areas, GNP prevention and control efforts may be improved. The model simply projected past invasive vegetation trends into the future. Without extensive campaigns to control or eradicate them, their expansion is projected to continue undisturbed, especially in protected areas. This would lead to an expansion of 11.0 km² of invasive vegetation cover in protected areas.

As general remark we want to note that our remote sensing classification for SC seems to suggest that in 2019, 76.3 km² (7.8%) of the island was degraded due to human activities either directly (agriculture, infrastructure expansion) or indirectly (invasive vegetation spread). This differs from what was previously reported by Watson et al. (2010), who found that 14% of the surface of SC was degraded in 2000. This difference was not due to a reduced human disturbance in SC but rather to the fact that Watson et al. considered the entire rural area to be degraded by human presence regardless of LCLU while we did not consider rural areas covered with native vegetation as degraded. Considering the future trends of LCLUC described by our SD model we found that under the moderate and no growth tourism scenarios, degraded areas would increase from 7.8% to 10.1% and 9.7% of SC Island area, respectively, while the high growth scenario projected almost twice as much (12.5%) total degraded surface.

Conclusions

In this paper we described the past and future trends of LCLUC in the Galapagos Island of Santa Cruz. We identified the most relevant drivers of change in LCLU and integrated them in a System Dynamics model describing the human and natural dimensions of the island, projecting past LCLUC trends into the next 20 years. The model designated the processes of tourism growth and Ecuadorian regional economic development as the main exogenous drivers of change, which remarkably influenced SC demographic dynamics through strong impacts on immigration and emigration. In the model, we represented SC population dynamics as the main direct and indirect drivers of LCLUC and we described their impacts on four types of LCLU: built-up (including all types of human infrastructure), agriculture, invasive vegetation, and native land cover. We classified remote sensing images to analyze LCLUC in SC over a period of 30 years and connect the observed transformations with the corresponding drivers of change. We found that 7.8% of the island surface was degraded by direct and indirect human activities and that most of this degradation was concentrated in the humid vegetation zone of the SC rural areas. Moreover the SD model results indicated that, under different tourism growth scenarios, the degraded areas of the island will continue to expand to the point of doubling in size under exponential tourism growth. Further studies should attempt to quantify the effects of other important drivers on LCLUC in SC and the Galapagos archipelago more generally, such as climate change and other human interventions like investments in conservation or changes in GNP management.

This study proposed a methodological approach where remote sensing image classification data was integrated with demographic, social, and economic data to inform a model that described the relevant drivers of LCLUC in a selected area. By joining different layers of information about SC we were able to describe the island as an integrated system where many processes and factors interact at different temporal and space scales. This description was integrated into a quantitative model that can be an important tool for policy makers and stakeholders making decisions about the management of the fragile ecosystems of an island like SC and other archipelagos like Galapagos more generally.

References

Bassett, C. A. (2009). *Galápagos at the crossroads: Pirates, biologists, tourists, and creationists battle for Darwin's cradle of evolution*. Edited by National Geographic. Washington, DC.

Cooner, A. J., Shao, Y., & Campbell, J. B. (2016). Detection of urban damage using remote sensing and machine learning algorithms. Revisiting the 2010 Haiti earthquake. *Remote Sensing, 8*(10), 868.

Darwin, C. (1859). *On the origin of species*. John Murray.

Díaz, J., Settele, E. S., Brondizio, E. S., Ngo, H. T., Guèze, M., Agard, J., ... Zayas, C. N. (2019). *Report of the Plenary of the Intergovernmental Science-Policy Platform on Biodiversity and Ecosystem Services on the work of its seventh session*. Bonn, Germany.

Espin, P. A., Mena, C. F., & Pizzitutti, F. (2019). A model-based approach to study the tourism sustainability in an island environment: The case of Galapagos Islands. In T. Kvan & J. Karakiewicz (Eds.), *Urban Galapagos* (pp. 97–113). Springer.

Foley, J. A., DeFries, R., Asner, G. P., Barford, C., Bonan, G., Carpenter, S. R., ... Snyder, P. K. (2005). Global consequences of land use. *Science, 309*(5734), 570.

Forrester, J. W. (1961). *Industrial dynamics*. Cambridge, MA: The MIT Press.

Georgantzas, N. C. (2003). Tourism dynamics: Cyprus' hotel value chain and profitability. *System Dynamics Review, 19*(3), 175–212.

Gobierno Autonomo Descentralizado Municipal de Santa Cruz. (2012). Plan De Desarrollo y Ordenamiento Territorial Cantón Santa Cruz 2012–2027.

González, J. A., Montes, C., Rodríguez, J., & Tapia, W. (2008). Rethinking the Galapagos Islands as a complex social-ecological system: Implications for conservation and management. *Ecology and Society, 13*(2), 13.

Graham, N. R., Gruner, D. S., Lim, J. Y., & Gillespie, R. G. (2017). Island ecology and evolution: Challenges in the Anthropocene. *Environmental Conservation, 44*(4), 323–335.

Hernández, J. M., & León, C. J. (2007). The interactions between natural and physical capitals in the tourist lifecycle model. *Ecological Economics, 62*(1), 184–193.

INEC, National Institute of Statistics and Census, Ecuador. (2019). Población y Demografía. Retrieved from https://www.ecuadorencifras.gob.ec/censo-de-poblacion-y-vivienda/.

Kennedy, R. E., Townsend, P. A., Gross, J. E., Cohen, W. B., Bolstad, P., Wang, Y. Q., & Adams, P. (2009). Remote sensing change detection tools for natural resource managers: Understanding concepts and tradeoffs in the design of landscape monitoring projects. *Remote Sensing of Environment, 113*(7), 1382–1396.

Kerr, S., Cardenas, S., & Hendy, J.. (2004). Migration and the environment in the Galapagos: An analysis of economic and policy incentives driving migration, potential impacts from migration control, and potential policies to reduce migration pressure. *Motu Working Paper Series No. 03-17*. Vol. 3–17.

Kerr, S. A. (2005). What is small island sustainable development about? *Ocean and Coastal Management, 48*(7–8), 503–524.

Kramer, P. (1983). The Galapagos Islands under siege. *Ambio, 12*(3/4), 186–190.

Lambin, E. F., & Meyfroidt, P. (2011). Global land use change, economic globalization, and the looming land scarcity. *Proceedings of the National Academy of Sciences, 108*(9), 3465–3472.

Lambin, E. F., Turner, B. L., Geist, H. J., Agbola, B., Angelsen, A., Bruce, J. W., … Xu, J. (2001). The causes of land-use and land-cover change: Moving beyond the myths. *Global Environmental Change, 11*, 261–269.

Law, A., De Lacy, T., McGrath, G. M., Whitelaw, P. A., Lipman, G., & Buckley, G. (2012). Towards a green economy decision support system for tourism destinations. *Journal of Sustainable Tourism, 20*(6), 823–843.

Lim, C. C., & Cooper, C. (2009). Beyond sustainability: Optimising island tourism development. *International Journal of Tourism, 11*, 89–103.

Lougheed, L. W., Edgar, G. J., & Snell, H. L.. (2002). Biological impacts of the Jessica oil spill on the Galapagos environment. *Final Report*. Puerto Ayora, Galapapgos.

Lu, F., Valdivia, G., & Wolford, W. (2013). Social dimensions of 'nature at risk' in the Galápagos Islands, Ecuador. *Conservation and Society, 11*(1), 83.

Lundh, J. P. (1995). A brief account of some early inhabitants of Santa Cruz Island. *Noticias de Galapagos*, no. 55.

Macdonald, L. H., Anderson, D. M., & Dietrich, W. E. (1997). Paradise threatened: Land use and erosion on St. John, US Virgin Islands. *Environmental Management, 21*(6), 851–863.

Mccleary, A. L. (2013). Characterizing contemporary land use\cover change on Isabela Island, Galápagos. In S. J. Walsh & C. F. Mena (Eds.), *Science and conservation in the Galapagos Islands* (pp. 150, 161). New York: Springer.

McGrath, G. M. (2010). Towards improved event evaluation and decision support: A systems-based tool. *Proceedings of the Annual Hawaii International Conference on System Sciences*, pp. 1–10.

Ministerio de Agricultura y Ganaderia, Programa Nacional de Regionalizacion Agraria (PRONAREG), Institut Francais de Recherche Scientifique Pour le Developpement en Cooperation (ORSTOM), & Instituto Nacional Galápagos (INGALA). (1987). Islas Galápagos: Mapa de Formaciones Vegetales (1:100000).

Orams, M. B. (2002). Feeding wildlife as a tourism attraction. *Tourism Management, 23*, 281–293.

Oreja Rodríguez, J. R., Parra-López, E., & Yanes-Estévez, V. (2008). The sustainability of island destinations: Tourism area life cycle and teleological perspectives. The case of Tenerife. *Tourism Management, 29*(1), 53–65.

Parque Nacional Galápagos, & Observatorio de Turismo de Galápagos. (2018). Informe Anual de Visitantes a Las Áreas Protegidas de Galápagos Del Año 2018. Galapagos, Ecuador.

Pizzitutti, F., Walsh, S. J., Rindfuss, R. R., Reck, G., Quiroga, D., Tippett, R., & Mena, C. F. (2016). Scenario planning for tourism management: A participatory and system dynamics model applied to the Galapagos Islands of Ecuador. *Journal of Sustainable Tourism*. https://doi.org/1 0.1080/09669582.2016.1257011

Rivas-Torres, G. F., Benítez, F. L., Rueda, D., Sevilla, C., & Mena, C. F. (2018). A methodology for mapping native and invasive vegetation coverage in archipelagos: An example from the Galápagos Islands. *Progress in Physical Geography, 42*(1), 83–111.

Sampedro, C., Pizzitutti, F., Quiroga, D., Walsh, S. J., & Mena, C. F. (2018). System dynamics in food security: Agriculture, livestock, and imports in the Galapagos Islands. *Renewable Agriculture and Food Systems, 1*(15), 1–15.

Shao, Y., Wan, H., Rosenman, A., Laso, F., & Kennedy, L. M.. (2020). Evaluating land cover change on the island of Santa Cruz through cloud-gap filling and multi-sensor analysis (Forthcoming). In *Springer book on islands*.

Steffen, W., Sanderson, R. A., Tyson, P. D., Jäger, J., Matson, P. A., Moore, B., III, … Wasson, R. J. (2004). *Global change and the Earth system*. Berlin: Springer.

The Nature Conservancy (TNC), & Centro de Levantamientos Integrados de Recursos Naturales por Sensores Remotos (CLIRSEN). (2006). Cartografía Galápagos 2006 Conservación En Otra Dimensión.

Tye, A., Atkinson, R., & Carrion, V.. (2008). Increase in the number of introduced plant species in Galapagos. *Informe Galápagos 2006–2007* (pp. 111–117).

UNESCO. 2019. Galapagos biosphere reserve, Ecuador. Retrieved from https://en.unesco.org/biosphere/lac/galapagos.

Villa, A., & Segarra, P.. (2011). Changes in land use and vegetative cover in the rural areas of Santa Cruz and San Cristóbal. *Galapagos Report 2009–2010* (pp. 85–91).

Vitousek, P. M., Mooney, H. A., Lubchenco, J., & Melillo, J. M. (1997). Human domination of Earth's ecosystems. *Science, 277*(5325), 494–499.

Walker, P. A., Greiner, R., McDonald, D., & Lyne, V. (1998). The tourism futures simulator: A systems thinking approach. *Environmental Modelling and Software, 14*(1), 59–67.

Walsh, S. J., Engie, K., Page, P. H., & Frizzelle, B. G. (2019). Demographics of change: Modeling the transition of fishers to tourism in the Galapagos Islands. In T. Kvan & J. Karakiewicz (Eds.), *Urban Galapagos*. Springer Nature.

Walsh, S. J., McCleary, A. L., Mena, C. F., Shao, Y., Tuttle, J. P., González, A., & Atkinson, R. (2008). QuickBird and Hyperion data analysis of an invasive plant species in the Galapagos Islands of Ecuador: Implications for control and land use management. *Remote Sensing of Environment, 112*(5), 1927–1941.

Walsh, S. J., & Mena, C. F. (2013). Perspectives for the study of the Galapagos Islands: Complex systems and human–environment interactions. In S. J. Walsh & C. F. Mena (Eds.), *Science and conservation in the Galapagos Islands* (p. 243). New York: Springer.

Watson, J., Trueman, M., Tufet, M., Henderson, S., & Atkinson, R. (2010). Mapping terrestrial anthropogenic degradation on the inhabited islands of the Galapagos Archipelago. *Oryx, 44*(1), 79–82.

Wikelski, M., Wong, V., Chevalier, B., Rattenborg, N., & Snell, H. L. (2002). Marine iguanas die from trace oil pollution. *Nature, 417*(6889), 607–608.

Zhao, B., Kreuter, U., Li, B., Ma, Z., Chen, J., & Nakagoshi, N. (2004). An ecosystem service value assessment of land-use change on Chongming Island, China. *Land Use Policy, 21*(2), 139–148.

Zhu, Z., Wang, S., & Woodcock, C. E. (2015). Improvement and expansion of the Fmask algorithm: Cloud, cloud shadow, and snow detection. *Statewide Agricultural Land Use Baseline, 2015*(1), 4388–4392.

How Do Non-Native Plants Influence Soil Nutrients Along a Hydroclimate Gradient on San Cristobal Island?

Jia Hu, Claire Qubain, and Diego Riveros-Iregui

Introduction

The introduction of non-native plants into islands across the world has led to cata-strophic consequences, with many islands now having double the number of natu-ralized exotic species compared to native plant species (Sax, Gaines, & Brown, 2002). The Galapagos Islands are no exception and studies have documented over 880 introduced plants on the island archipelago (Gardener et al., 2013; Gardener, Atkinson, & Renteria, 2010), with the majority of invasive species introduced since 1950 (Tye, 2006). However, of the 880 non-native plants, about 10 non-native plant species dominate across the archipelago. Most of the non-native plants were intro-duced for use by human settlers for a variety of reasons, including food, timber, and other cash crops. Some of the most abundant non-native plants include *Chinchona pubescens* (red quinine), *Cedrela ordorata* (Spanish cedar), *Rubus niveus* (black-berry), and *Psidium guayava* (guava fruit) (Gardener et al., 2013). Many studies have identified common mechanisms through which non-native plants can outcom-pete native plants; these include lack of natural enemies, lack or herbivore pressure, allelopathy, higher resource availability, and ability to transform abiotic-biotic con-ditions to favor invasive plants. For example, it has been proposed that the higher nutrient content of non-native plants can alter soils nutrient dynamics, creating a positive feedback in which the altered soils favor non-native plants (Jäger, Alencastro, Kaupenjohann, & Kowarik, 2013).

J. Hu (✉)
University of Arizona, Tucson, AZ, USA
e-mail: jiahu@email.arizona.edu

C. Qubain
Montana State University, Bozeman, MT, USA

D. Riveros-Iregui
University of North Carolina-Chapel Hill, Chapel Hill, NC, USA

© Springer Nature Switzerland AG 2020
S. J. Walsh et al. (eds.), *Land Cover and Land Use Change on Islands*, Social and Ecological Interactions in the Galapagos Islands, https://doi.org/10.1007/978-3-030-43973-6_9

Plant invasions can influence nutrient cycling in tropical island ecosystems, but these patterns are not always consistent. In some cases, plant invasion can increase N availability, mineralization rates, and litter decomposition rates (Ehrenfeld, 2003). Invasive plants not only directly affect nutrient cycling, but they also trigger indirect positive feedbacks in soil nutrient dynamics that accelerated biogeochemical cycling (Ostertag & Verville, 2002). While the majority of research suggests that invasive plants increase soil nutrient availability and decomposition rates, some studies find declines or no change in soil nutrient availability after plant invasion (Ehrenfeld, Kourtev, & Huang, 2001). To date, only a handful of studies have focused on characterizing soil N or P concentrations in the Galápagos (de la Torre, 2013; Jäger et al., 2013; Kitayama & Itow, 1999), and these studies have focused primarily on Santa Cruz Island. To our knowledge, no studies have examined soil N and P availability on San Cristóbal Island, especially across the strong elevation and hydroclimate gradient.

The processes of soil formation and biogeochemical cycling are driven by five state factors: topography, parent material, time, climate, and potential biota (Jenny, 1941). Topography influences nutrient dynamics through leaching and erosion (Weintraub et al., 2015), while different parent materials alter the nutrient composition of overlying soils (Yavitt, 2000). Phosphorus is most abundant in young soils, while nitrogen (N) transformation by microbes increases in older soils (Lambers, Raven, Shaver, & Smith, 2008). Climate influences nutrient dynamics both directly and indirectly through weathering, leaching and erosion as well as through influencing microbial activity and nitrogen transformation (Richardson, Peltzer, Allen, McGlone, & Parfitt, 2004; Vitousek & Matson, 1988). In addition to the influence of climate, plant community composition can also change biogeochemical cycling under vegetation canopies (Schlesinger et al., 1990; Zinke, 1962). Island ecosystems have served as natural laboratories to conduct studies of pedogenesis because each formation process can be more easily isolated. Islands present strong elevation gradients that drive climate variability across a small area, develop along chronosequences within archipelagos, and host different microbial and vegetation communities across space. While the Hawaiian Islands have served as a central site for many studies of soil formation, other archipelagos, such as the Galápagos, have similar characteristics but have received less attention. On San Cristóbal Island in the Galápagos Archipelago, climate and biota are two key state factors that influence nutrient dynamics.

Climate patterns in the Galápagos are quite variable. The archipelago experiences two seasons, the cool season (June–December), and the warm season (January–May), due to an inter-annual migration of the Inter-Tropical Convergence Zone. Temperature and precipitation vary along the elevation gradient and between the leeward and windward sides of the islands (Snell & Rea, 1999). The climosequence is divided between the very arid to arid zone at low elevations, the transition zone, and the humid to very humid zone at high elevations (Percy, Schmitt, Riveros-Iregui, & Mirus, 2016). Higher elevations are cooler and receive more precipitation in the form of fog (Violette et al., 2014), while lower elevations are dryer and typically only receive rainfall. However, the climate regime in the Galápagos is predicted to change due to intensification of the El Niño Southern Oscillation (ENSO) (Trueman & d'Ozouville, 2010). The Galápagos will experience more intense drought in La Niña years followed by prolonged rainfall events during El Niño years. Longer droughts and prolific rainstorms will likely alter rates of parent material weathering and soil development (White & Blum, 1995), but we lack thorough knowledge of soil processes on the islands to detect those changes. Therefore, establishing basic knowledge of soil N and P chemistry on the islands, as well as understanding if nutrient dynamics do contribute to non-native plant success, will become more important as the climate regime intensifies.

The main objective of this study was to characterize seasonal soil N and P concentrations on San Cristóbal, Galápagos in the context of native and non-native plant interactions. We evaluated how climate controls seasonal nutrient concentrations through the influence of temperature, precipitation and soil moisture, and we characterized how non-native plant species mediated changes in soil N and P concentrations across the island's elevation gradient. Few studies have examined these processes thoroughly in the Galápagos, and further investigation is necessary to inform the management of ecosystem change in the Galápagos.

Methods

Site Description

The Galápagos Archipelago is located 1000 km off the coast of Ecuador and is comprised of 13 main islands. They formed when the Nazca Plate passed over a hot spot in the Pacific Ocean, causing volcanic eruptions, and eventually islands (Geist, Snell, Snell, Goddard, & Kurz, 2014). Of the islands in the archipelago, this study was performed on San Cristóbal Island, one of the oldest, dated to 2.35 ma (Geist et al., 2014). On the leeward, drier side of the island, we sampled soils along an elevation gradient at low (300 m above sea level (a.s.l.)), mid (500 m a.s.l.) and high (650 m a.s.l.) elevation sites under both native and invasive plant canopies (Fig. 1). *Bursera graveolens* (torchwood), *Zanthoxylum fagara* (cat's claw), and *Miconia robinsoniana* (miconia) are dominant native plant species at the low, mid, and high elevations, respectively. However, non-native species, such as *Psidium guajava* (guava) and *Rubus niveus* (blackberry), were dominant at all elevations sampled.

Across the three study sites, there were also different management strategies. The low and mid elevation sites was located on private land, part of a local farm in which active native plant conservation plans were underway. The high elevation site was located near El Junco Lake, which is managed by the Galapagos National Park. At the low elevation site, non-native plants were manually removed and native plants were introduced two years prior to our research project. Because the low elevation site was also close to active farming plots, some manure application had occurred close to the research site. While the mid-elevation site was also located on

Fig. 1 Map of study sites along the elevational gradient with photos demonstrating the difference in vegetation communities along the climosequence. *Zanthoxylum fagara* (left) and *Scalesia gordilloi* (right) pictured at the 300 m site. El Junco lake (left) and *Psidium guyaba* (right) at the high elevation site. Arrow on map inset shows location of the Galápagos Archipelago

the same farm, there were no active native planting or farming activities present. The high elevation site did experience consistent non-native eradication efforts from the Galapagos National Park, including removal or *R. niveus* and replanting of *M. ribsoniana* in the last 5–10 years.

Typical of many tropical islands, the Galápagos experience a warm and wet season (January–May) and a cool and dry season (June–December). From 1977 to 1983 on San Cristóbal at c. 6 m.a.s.l, the mean annual temperature was 24.8 °C, and the mean temperature during the cool season was 23.5 °C and was 26.5 °C during the warm season (Violette et al., 2014). The mean annual rainfall during this time period was 2961 mm/yr above 650 m.a.s.l., and the measured mean annual rainfall at c. 6 m a.s.l was 368 mm/yr (Pryet, 2011; Violette et al., 2014). Based on data from Santa Cruz island, an additional 26% of the median annual rainfall in the highlands occurs in the form of occult precipitation (fog), although fog is rarely present at the low elevations (Pryet et al., 2012).

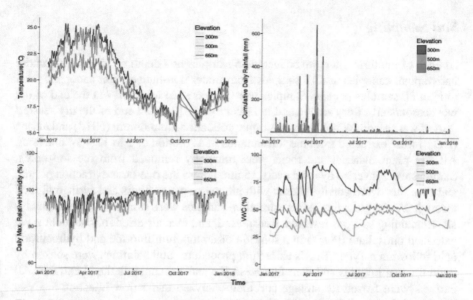

Fig. 2 Meteorological measurements of (clockwise) mean daily air temperature, cumulative daily rainfall, volumetric water content (VWC), and mean daily maximum relative humidity (RH). Note that the air temperature at the 650 m elevation site is modeled from the temperature inside the data logger. The VWC at the 300 m elevation site is from the 30 cm depth while the VWC from the 500 m and 650 m elevation sites are from the 20 cm depth

Meteorological Data

EM50 data loggers recorded air temperature, relative humidity, precipitation, and volumetric water content (VWC) at 15-min intervals at each of the three sites (METER Group, Pullman, WA). To measure VWC, EC-5 sensors were buried at 10, 20, and 50 cm at the mid and high elevation sites and at 10, 20, and 30 cm at the low elevation site. Due to sensor malfunctions at the high elevation site, air temperature and relative humidity were not recorded in 2017. However, temperature inside the data logger was recorded. In order to estimate the air temperature at the high elevation site, we modeled outside air temperature as a function the temperature inside the data loggers from both the low and mid elevation sites. We then used the model to predict air temperature at the high elevation site. Further, at the low elevation site, only values from the 30 cm depth were reported due to instrument malfunction. The average VWC from all three depths was estimated for the mid and high elevation sites (Fig. 2).

Soil Sampling

At each of the three sites, we collected 15 samples at a depth of 15 cm from under native plant canopies and 15 samples from under non-native plant canopies, for a total of 30 samples per site. Samples were collected in May 2017 at the end of the wet season/start of dry season and in December 2017 at the end of the dry season/ start of wet season. Within 24 hr of sample collection, ammonium (NH_4^+) and nitrate (NO_3^-) were extracted from the soil samples in a solution of 2 M KCl for one hour. All live plant material and rocks were manually removed from the soil cores. Samples were inverted by hand every 15 min across the one-hour extraction period, and then they were gravity filtered with p8 coarse paper filters and syringe filtered through 0.7 μm glass tips. After extraction, samples were stored frozen until analysis. Remaining soil was reserved, transported, and then air-dried for one month. We extracted phosphate (PO_4^{3-}) in a solution of ammonium fluoride and hydrochloric acid following a Type I Bray's extraction procedure. Soil solutions were shaken by hand for five minutes and then filtered in the same manner as the N extracts. All extracts were frozen for storage and then analyzed using flow injection analysis (Lachat Quik-Chem Series 2400, Colorado).

Plant Sampling

During each sampling period, we collected leaf samples from the vegetation we sampled soils under. All samples were transported back to the lab and dried at 60 °C for 48 hr. Leaves were hand ground using liquid nitrogen and then weighed into tin capsules. Then the samples were analyzed for %N and %C using a CHNOS Elemental Analyzer.

Statistical Analysis

In order to examine spatial and temporal patterns of NH_4^+, and NO3, and PO_4^{3-} concentrations, we fit linear mixed effect models using the lme function in R (Pinheiro, Bates, DebRoy, & Sarkar, 2014). To select each model, we used a stepwise Akaike information criterion (AIC) process (Table 1). We started with a full model testing a three-way interaction between elevation, plant canopy type, and season, after accounting for the random effect of sampling location. We removed variables or interactions one at a time and selected the model with the lowest AIC score. After visually assessing if each model distribution met the assumptions of normality and constant variance using residual and normal-QQ plots, the concentrations of each nutrient were log transformed.

Table 1 Results from the stepwise AIC model selection processes

Analyte	Fixed effects	AIC
NH_4^+	Elevation × Season × Plant Type	355.5612
	Elevation × Season + Plant Type × Season	345.837*
	Elevation × Season + Plant Type	346.826
	Elevation + Season + Plant Type	346.808
	Elevation × Season	346.758
	Elevation + Season	346.755
NO_3^-	Elevation × Season × Plant Type	375.051
	Elevation × Season + Plant Type × Season	370.6681
	Elevation × Season + Plant Type	367.063*
	Elevation + Season + Plant Type	483.803
	Elevation × Season	390.261
	Elevation + Season	497.262
PO_4^{3-}	Elevation × Season × Plant Type	390.238
	Elevation × Season + Plant Type × Season	387.382
	Elevation × Season + Plant Type	387.164
	Elevation × Season	382.595*
	Elevation + Season	384.67

We selected the models with AIC scores listed with*

Results

Soil Sampling

We examined differences in NH_4^+, NO_3^-, and PO_4^{3-} across three elevations (300 m, 500 m, and 650 m), between the wet and the dry season, and under native and non-native plant canopies. For each analysis, we first fit a full linear mixed effects model including a three-way interaction between elevation, season, and plant type (native or non-native), after accounting for a random effect of sampling location. We then used a stepwise AIC process to select each model (Table 1). For NH_4^+, we selected the following model:

$$\text{Log}(NH_4^+) = \beta_0 + \beta_1 I_{300m} + \beta_2 I_{500m} + \beta_3 I_{650m} + \beta_4 I_w + \beta_5 I_d + \beta_6 I_n + \beta_7 I_{in} + \beta_8 I_{300m} * I_w + \beta_9 I_{500m} * I_w + \beta_{10} I_{650m} * I_w + \beta_{11} I_{300m} * I_d + \beta_{12} I_{500m} * Id + \beta_{13} I_{650m} * I_d + \beta_{14} I_w * I_n + \beta_{15} I_w * I_{in} + \beta_{16} I_d * I_n + \beta_{17} I_d * I_{in} + \text{location}_i + \varepsilon_{ij}, \text{location}_i \sim N(0, \sigma 2), \varepsilon_{ij} \sim N(0, \sigma 2_\varepsilon)$$

where 300 m, 500 m, and 650 m refer to the site elevation expressed in meters; w and d refer to wet or dry season; n and in refer to native or invasive vegetation, ε_{ij} = residual variability of the jth occasion for the ith subject, and $N(0, \sigma 2)$ denotes that errors are normally distributed with a mean of 0 and a variance of σ^2.

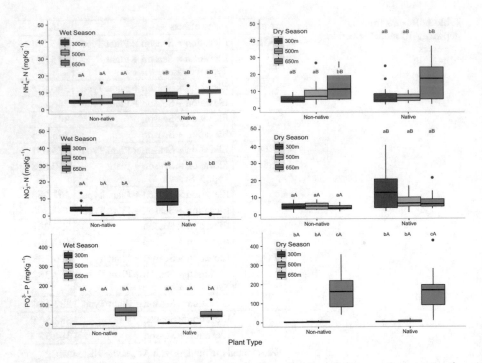

Fig. 3 Soil NH_4^+, NO_3^-, and PO_4^{3-} concentrations across elevations, between native or non-native plant canopies, and between the warm wet season (left) and cool dry season (right). Lower cases letters denote differences in nutrient pools between elevation and between season and upper case letter denote differences between pools under native and non-native canopies

At the mid and high elevation sites, median NH_4^+ was higher during the dry season, but at the low elevation site, median NH_4^+ was higher during the wet season (Fig. 3). In the dry season, NH_4^+ increased along the elevation gradient from 5.47 mg kg^{-1} (95% confidence interval (CI) of 4.48 to 6.68 mg kg^{-1}) at the low elevation to 11.59 mg kg^{-1} at the high elevation (95% CI of 9.02 to 14.15 mg kg^{-1}). However, during the wet season, NH_4^+ did not vary across the elevations. During the dry season, there was no difference in NH_4^+ concentrations under native or non-native plant canopies. Yet, during the wet season, NH_4^+ concentrations were 2.69 mg kg^{-1} greater under native versus non-native plant canopies (95% CI of 2.54 to 3.28). We also found an interaction between elevation and season (χ^2 (2,81 df) = 3.70, p-value = 0.03), suggesting that elevational differences in median NH_4^+ concentrations depended on the season sampled. Furthermore, we detected evidence for an interaction between NH_4^+ concentrations and vegetation type (χ^2 (1, 81 df) = 4.86, p-value = 0.03), suggesting differences in soil median NH_4^+ concentrations between plant types also depended on the season sampled.

To examine differences in NO_3^- concentrations between elevations, seasons, and plant type, we selected the following model:

$Log(NO_3^-) = ß_0 + ß_1I_{300m} + ß_2I_{500m} + ß_3I_{650m} + ß_4I_w + ß_5I_d + ß_6I_n + ß_7I_{in} + ß_8I_{300m} * I_w + ß_9I_{500m} * I_w + ß_{10}I_{650m} * I_w + ß_{11}I_{300m} * I_d + ß_{12}I_{500m} * I_d + ß_{13}I_{650m} * I_d + location_i + \varepsilon_{ij}, location_i \sim N(0, \sigma2), \varepsilon_{ij} \sim N(0, \sigma^2_\varepsilon)$

During the dry season, NO_3^- did not vary across elevations and ranged from 4.9 mg kg^{-1} at the low elevation (95% CI of 4.05 to 6.04 mg kg^{-1}) to 6.0 mg kg^{-1} at the high elevation (95% CI 4.95 to 8.17 mg kg^{-1})(Fig. 3). However, during the wet season, the median NO_3^- concentration at the low elevation was 6.04 mg kg^{-1} (95% CI of 4.95 to 7.39 mg kg^{-1}), which was 17 times greater than at the high elevation and 35 times greater than that at the mid-elevation. The median NO_3^- concentration under native plant canopies was 0.59 mg kg^{-1} greater than under non-native plant canopies (χ^2 (1, 86 df) = 32.10, p-value <0.0001). We detected an interaction between NO_3^- and elevation, suggesting that differences in median NO_3^- concentrations between elevations depended on the season sampled (χ^2 (2, 82 df) = 103.06, p-value <0.0001).

We selected the following model to examine differences in PO_4^{3-} concentrations between elevations and seasons:

$Log(PO_4^{3-}) = ß_0 + ß_1I_{300m} + ß_2I_{500m} + ß_3I_{650m} + ß_4I_w + ß_5I_d + ß_6I_{300m} * I_w + ß_7I_{500m} * I_w + ß_8I_{650m} * I_w + ß_9I_{300m} * I_d + ß_{10}I_{500m} * I_d + ß_{11}I_{650m} * I_d + location_i + \varepsilon_{ij}, location_i \sim N(0, \sigma2), \varepsilon_{ij} \sim N(0, \sigma^2_\varepsilon),$

For PO_4^{3-}, we found that plant type did not influence PO_4^{3-} concentrations (p-value >0.05), so it was not included in the final model. We still detected that the elevational differences in median PO_4^{3-} concentrations depended on the season sampled (χ^2 (2, 87 df) = 4.57, p-value = 0.01) (Fig. 3). During both seasons, PO_4^{3-} was most concentrated at the high elevation site but did not differ between the low and mid-elevations. The median PO_4^{3-} concentration at the high elevation during the wet season was 51.93 mg kg^{-1} (95% CI of 40.44 to 66.69 mg kg^{-1}) and was 134.30 mg kg^{-1} during the dry season (95% CI of 99.48 to 148.41 mg kg^{-1}). The seasonal differences at the low and mid-elevations each differed by less than 3 mg kg^{-1}.

Foliar N in Natives Versus Non-natives

Preliminary examinations showed that foliar %N was higher in native compared to non-native species at two of the three elevations sampled (Table 2). At the low elevation, the mean foliar %N of three native species, *Bursera graveolens*, *Chiocacca alba*, and *Leocarpus darwinii*, was 0.35% greater than that of the non-native species *Psidium guayaba*. At the mid elevation, the foliar %N in the native species *Zanthoxylum fagara* was 1.0% greater than in the mean %N of two non-native individuals, *Psidium guayaba* and *Pennisetum purpureum*. At the high elevation, leaf % N in the dominant native specie, *Miconia ribsoniana* was 0.55 % greater than the non-native specie, *Psidium guayaba*.

Table 2 Site descriptors across San Cristobal

Site name	Site elevation	Soil type*	Soil texture*	Mean soil pH*	Native foliar %N	Non-native foliar %N
Mirador	300 m	Ultisol	Sandy loam, Sandy clay loam, and Clay	6.06	2.22	1.87
Cerro Alto	500 m	Ultisol	Loamy sand and clay	6.02	3.28	2.16
El Junco	650 m	Oxisol	Sandy loam	4.05	1.59	2.14

Metrics with* are described by Percy et al. (2016) (Unpublished data)

Discussion

Seasonal, elevational changes in climate, and the presence of non-native plants influenced soil N and P pools in the tropical montane forests on San Cristóbal Island, Galápagos. Soil N and P varied between elevations but depended on the season sampled. Similar to other studies in volcanic tropical islands, we found that precipitation, temperature, and soil moisture influencing soil nutrient dynamics (Jenny, 1941; Vitousek & Chadwick, 2013; Vitousek & Matson, 1988). On the other hand, the influence of non-native plants on soil N and P pools in this study did not agree with the general consensus that nutrient pools increase under non-native canopies (Liao et al., 2008). Instead, we discovered that soil N pools decreased under non-native canopies while available P did not change between the two plant types.

Nitrogen Pools

Climate controlled nutrient dynamics on San Cristóbal Island, but the effect sizes depended on the season sampled. This result is expected, since rainfall and temperature gradients also determine thresholds of soil development in other ecosystems (Chadwick et al., 2003; Ewing et al., 2006; Vitousek & Chadwick, 2013). Here, NH_4^+ increased with elevation in the dry season but did not vary with elevation during the wet season. NO_3^- concentrations decreased with elevation during the dry season. During the wet season, NO_3^- was most concentrated at the low elevation and least concentrated at the mid elevation. Leaching and denitrification likely caused export of NO_3^-, a highly mobile ion, out of the system during the wet season and at wetter elevations (Vitousek & Matson, 1988; Weintraub et al., 2015), but $NH4+$ was less variable across the elevation gradient and between seasons, presumably because it is not as prone to leaching and denitrification as NO_3^-. Interestingly, Chacón (2010) observed that both NH_4^+ and NO_3^-, not just NH_4^+, increased along the elevation gradient on neighboring Santa Cruz Island. However, N concentrations in our study still fell within the N concentration ranges observed in other soils in the Galápagos (Jäger et al., 2013; Kitayama & Itow, 1999). The differences in N pools across elevation found between this study and others could have been due to the age of the islands where observations were made, to the sampling season, or to varying levels of disturbance or land use where sampling took place. Some of the studies focused their sampling on one island zone or during one season. Further, sampling could have occurred in sites with greater or lesser disturbance or land use change, but there is little way to compare between studies without knowing specific sites.

We detected that NH_4^+ and NO_3^- were more concentrated under native than under non-native canopies. However, most studies examining how invasion influences nutrient dynamics find that soil nutrient pools are more concentrated under non-native canopies (Ehrenfeld, 2003; Liao et al., 2008; Scott, Saggar, & McIntosh, 2001; Vila et al., 2011; Vitousek, Walker, Whiteaker, Muellerdombois, & Matson, 1987). Because the magnitude and direction of changes in nutrient pools in invaded systems varies greatly depending on plant functional traits, ecosystem types, and N-fixation status of invaders, Liao et al. performed a meta-analysis to generalize how invasion influences ecosystem properties. Based on 94 studies, they determined that NH_4^+ and NO_3^- pools under non-native plant canopies were 30% and 17% greater, respectively, compared to under native plant canopies (Liao et al., 2008). In our study, however, NH_4^+ and NO_3^- concentrations were 33% and 42% lower, respectively, under non-native canopies than native canopies.

In other studies conducted in the Galápagos, non-native plants typically increased soil N pools. In the arid zone on San Cristóbal Island, percent N (%N) in bulk soil was highest in a pasture restored with native species and lowest in actively managed agricultural lands containing non-natives (de la Torre, 2013). In the humid zone on Santa Cruz, NH_4^+ was more concentrated under non-native canopies, but they detected no differences in NO_3^- under different canopy types (Jäger et al., 2013). Whereas we examined numerous non-native and native species growing in all climate zones on San Cristóbal, other studies carried out on the Galápagos either followed the influence of a single non-native species or of plant communities within single climatic zones, which may have influenced the magnitude of non-natives' effect on soil chemistry. Increases in soil N pools after invasion are far from universal, however, and decreases or no differences, as we observed in this study, has also been observed (Ehrenfeld, 2010; Ehrenfeld & Scott, 2001; Martin, Tipping, & Sickman, 2009; Scharfy, Eggenschwiler, Venterink, Edwards, & Gusewell, 2009; Svejcar & Sheley, 2001).

Phosphorus Pools

Climate was the strongest control of PO_4^{3-} dynamics in the Galápagos. PO_4^{3-} concentrations were significantly higher at the high elevation site, suggesting that greater soil moisture at the high elevation could have increased the organic P turnover rates, in turn elevating the P pool (Turner & Engelbrecht, 2011). However, this is opposite to the result that Chacón (2010) observed on Santa Cruz, where PO_4^{3-} concentrations decreased with increasing elevation. PO_4^{3-} varied seasonally as well, suggesting that plant uptake controlled P pools along shorter time intervals.

After the cool, dry season, PO_4^{3-} was almost double the concentration of PO_4^{3-} after the warm, wet season. Under drier conditions, diffusion of PO_4^{3-} from the soil matrix to root depletion shells slow (Gahoonia, Raza, & Nielsen, 1994), and this process could have caused the build-up of PO_4^{3-} during the cooler, drier season. Further, cooler temperatures may have caused plant activity, including nutrient uptake, to decrease, leaving more P available in the soil pool (Lambers et al., 2008).

We did not observe differences in PO_4^{3-} concentrations in soils under native and non-native plant canopies, which in contradictory to findings by Jäger et al. (2013). However, the study by Jäger et al. (2013) was on Santa Cruz Island, and the dominant introduced species was *Cinchona pubescens* and not *Psidium guayaba*. Similar to our study, in the Seychelles Islands, Kueffer et al. (2008) found that three non-native tree species did not induce changes in total soil P under their canopies. However, even though soil nutrient pools may not change, non-natives may still impact litter concentrations or decomposition rates. For example, total P increased in litter mass from *Hieracium pilosella*, a non-native forb growing in New Zealand (Scott et al., 2001), and litter P and litter decomposition rates increased in invaded plots in Hawaii (Allison & Vitousek, 2004). Similarly, in another tropical forest in Panama, there was a negative relationship between litter-quality (measured as ratio of leaf lignin to P) to soil P, suggesting that litter could influence P availability in soils (Santiago, Schuur, & Silvera, 2005). Although we did not detect changes in soil P pools, changes in plant litter or alteration of decomposition rates may still be apparent if we were to observe those processes.

Conclusions

Even though non-native species do not benefit from increasing soil nutrient pools on the island, they are still far more successful than native plants. Both soil and plant processes observed here suggest that escaping physiological constraints is not what allows non-natives to succeed in this ecosystem. Instead, non-native plants are likely successful because have escaped dispersal limitations (Belyea & Lancaster, 1999). In the 1980s, the feral goat population on the islands grew to around 100,000 individuals, and as a result, rapid plant community changes took place (Coblentz, 1978). Goats acted as seed vectors after eating guava and blackberry fruits, and the plants were widely dispersed. This escape from dispersal limitation appears to be more important than escape from nutrient limitation on the island. In fact, our study suggests that the most important factors in influencing soil nutrients were not non-native or native plants, but seasonality (wet versus dry season) and elevation, which influences soil moisture and nutrient leaching dynamics.

References

Allison, S. D., & Vitousek, P. M. (2004). Rapid nutrient cycling in leaf litter from invasive plants in Hawai'i. *Oecologia, 141*, 612–619.

Belyea, L. R., & Lancaster, J. (1999). Assembly rules within a contingent ecology. *Oikos, 86*, 402–416.

Chacón, G. (2010). Variación de las propiedades químicas del suelo bajo vegetación nativa e introducida en las zonas altas de Santa Cruz, Galápagos.

Chadwick, O. A., Gavenda, R. T., Kelly, E. F., Ziegler, K., Olson, C. G., Elliott, W. C., & Hendricks, D. M. (2003). The impact of climate on the biogeochemical functioning of volcanic soils. *Chemical Geology, 202*, 195–223.

Coblentz, B. E. (1978). The effects of feral goats (*Capra hircus*) on island ecosystems. *Biological Conservation, 13*, 279–286.

de la Torre, S. (2013). Research in agricultural and urban areas in Galapagos: A biological perspective. In S. J. Walsh & C. F. Mena (Eds.), *Science and conservation in the Galapagos Islands* (pp. 185–198). New York: Springer.

Ehrenfeld, J. G. (2003). Effects of exotic plant invasions on soil nutrient cycling processes. *Ecosystems, 6*, 503–523.

Ehrenfeld, J. G. (2010). Ecosystem consequences of biological invasions. *Annual Review of Ecology, Evolution, and Systematics, 41*, 59–80.

Ehrenfeld, J. G., Kourtev, P., & Huang, W. Z. (2001). Changes in soil functions following invasions of exotic understory plants in deciduous forests. *Ecological Applications, 11*, 1287–1300.

Ehrenfeld, J. G., & Scott, N. (2001). Invasive species and the soil: Effects on organisms and ecosystem processes. *Ecological Applications, 11*, 1259–1260.

Ewing, S. A., Sutter, B., Owen, J., Nishiizumi, K., Sharp, W., Cliff, S. S., ... Amundson, R. (2006). A threshold in soil formation at earth's arid-hyperarid transition. *Geochimica et Cosmochimica Acta, 70*, 5293–5322.

Gahoonia, T. S., Raza, S., & Nielsen, N. E. (1994). Phosphorus depletion in the rhizosphere as influenced by soil-moisture. *Plant and Soil, 159*, 213–218.

Gardener, M. R., Atkinson, R., & Renteria, J. L. (2010). Eradications and people: Lessons from the plant eradication program in Galapagos. *Restoration Ecology, 18*, 20–29.

Gardener, M. R., Trueman, M., Buddenhagen, C., Heleno, R., Jäger, H., Atkinson, R., & Tye, A. (2013). A pragmatic approach to the management of plant invasions in Galapagos. In L. Foxcroft, P. Pysek, D. Richardson, & P. Genovesi (Eds.), *Plant invasions in protected areas*. Dordrecht: Springer.

Geist, D., Snell, H., Snell, H., Goddard, C., & Kurz, M. (2014). Paleogeography of the Galápagos Islands and biogeographical implications. In *The Galapagos: A natural laboratory for the Earth Sciences* (pp. 135–166). Hoboken, NJ: John Wiley & Sons.

Jäger, H., Alencastro, M. J., Kaupenjohann, M., & Kowarik, I. (2013). Ecosystem changes in Galapagos highlands by the invasive tree Cinchona pubescens. *Plant and Soil, 371*, 629–640.

Jenny, H. (1941). *Factors of soil formation: A system of quantitative eastern area*. Radnor, PA: State and Private Forestry.

Kitayama, K., & Itow, S. (1999). Aboveground biomass and soil nutrient pools on a *Scalesia pedunculata* montane forest on Santa Cruz, Galapagos. *Ecological Research, 14*, 405–408.

Kueffer, C., Klingler, G., Zirfass, K., Schumacher, E., Edwards, P. J., & Gusewell, S. (2008). Invasive trees show only weak potential to impact nutrient dynamics in phosphorus-poor tropical forests in the Seychelles. *Functional Ecology, 22*, 359–366.

Lambers, H., Raven, J. A., Shaver, G. R., & Smith, S. E. (2008). Plant nutrient-acquisition strategies change with soil age. *Trends in Ecology & Evolution, 23*, 95–103.

Liao, C. Z., Peng, R. H., Luo, Y. Q., Zhou, X. H., Wu, X. W., Fang, C. M., ... Li, B. (2008). Altered ecosystem carbon and nitrogen cycles by plant invasion: A meta-analysis. *New Phytologist, 177*, 706–714.

Martin, M. R., Tipping, P. W., & Sickman, J. O. (2009). Invasion by an exotic tree alters above and belowground ecosystem components. *Biological Invasions, 11*, 1883–1894.

Ostertag, R., & Verville, J. H. (2002). Fertilization with nitrogen and phosphorus increases abundance of non-native species in Hawaiian montane forests. *Plant Ecology, 162*, 77–90.

Percy, M. S., Schmitt, S. R., Riveros-Iregui, D. A., & Mirus, B. B. (2016). The Galapagos archipelago: A natural laboratory to examine sharp hydroclimatic, geologic and anthropogenic gradients. *Wiley Interdisciplinary Reviews-Water, 3*, 587–600.

Pinheiro, J., Bates, D., DebRoy, S., & Sarkar, D. (2014). *R Core Team (2014) nlme: Linear and nonlinear mixed effects models*. R package version 3.1-117. Retrieved from http://CRAN.R-project.org/package=nlme

Pryet, A. (2011). *Hydrogeology of volcanic islands: A case-study in the Galapagos Archipelago (Ecuador)*. l'Universite Pierre et Marie Curie.

Pryet, A., Dominguez, C., Tomai, P. F., Chaumont, C., d'Ozouville, N., Villacis, M., & Violette, S. (2012). Quantification of cloud water interception along the windward slope of Santa Cruz Island, Galapagos (Ecuador). *Agricultural and Forest Meteorology, 161*, 94–106.

Richardson, S. J., Peltzer, D. A., Allen, R. B., McGlone, M. S., & Parfitt, R. L. (2004). Rapid development of phosphorus limitation in temperate rainforest along the Franz Josef soil chronosequence. *Oecologia, 139*, 267–276.

Santiago, L. S., Schuur, E. A. G., & Silvera, K. (2005). Nutrient cycling and plant-soil feedbacks along a precipitation gradient in lowland Panama. *Journal of Tropical Ecology, 21*, 461–470.

Sax, D. F., Gaines, S. D., & Brown, J. H. (2002). Species invasions exceed extinctions on islands worldwide: A comparative study of plants and birds. *American Naturalist, 160*, 766–783.

Scharfy, D., Eggenschwiler, H., Venterink, H. O., Edwards, P. J., & Gusewell, S. (2009). The invasive alien plant species Solidago gigantea alters ecosystem properties across habitats with differing fertility. *Journal of Vegetation Science, 20*, 1072–1085.

Schlesinger, W. H., Reynolds, J. F., Cunningham, G. L., Huenneke, L. F., Jarrell, W. M., Virginia, R. A., & Whitford, W. G. (1990). Biological feedbacks in global desertification. *Science, 247*, 1043–1048.

Scott, N. A., Saggar, S., & McIntosh, P. D. (2001). Biogeochemical impact of Hieracium invasion in New Zealand's grazed tussock grasslands: Sustainability implications. *Ecological Applications, 11*, 1311–1322.

Snell, H., & Rea, S. (1999). The 1997–98 El Niño in Galápagos: Can 34 years of data estimate 120 years of pattern? *Noticias de Galápagos, 60*, 111–120.

Svejcar, T., & Sheley, R. (2001). Nitrogen dynamics in perennial- and annual-dominated arid rangeland. *Journal of Arid Environments, 47*, 33–46.

Trueman, M., & d'Ozouville, N. (2010). Characterizing the Galapagos terrestrial climate in the face of global climate change. *Galapagos Report, 67*, 26–37.

Turner, B. L., & Engelbrecht, B. M. J. (2011). Soil organic phosphorus in lowland tropical rain forests. *Biogeochemistry, 103*, 297–315.

Tye, A. (2006). Can we infer island introduction and naturalization rates from inventory data? Evidence from introduced plants in Galapagos. *Biological Invasions, 8*, 201–215.

Vila, M., Espinar, J. L., Hejda, M., Hulme, P. E., Jarosik, V., Maron, J. L., … Pysek, P. (2011). Ecological impacts of invasive alien plants: A meta-analysis of their effects on species, communities and ecosystems. *Ecology Letters, 14*, 702–708.

Violette, S., d'Ozouville, N., Pryet, A., Deffontaines, B., Fortin, J., & Adelinet, M. (2014). Hydrogeology of the Galapagos Archipelago: An integrative and comparative approach between islands. In K. S. Harpp, E. Mittelstaedt, N. d'Ozouville, & D. W. Graham (Eds.), *The Galapagos: A natural laboratory for the earth sciences* (pp. 167–183). Hoboken, NJ: John Wiley & Sons.

Vitousek, P. M., & Chadwick, O. A. (2013). Pedogenic thresholds and soil process domains in basalt-derived soils. *Ecosystems, 16*, 1379–1395.

Vitousek, P. M., & Matson, P. A. (1988). Nitrogen transformations in a range of tropical forest soils. *Soil Biology & Biochemistry, 20*, 361–367.

Vitousek, P. M., Walker, L. R., Whiteaker, L. D., Muellerdombois, D., & Matson, P. A. (1987). Biological invasion by Myrica-Faya alters ecosystem development in Hawaii. *Science, 238,* 802–804.

Weintraub, S. R., Taylor, P. G., Porder, S., Cleveland, C. C., Asner, G. P., & Townsend, A. R. (2015). Topographic controls on soil nitrogen availability in a lowland tropical forest. *Ecology, 96,* 1561–1574.

White, A. F., & Blum, A. E. (1995). Effects of climate on chemical-weathering in watersheds. *Geochimica et Cosmochimica Acta, 59,* 1729–1747.

Yavitt, J. B. (2000). Nutrient dynamics of soil derived from different parent material on Barro Colorado Island, Panama. *Biotropica, 32,* 198–207.

Zinke, P. J. (1962). Pattern of influence of individual forest trees on soil properties. *Ecology, 43,* 130.

A Critical Physical Geography of Landscape Changes in Southeast Sulawesi, Indonesia, 1950s–2005

Lisa C. Kelley

Introduction

Remote sensing has been increasingly utilized to study socio-natural change and tracking land use and cover change has been a particularly important application of remotely sensed data. Since 2008, open-access to the entire Landsat imagery archive has created the world's largest openly accessible library of geospatial information at 30-m resolution (Wulder, Masek, Cohen, Loveland, & Woodcock, 2012). The development of Google Earth Engine and other cloud-based computing techniques has also enabled this imagery to be processed and analyzed on the fly (Gorelick et al., 2017). Such approaches facilitate increasingly precise assessments of land use and cover change, enabling wall-to-wall land cover change maps for most parts of the world. New technologies for mapping land change also inform new environmental politics; shaping, reworking, and/or reinforcing prior accounts of change and transformation.

The current moment is thus a good time in which to highlight the histories poorly captured by remotely sensed data, and arguably, invisible to these techniques. These questions are pressing in tropical islands, which, like other tropical regions and ecosystems, have often been understood as places without history. In such locations, remotely sensed data not only appear to give new visibility where it was formerly lacking. They also orient a conversation of landscape change around those dynamics most easily captured by remotely sensed mapping techniques, in particular, forest loss. This chapter thus focuses on reconstructing relatively more invisible histories of landscape change in Southeast Sulawesi, Indonesia, here integrating

L. C. Kelley (✉)
Department of Geography and Environment, College of Social Sciences, University of Hawaii at Manoa, Honolulu, Hawaii, USA
e-mail: lckelley@hawaii.edu

© Springer Nature Switzerland AG 2020
S. J. Walsh et al. (eds.), *Land Cover and Land Use Change on Islands*, Social and Ecological Interactions in the Galapagos Islands,
https://doi.org/10.1007/978-3-030-43973-6_10

ethnographic, archival, and remotely sensed data. It also explores the relative contributions of remotely sensed data to resulting accounts of change.

The Southeast Sulawesi site I focus on, like the insular tropics more generally, has long been assumed to be isolated by both choice and geographical circumstance, remote from trade and modernity. Often implicit in this framing, however, is the assumption that history first began with the remarkable boom in smallholder cacao production that took place from the late 1970s onward. As this chapter will show, however, while the landscape and market changes ushered in by cacao were significant, these changes are best understood as a product of earlier resource governance histories: from insurgency and counter-insurgency through the 1950s to the establishment of "political forests" in the 1960s to the elaboration of timber and rattan industries in state-claimed forests through the 1970s. These dynamics worked in and through localized politics and power relations to significantly reshape rural livelihoods, property relations and land use practices all before cacao was first adopted in many areas.

The corollary of these complexities is the partiality of remotely sensed records in shedding light on the varied processes that have shaped contemporary landscapes in Southeast Sulawesi. Not only are remotely sensed data limited in their historical reach, dating in this case to the first available Landsat images in 1972. They are also unable to untangle or reveal the complex on-the-ground relations shaping landscape change on their own (Dennis et al., 2005; Kelley, 2018; Kelley, Evans, & Potts, 2017; Turner, 2003). To supplement remotely sensed data in this case, I thus rely on both archival and ethnographic research. This data includes more than 50 oral histories and 40 in-depth interviews collected from two case study villages over roughly 18 months between 2014–2018. Wherever possible, oral histories and interviews were conducted during field visits, allowing for qualitative findings to be cross-referenced with remotely sensed data. Most oral histories and interviews were also taped and transcribed for subsequent analysis.

In bringing together these approaches, this chapter is thus inspired by emerging work within the field of Critical Physical Geography (CPG). Work in CPG draws closely on work from Political Ecology, sharing a strong interest in the dynamics of resource change and access that structure human-environment relations (Blaikie, 1985; Carney, 2001; Fairhead & Leach, 1996; Watts, 1987). Approaches within CPG, however, also commonly draw on techniques from physical geography and environmental science to enable a more explicit assessment of the environmental changes and conditions structuring human decisions and responses (Lave et al., 2014; Lave, Biermann, & Lane, 2018). Past work suggests that integrated approaches such as this can enrich understandings based solely on remote sensing or political ecology, unsettling, for example, the apparent stability of cartographic categories (e.g., Robbins, 2001) and assumed but unexamined linkages between state engagement and landscape change (e.g., Lukas, 2014).

Below, I first begin by reviewing known and unknown dimensions of landscape change in Southeast Sulawesi. I then turn to the re-reading of landscape change that is made possible by looking beyond the existing emphasis on cacao-deforestation linkages.

Visibilities and Invisibilities of Prior Landscape Changes in Sulawesi

Southeast Sulawesi is often understood as a place without history; "a small undeveloped province which has struggled to attract much outside attention" (Midgley, Rimbawanto, Mahfudz, Anies Fuazi, & Brown, 2007). This representation reflects a thin archival record. Unlike Java, Bali, and many coastal regions in Indonesia, Southeast Sulawesi was not the site of a powerful state that has attracted historians' attention but was inhabited by dispersed groups of swidden cultivators with strong kinship systems and weak traditions of kingship (Reid, 2011: 5). Most seventeenth and eighteenth century traders also avoided the eastern shore of the province entirely, scared off by coastal bluffs and severe storms. The few surviving accounts from the nineteenth century recount relatively barren coastlines.[1] The lack of information regarding the province in earlier centuries is reflected in the distorted geography of the province and lack of place information for Southeast Sulawesi in maps of Indonesia produced before the twentieth century (Fig. 1).

Among all the islands in Indonesia and Southeast Asia, Sulawesi has also been noted for particularly high levels of endemism, particularly of fauna; a fact which is all the more notable given as-yet limited taxonomic attention on the island (Wallace, 1869; Whitmore & Burnham, 1975). Southeast Sulawesi's biological diversity is contained within various ecotypes, including mangrove, lowland, hill, karst, and montane forests, grasslands, and peat swamps and marshes fed by the Lalinda, Lasolo and Sampara (Konawe'eha) rivers (Whitten et al., 1987). This biological diversity, rather than being understood in relation to histories of human habitation,

Fig. 1 Historical maps from Southeast Sulawesi. Early maps of Southeast Sulawesi illustrate how poorly known the province was prior to the twentieth century, even relative to other regions within Sulawesi

---

[1] As a comprehensive regional history by Will de Jong notes, merchant Jacques Nicolas Vosmaer noted deserted coasts when entering the Bay of Kendari (now the capital of the province) on May 9, 1831. The crew of a second ship entering in 1836 also noted few inhabitants; Buginese and Wajo traders rather than the Tolaki or Tomoronene peoples indigenous to the inland mountain regions. Near the Bay of Kolaka on the western side of the province, one traveler noted, "The East Indies fjords are beautiful, yet over it all hangs a shroud of gloom, mystery and mysticism, just as in the forests above on the black mountain ridges" (de Jong, 2017: 16).

has often been understood as part of the natural history of the area, with many forests considered to be 'primary' (Cannon, Summers, Harting, & Kessler, 2007).

These ahistorical presentations of Southeast Sulawesi closely inform existing stories of landscape change, most of which are oriented around a substantial expansion of cacao tree cropping (*Theobroma cacao* L.) into 'virgin' or primary forests (Ruf & Schroth, 2004). Up from almost no cacao production in 1979 when boom began, Indonesia produced more than 600,000 tons of cacao annually by 2002, with more than 90% of this production generated by smallholder producers on the island of Sulawesi managing under two hectares of land (BPS, 2018). This remarkable boom—considered to be "*one of the most spectacularly efficient* [smallholder cacao booms] *in the world*" (Ruf & Siswoputranto, 1995)—was widely celebrated as a development success. As is true in other key sites of cacao production globally (Nieston et al., 2004; Ruf & Schroth, 2004), however, Sulawesi's smallholder cacao boom has also been seen as productive of deforestation, an understanding reinforced by accounts reliant on remote sensing (Erasmi & Twele, 2009; Erasmi, Twele, Ardiansyah, Malik, & Kappas, 2004; Steffan-Dewenter et al., 2007).

Cacao-deforestation linkages have also commonly been explained with reference to the ecological-come-economic yield advantages growers can obtain by planting cacao in tropical forests. Francois Ruf's forest rent theory of cacao expansion, for example, argues that the Ricardian rents afforded by freshly cleared forested land—high soil fertility, low disease levels and protective shade for newly planted seedlings—combine with rapid in-migration and moments of opportunity in world markets to propel cacao expansion into forested lands. Forest rents also beget what Clarence-Smith and Ruf (1996: 1) suggest is the basic problem of cacao economies: "their need for fresh supplies of primary forest to maintain themselves." Because forest rents eventually expire under intensive production, growing material and labor inputs are considered necessary to maintain the same yields. Assuming no premium to offset these costs, forest rent theory posits that forested land will outcompete existing sites of production.

Consistent with the forest rent theory of expansion, most work to explain smallholder cacao boom in Sulawesi has thus emphasized three key factors: first, Sulawesi's ample forest reserves, particularly in the fertile alluvial lowlands and in sites of abundant rainfall (Durand, 1995; Ruf & Siswoputranto, 1995; Ruf & Yoddang, 1996); second, the in-migration of people of Bugis ethnicity into forested regions from the 1970s onward, many of whom had experience with cacao from an earlier regional boom in Malaysia (Jamal & Pomp, 1993; Pomp & Burger, 1995; Steffan-Dewenter et al., 2007); and third, the lack of trade or taxation policies in the sector, which allowed smallholders to benefit from some of the highest free-on-board prices globally (Akiyama & Nishio, 1997; Neilson, 2007; Ruf & Yoddang, 1996).

In other words, much work to explain cacao expansion has assumed that the historical and socio-spatial relations pre-dating cacao were largely contingencies (Leiter & Harding, 2004); a reduction compatible with more ahistorical remotely sensed analyses of forest loss due to cacao. More economistic accounts of cacao expansion have also informed policy propositions in the sector, including claims surrounding the benefits of yield intensification as a means of "sparing" further forest from clearance into cacao (Clough, Faust, & Tscharntke, 2009; Neilson, 2007). Clough et al. (2009), for example, has argued that: "a failure to sustain production in current cacao-growing areas in Sulawesi likely means a shift to the forest margins" while Neilson (2007: 2) has noted that "[l]eft to market forces alone, the 'mining' of cocoa regions will in all likelihood continue unabated across tropical frontiers until all potential cocoa lands have been physically exhausted."

Coupled to an emphasis on cacao as the primary driver of landscape change in this region, these arguments shape a largely essentialized and economistic understanding of prior landscape changes. The next two sections expand these accounts by exploring the various land and livelihood transformations that pre-dated and conditioned the eventual shift into cacao.

Alternative Accounts and Earlier Histories of Landscape Production

One of the most notable early accounts of landscape change in this region (though not originally written as such) is an ethnography of Tolaki people produced by Tarimana in 1989. Tarimana's ethnography highlights the diversity of livelihood pursuits characteristic of indigenous Tolaki people in Southeast Sulawesi, including nickel and gold mining; swidden rice (*ladang*), rain-fed rice (*sawah*), and tree-crop production (*berkebun tanaman jangka panjang*); buffalo raising (*beternak kerbau*); swamp and river fishing (*menangkap ikan di rawa-rawa dan di sungai*); and forest product collection, including varied forms of hunting (such as for deer and anoa) and timber, rubber, and rattan harvest. As Tarimana shows, these diverse practices involved considerable landscape mobility as well as strategic and collective forms of landscape manipulation.[2]

While waiting for rice to yield, most people relied on corn as well as planted and wild cassava (*ubi*). Sago (*Metroxylon spp.*) was also eaten in the 1–2 months before the rice harvest when prior stocks of rice and corn had run out. An important subsistence staple in Eastern Indonesia, sago is also integral to Tolaki diet and food systems; planted along the banks of a river, along the edge of a swamp, or in marshy plains, and possible to leave untended for as long as 25 years before trunks are ready to be cut and harvested (Melamba, 2014). Many households also managed other tree crops in home gardens and in swidden plots (here, trees also served as a claim over lands). As Tarimana notes (1989: 84–85), this "*planting [was] not done in separate*

---

[2] Temporary houses (*rumah kebun*) were inhabited during the swidden season, located at the borders of the larger complex of all households' adjoining swidden plots. Fields were opened and planted side-by-side both to facilitate tasks performed collectively (such as field burning) and to saturate a given area, delimiting pest pressure on any single field. Some settlements instead practiced wet rice production in the lowlands, particularly near swamp forests in Lambuya, Rate-Rate, Tinondo, and Mowewe.

areas for each tree species but all [were] *planted in mixed systems ...* [managed] *during breaks in the management of swidden rice plots*." These practices, in turn, produced many of the agrarian and forested landscapes later converted to cacao.

Tarimana's ethnography also recounts the complex forms of fish and wildlife harvest that complemented these practices, particularly during the rainy season before or immediately following swidden rice harvest. Fish were caught from streams, rivers, or swamps using hooks (obtained via trade) or with traps (made by women from rattan collected in the forest). During long dry seasons, fish could also be collected by drumming bamboo against the surface of the water in larger puddles to force fish into adjacent puddles, where earthen walls could be built about a foot high to prevent the fish from escaping. Fish could subsequently be caught by hand. Buffalo were also raised in grasslands adjacent to lowland or forest settlements, typically by elites. *Imperata* grasslands were either created through fire or could result from a state-shift in long-utilized swidden lands (see also Garrity et al., 1996 and Henley, 2002 for a discussion of these dynamics of grasslands elsewhere in the region).

Tarimana's account is thus important in depicting a landscape marked by use and manipulation; *cultivated* and *inhabited* localities rather than the primary or undisturbed forest systems believed affected by cacao expansion. A regional history by the scholar Will de Jong reiterates this point through a discussion of the multiple phases of colonization that contributed to landscape change prior to the 1970s, including that associated with the arrival of Buginese, Makassarese, Butonese, and sporadic English and Dutch traders from regional trade ports that brought ships to Kendari prior to the twentieth century (de Jong, 2017: 10). This trade activity seems to have grown around 1839 when the "narrow entrance" into Kendari Bay was discovered (Whitten et al., 1987), pre-dating the establishment of a Dutch colonial presence in the province in the early 1900s. Suggestive of the diverse forms of landscape manipulation that existed, de Jong notes (2012: 4–5) that items traded to and from Kendari likely included: "*poultry, fish, rice, cane sugar, salt, resins for lamps and torches, coconut oil, nutmeg, and its by-product mace, tubers (ubi in Malay), sago, peas, corn, beans, mangoes, coconuts, betel nuts and big and smaller boats, pottery, rattan, woven mats and hats, honey and beeswax, fishing lines, pigs, goats, cattle, poultry and hides*."

Important to the more contemporary history of landscape change that follows, de Jong also suggests how dynamics of landscape use and manipulation would have shifted following the expansion of Dutch colonial interest in the region in 1906 when "the [Dutch] colonial state spread its wings in the eastern part of the archipelago and with its officials, rules, procedures, courts, taxes and reform plans penetrated to the farthest corners" (de Jong, 2017: 10). In Southeast Sulawesi, for instance, the Dutch widened and solidified road networks and built bridges to cross the diverse marshes, coastal flood forests, and river networks, focusing on reinforcing the east-west road between the port towns of Kendari and Kolaka using *herendienst* (enforced labor). The resulting network of gravel roads and paths subsequently became the basis for the formal establishment of new settlements (*kampong*), with households given numbers and all tax-paying adult men given a *kampong*-identity card.

Despite the dynamics that limited colonial ambitions,[3] these engagements would have significantly altered landscapes and livelihoods through the first several decades of the 1900s: reworking demographics through the rapid spread of malaria, scabies, syphilis and other venereal diseases (18) and altering and intensifying resource production the introduction of trucks for forest product extraction, irrigation for wet rice production (22), and intensified mining operations for nickel, magnesium and chromium in sites such as Lapao-Pao, Oko-Oko and Pomala'a (142). Subsequent Japanese occupation during World War II (1942–1945) further elaborated these pursuits. While no aerial imagery or remotely sensed data are available for this period, early maps provide some indication of the resulting landscape formations (Fig. 2).

De Jong's account stops short of describing an important "fourth" phase of colonization on the island: that associated with the declaration of Indonesian independence in 1945 and formal acknowledgement of Southeast Sulawesi as an independent province in 1964 (Potter & Lee, 1998). This moment in time, however, was crucially important to subsequent landscape changes through the property and livelihood reconfigurations it shaped. These changes, as I discuss below, included the violence shaped by the Darul Islam insurgency (DI/TII) and associated counterinsurgency through the 1950s[4] and the shifts in land control produced by the 1967 Basic Forestry Law, which provided the basis for deeming many prior settlements and land holdings 'under-utilized' or 'unused,' and as such, property of the state (Peluso & Vandergeest, 2011). These two developments worked in tandem to reorder agrarian society, with many settlements that had been disrupted by political violence in the 1950s placed under state control by the 1960s.

As I show in the following chapter, development policies and processes enabled by the establishment of state control would shape much of the deforestation that has more commonly been attributed to the cacao sector. State land control also informed the abandonment of swidden fallows and adoption of tree crops in grassland ecosystems less subject to expropriation, processes productive of substantial tree cover gain over a four decade period. These processes did not unfold evenly across either

---

[3] As de Jong notes, the building of roads, culverts, dikes, and bridges often "progressed slowly, not only because of financial constraints, but also because many *kampongs* existed only on paper, or were inhabited for just a few months of the year and therefore few or no laborers were available" (de Jong, 2017: 25). Even after the opening of the road network in 1930, "[o]ver land the peninsula of Southeast Celebes was not accessible by car from South and Central Celebes. In this respect it had the characteristics of an island" (de Jong, 2017: 24). Travel in the province also remained difficult by motorized vehicle or horse, particularly through the low-lying and muddy floodplains adjoining the arms and branches of major rivers in the province. These dynamics would also affect the viability of development schemes pursued by the Indonesian government following independence.

[4] This insurgency, which fought for the establishment of an Islamist state in Indonesia, involved guerrilla warfare from forested hillsides, a violence amplified by the arrival of counter-insurgency troops in the province.

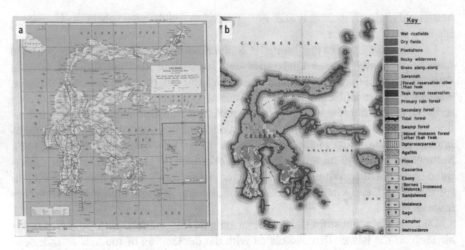

Fig. 2 Mid-twentieth century maps from Southeast Sulawesi. Despite their limited value in conveying the complexity of use arrangements that existed, these maps from the mid-twentieth century show the limited road network reinforced by the Dutch in the early twentieth century as well as the extent of cultivated land that existed by this point in time. The "Special Strategic Map" made by the US Army at the end of the Dutch colonial period in 1943 (**a**) for instance, shows the reinforced road network that would have facilitated accelerated trade, particularly during flood months when the river became dangerous to navigate by bamboo raft. A vegetation map composed by the Planning Department of the Forestry Service in 1950 (**b**) also suggests the significance of land held in either secondary forest or grassland (*alang-alang*) by this point, particularly in the lowlands of the Mekong and Tanggeasinua mountain ranges

space or time, however, and were closely shaped by idiosyncrasies in settlement practices, the specificity of local resistances, and the exact configuration of institutions, actors, and interests present. The next section discusses what these distinctions implied for landscape change, and in particular, for the uneven ways cacao tree cropping spread.

Reconstructing Landscape Change in the Konawe Lowlands, 1950–Onwards

This section builds on the landscape history presented above by situating histories of landscape change in two lowland villages from the 1950s onward using oral histories, in-depth interviews, and remotely sensed data. I focus on the villages of Besulutu and Unahoa for comparative analysis because although they exhibit certain shared demographic or agro-ecological characteristics (Table 1, Fig. 3), they present contrasting histories of land and resource governance as well as road and market access. For example, though most land in Besulutu was nominally held by the state, no forestry officials would visit the area until 1981 given the muddy, watery plains and seasonal floods that affected mobility in and out of the area. In

Table 1 Agro-ecological and demographic overview of the two villages selected for comparative analysis

	Unahoa	*Besulutu*
District	Kolaka Timur	Konawe
Population (2012)	769	731
Area (km²)	13.87	9.91
Pop. Density (ppl/km²)	55	74
First in-migration	1970s	1970s
In-migrants (%)	43%	57%
Avg. elevation (m)	170	50

An overview of basic demographic and agro-ecological characteristics of the two village case studies compiled with data obtained from Cannon et al. (2007). Village names are pseudonyms.

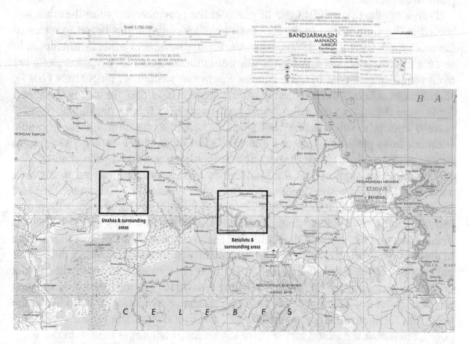

Fig. 3 Map prepared by the Army Map Service, Corps of Engineers, 1965. This map, subset to the two focal study areas, illustrates settlements which had been formally established along the main road by the mid-1960s. It also provides a sense of regional topography

contrast, the areas around Unahoa became a base for provincial forestry and police departments, building on earlier colonial engagements and infrastructure. These differences were central to subsequent landscape changes.

Besulutu

Prior to the 1950s, most people were formally settled outside the reach of what would become Besulutu settlement along the main road through the province. What would become the administrative area of Besulutu was a seasonal resource: exploited for the lands along the Konawe'eha River and for occasional swidden production in the hills. Riparian lands were used for vegetable cultivation and as a fishing grounds during flood months at the end of the rainy season. This remained true until counter-insurgency troops sent to contain the DI/TII insurgency burned homes in nearby Wawotobi in 1951, at which point, many individuals located along the main provincial road network fled inland. People fleeing to what is now Besulutu settled in a site they called Tundu Puri Wuta ('the end of the land'), a small "island" flanked by the Konawe'eha River on both sides and thus difficult for either army troops or insurgents to easily reach.

Following the re-establishment of peace in the province, most families that had settled in Besulutu stayed in the area even as other communities began to return to the main road. By the late 1960s, however, local government officials attempted to resettle these households to integrate Besulutu into administrative regions along the main road. Reminiscent of what Nathan Sayre has written on rangelands (2017), resettlement schemes further derived from a perception that the newly inhabited lands were difficult to control, understand, manage, or render profitable: "muck" moreso than productive agricultural lands. Schemes to develop and territorialize Besulutu from the 1960s onwards thus tended to focus either on ways to relocate people or to establish wet rice production in low-lying and flood-prone areas.

Oral histories suggest that the four family groups that had moved to the area amid conflict at first resisted relocation. During the years of displacement, they had claimed and managed flooded lands along the river for seasonal vegetable crops, including corn, soybean, eggplant, long beans, and peanuts. They had also claimed higher lands in the middle of the island and in the forested hillsides for swidden rice in times of rising waters, re-activating older family claims on land. When the district government ordered families in Tundu Puri Wuta to resettle to more established areas, they instead attempted to stay, with one woman recounting how they embraced an old coconut tree, holding hands with one another, and vowing to build life in the area. The families were eventually moved, not to a distant site, as had been proposed, but across the island into slightly higher lands adjoining the swamp.

By the late 1960s, governments efforts shifted to a strategy of further developing the area through local transmigration schemes, or by bringing other Tolaki people into the area while simultaneously converting swamplands into wet rice. Two specific in-migration schemes were pursued, a first in 1969 and a second in 1978.[5]

---

[5] The first seems to have involved at least fifty households and the second 100 households, sponsored by the Social Welfare Department. Though the resettled households were provided with sugar, oil, dried fish, and other supplies for several years, support for irrigation was provided, making wet rice production a difficult prospect.

Through these schemes, each relocated household was allocated a 50×100 m^2 plot of land intended for wet rice production and settlement, with houses arranged in blocks mirroring villagization schemes elsewhere. Both schemes, however, failed to establish wet rice production in the area. People lacked expertise with wet or swamp rice production and regularly lost their crop to the annual floods. Many Tolaki people also seasonally abandoned wet rice plots for hillside swidden rice production. Most resettled households returned to their former settlements after just a few years.[6]

Despite their limited traction, the two schemes played an important role in structuring landscape change, including through their role in driving secondary forest and swamp conversion. Oral histories suggest an estimated 74 hectares of land were cleared between the two schemes, including land previously dense with timber species, bamboo grasses, and other cultivated vegetation, including planted sago palms and fruit trees. Simultaneously, because of their limited efficacy in effecting sedentary wet rice production, landscape mobility for swidden agriculture in the forested hillsides persisted, resulting in a mosaic of rotationally opened and fallowed lands. Such cultivations were particularly necessary in years of heavy rainfall when longer floods precluded early dry season vegetable cropping. During these years, tree growth was likely as common as tree loss; swidden lands were typically harvested for 1–2 years before being fallowed.

The 1979 transmigration scheme also transformed land uses in the area indirectly by brokering an important connection between a local landholder, Pak Sidik, and one of the wealthiest individuals in the province, Pak Wahab. Pak Wahab worked as a contractor on the construction of the four housing rows (*lorong*) for in-migrant households in 1979. During his time in Besulutu, and on the basis of his wealth and government connections, Pak Sidik invited him to take part in a scheme to plant cashew (*jambu mete*) in the forests of Sambarapa, an area in the eastern half of Lawonua. Pak Wahab quickly acquired much of the land previously managed for swidden agriculture,[7] later extricating it officially from state control with an HGU obtained in 1999 through government connections in Kendari. Many of these land transactions remain contentious decades later. In 2010, the initial use permit (*Hak Guna Usaha*, HGU) he had obtained would also become the basis for an industrial oil palm concession.

---

[6] The in-migration schemes also resulted in the official administrative designation of the area as a village; a shift that implied the need to maintain local populations even after local transmigrants had departed. The village head thus opened up the previously established resettlement lands to Javanese transmigrants from elsewhere in the province seeking new lands. As before, the intention was to develop rice fields in the eastern swamps of Besulutu. Because of the absence of paddy fields and irrigation, however, the Javanese families also lasted only several years. As time went on, some of the houses established as part of the resettlement scheme began to break down while others were claimed by local residents or entered into village holdings.

[7] Some accounts suggest that many lands claimed and sold by Pak Sidik were in fact held by other claimants unacknowledged in these tractions. Other people suggest that once the initial partnership was struck, cash-strapped families began to seek Wahab as a land purchaser. It is likely both accounts are true to a degree.

Pak Wahab's land acquisitions shifted land use practices by effectively enclosing many lands previously used for swidden agriculture. Throughout the 1980s and 1990s, Pak Wahab hired people from Besulutu and adjoining villages to clear the acquired lands for various commodity ventures. After a failed cashew scheme, for instance, more than 50,000 *sengon* seedlings were planted. Around 20 hectares of rice were also planted in 1994 with plans to expand a small existing cattle ranch. These dynamics largely resulted in tree clearance as the *sengon* trees that had been planted were mostly eaten by the handful of cattle that had been brought onto the property at an earlier point in time. Individuals' land use practices also continued to contribute to land change. Many land claimants continued to visit previously planted sago palms and fruit trees seasonally. Specialized forest product collectors and traders also continued to trade timber and rattan from lands acquired by Pak Wahab.

In 1997, Pak Wahab partnered with a parastatal cacao plantation that provided seeds and other inputs for production. Subsequently, 25 Buginese families were recruited into the area on promises of a *bagi tanah* or land sharing scheme that would return half of the opened land to Pak Wahab after cacao began to harvest. Cacao expansion further accelerated in secondary forested lands and former swidden fallows in the late 1990s amid administrative confusion following the fall of the Suharto dictatorship between 1997–1998 (McCarthy, 2004). Though the Village Head had prevented land sales for cacao in the hillsides more proximate the primary settlement in earlier years, a land broker took advantage of administrative confusion to approach the Sub-District Head (*Camat*), obtaining permission for the land sales from him. Land sales quickly emerged, enabling further in-migration, including among people learning of the program from their kin already established in lands controlled by Pak Wahab.

By 1998, the village had also been selected to participate in the Sulawesi Rain-Fed Agricultural Development Program (SRADP)—a partnership between the Asian Development Bank and the provincial Plantation Department that provided a staggering $43.8 million dollar budget to support the expansion of cacao, peppercorn and coconut, then the most important estate crops in the region (ADB, 2004). SRADP provided an initial 250,000 Rupiah to compensate growers for opening forested lands for production. It also provided title to the land, salary for planting the land in cacao, and seedlings, fertilizers and pesticides. While the availability of SRADP support further encouraged Buginese in-migration, it also led to the adoption of cacao among Tolaki people, including those who had previously left the area or left the area for temporary work elsewhere. By the early 2000s, these livelihood shifts would complete the reorientation of most Tolaki agricultural practices away from swidden production, in addition to driving the conversion of diverse forest fallows. Much more forest loss, however, was driven by earlier openings and by the eventual transfer of lands held by Pak Wahab to a corporate oil palm plantation in 2010, conversions which explain the major uptick in tree cover loss after this point in time (Fig. 4a). In other words, much of what presents as deforestation, even over just the past two decades, has had less to do with forest rents than it has with the shifting politics of access to and control over land.

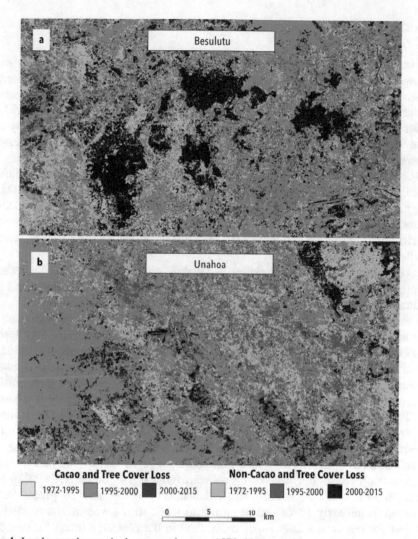

Fig. 4 Landscape changes in the two study areas, 1972–2015. **a** and **b** present data on the overlap between current cacao and non-cacao land covers and tree cover loss in Besulutu and Unahoa over the period 1972–2015. These images depict the centrality of non-cacao land changes to the story of landscape change in these areas. They also highlight that many cacao plantings took place in lands that had been cleared at an earlier point in time (including in lands cleared of trees at some point between 1972–1995). The methods underlying these data are detailed in Kelley et al. (2017) and Kelley (2018) and are reliant on tree cover loss data provided by Hansen et al. (2013) and Kelley et al. (2017)). Non-cacao land covers mapped through this approach include wet rice, built surfaces, other forms of tree crop production, and forested land, and the uneven time intervals presented reflect a lack of cloud-free imagery prior to 1995

Unahoa

In contrast with Besulutu, Unahoa had been well-connected to the primary road network since the 1930s. Correspondingly, the area was well-connected to nearby towns and to an ever-growing network of traders. The areas surrounding Unahoa were also the site of repeat colonization initiatives beginning in the early 1970s, with growers in Unahoa gaining access to seedlings for cacao and cashew by the late 1980s as part of a smallholder tree crop development initiative. In Unahoa, however, well-enforced state claims over forested lands successfully suppressed most swidden practices by the 1970s. State land enclosures also inhibited commodity-oriented tree crop plantings in protected forests until the early 2000s. Significant landscape changes nonetheless took place in the decades pre-dating cacao expansion, including forest loss linked to timber and rattan concessions and agricultural development schemes and forest regrowth in former forest settlements.

The strength of state land claims in Unahoa was linked to earlier Dutch and Japanese occupations. Because of the direct connections between the area and the primary road network, Unahoa and adjacent areas had become a key site of administrative activity throughout the twentieth century. Administrative infrastructure in the area was then used by the Indonesian Army to establish a base in the area during the DI/TII occupations. This meant that resource use practices repeatedly shifted in response to political violence. One woman that was 16 years old when the Japanese invaded, for instance, recounts that villagers were forcibly relocated from hillside settlements into the lowlands as part of a sedentarization and villagization scheme.

Tolaki households were prohibited from using the forest during the DI/TII insurgency and remained so after the reestablishment of state control in the area. As in Besulutu, efforts to suppress swidden agriculture largely reflected the goal of administratively developing state-designated villages along the main road. Settlement practices in Unahoa however were more strictly enforced than they had been in Besulutu. One woman, for instance, recounts around 20 people being arrested in the early 1970s and being put to work on a government construction project for one month after found using land in the protected forest for swidden agriculture. Respondents also recounted the time when many people relocated to former forest settlements during a protracted drought in the 1970s only to have police burn their forest dwellings (*rumah kebun*). As one person described: "*At that point, the village nearly died. … It was as though the village didn't have inhabitants. Finally so that they [the swiddeners] would come down (turun, i.e. to the flatlands), the government burnt their houses.*"

In this context, fallowed swidden fallows began to regenerate, shaping a growth in vegetative cover in many former settlement sites since the 1950s and 1960s invisible to the remotely sensed record. Simultaneously, the area became a key site of transmigration initiatives.[8] These schemes encompassed more area and many more

---

[8] Transmigration schemes generally involve the movement of poor, landless, and other urban and rural people from more densely populated areas of Indonesia to less populated areas.

people than had approaches in Besulutu: between 1975–1984, more than 3000 Javanese and Balinese individuals were settled in the areas surrounding Unahoa adjacent to the Konawe'eha River (Departemen Tenaga Kerja dan Transmigrasi, 2014). Irrigation infrastructure was also established to ensure wet rice production. These dynamics produced bifurcated processes of landscape change from the 1960s onward. While satellite imagery captures the clearance of hundreds of hectares of lowland swamp forest in and around Unahoa where transmigration schemes developed, for instance, many swidden fields would have already regrown by the 1972 date of the earliest satellite imagery in the area.

Many Tolaki people continued to rely on forest resources despite the growth of wet rice in the region and despite risks of arrest or property confiscation, in part as a response to irregularities in irrigation water and challenges with pests in wet rice fields. Forest prohibitions, however, continued to gain strength throughout the 1970s and 1980s as the capacity of the Forestry Department to enforce state lands progressively improved. By the late 1980s, a state water company had also been established in the foothills proximate to former swidden lands. This led to the formal demarcation of state-claimed lands. It also led the Forestry Department to more carefully police state-claimed lands for swidden agriculture or other seasonal crops, perceiving such lands to be more vulnerable to erosion on the steep slopes. By the early 1990s, only a handful of swidden agricultural plots remained.

Two concurrent developments worked alongside these dynamics to reinforce forest gain and loss. The first was the de facto establishment of a concession within the forestry reserve for rattan extraction. The second was a growing emphasis on establishing tree crop production, with villagers granted permission to cultivate these crops on grasslands and some former swidden lands by the 1980s. These tree crop plantings produced substantial tree cover gain, particularly in former swidden lands or grassland areas. When the politics of access to forested land momentarily shifted with decentralization, local elites capitalized on the administrative confusion to sell 20 hectares of forests land to ten Buginese migrants seeking "cheap land" for cacao. These transactions would provide later basis for provisional land use rights following a shift in district leadership in 2004. From 1999–2004, however, their access to land was blocked by forestry department officials.

Though cacao production became possible in forested areas by 2004, eventual expansion was muted because the ecology and economics of cacao had eroded by the late 1990s. Pests and pathogens (the Cacao Pod Borer in particular) were widespread by the early 2000s and many growers perceived that expensive synthetic inputs were necessary to ensure the success of their crop. Simultaneously, the price of fertilizers rose by 100–200% between 1997 and 1998 with the removal of Green Revolution-era subsidies (Ruf & Yoddang, 1999). In this context, far less forest was cleared for cacao production than was cleared for other development opportunities and programs, including the timber concessions and colonization schemes in and around Unahoa (Fig. 4b). Building on the case of Besulutu, then, the case of Unahoa shows the processes of landscape change associated with cacao but largely invisibilized to date, in particular the role of tree crop plantings in "reforesting" grasslands and former swidden lands.

Conclusions: Humility and History in Reconstructing Landscape Change

The years preceding cacao expansion saw tremendous, if highly variable, dynamics of landscape change. Landscape changes were produced not only by Tolaki livelihood systems but by successive waves of colonization, including that associated with early trade networks, Dutch colonization, and the establishment of Suharto's New Order regime and associated claims over forested lands. These changes were particularly notable in the lowland forests that would become a key site of cacao expansion through the 1980s, 1990s and early 2000s.

Overly emphasizing remotely sensed data in reconstructing these changes, however, would mask much of the complex environmental history with which cacao expansion intersected. Though remotely sensed data serve as an important tool in landscape reconstruction, supplementing these records with archival and ethnographic data here was essential in shaping three critical insights. First, that what has often been described in remotely sensed records as deforestation for cacao has often been the product of earlier development regimes. Second, that cacao expansion, tree crop plantings, and forest displacement have occurred alongside substantive tree cover gain and arguably, a story of forest regrowth. And third, that the forests cleared for cacao and other commodity or development ventures were not 'primary' or 'untouched' in any meaningful sense, but former logging lands, swidden lands, and existing agro-forests. Collectively, these insights compel a broader re-reading of cacao expansion, and help to understand deforestation not as a product of forest rents or in-migrants (per se) but as a product of the shifting and often path-dependent politics of accessing land and capital.

This case is thus suggestive of how historical understandings enabled by ethnographic and archival research can work alongside remotely sensed data to both supplement and shift dominant accounts of land use change (and the policy propositions that emerge out of such accounts). Such reconstructive projects are particularly important in island locales where an overemphasis on remotely sensed data or mapping discrete shifts in land change has often allowed for complex socio-spatial relations to be condensed into essentialized visions of nature and nativity; people and lands without history.

References

Akiyama, T., & Nishio, A. (1997). Sulawesi's cocoa boom: Lessons of smallholder dynamism and a hands-off policy. *Bulletin of Indonesian Economic Studies, 33*, 97–121. https://doi.org/10.108 0/00074919712331337145

Asian Development Bank. (2004). Project completion report on the Sulawesi Rain-Fed Agriculture Development Project in Indonesia.

Badan Pusat Statistik Indonesia. (2018). Statistik Perkebunan.

Blaikie, P. (1985). *The political economy of soil erosion in developing countries*. Routledge. https://doi.org/10.4324/9781315637556

Cannon, C. H., Summers, M., Harting, J. R., & Kessler, P. J. A. (2007). Developing conservation priorities based on forest type, condition, and threats in a poorly known ecoregion: Sulawesi, Indonesia. *Biotropica, 39*, 747–759. https://doi.org/10.1111/j.1744-7429.2007.00323.x

Carney, J. A. (2001). *Black rice: The African origins of rice cultivation in the Americas*. Harvard University Press.

Clarence-Smith, W. G., & Ruf, F. (1996). Cocoa pioneer fronts: The historical determinants. In W. G. Clarence-Smith (Ed.), *Cocoa pioneer fronts since 1800* (pp. 1–22). Palgrave Macmillan UK. https://doi.org/10.1007/978-1-349-24901-5_1

Clough, Y., Faust, H., & Tscharntke, T. (2009). Cacao boom and bust: Sustainability of agroforests and opportunities for biodiversity conservation. *Conservation Letters, 2*, 197–205. https://doi.org/10.1111/j.1755-263X.2009.00072.x

de Jong, C. (2012). *A footnote to the colonial history of the Dutch East Indies: The 'Little East' in the first half of the 19th century*. Historical and cultural notes on the Southeastern and Southwestern islands.

de Jong, C. G. F. (2017). New chiefs, new beliefs. A history of the Tolaki and the Tomoronene, two nations in Southeast Celebes (Indonesia) until ca. 1950.

Dennis, R. A., Mayer, J., Applegate, G., Chokkalingam, U., Colfer, C. F. P., Kurniawan, I., ... Stolle, F. (2005). Fire, people and pixles: Linking social science and remote sensing to understand underlying causes and impacts of fire in Indonesia. *Human Ecology, 33*, 465–504.

Departemen Tenaga Kerja dan Transmigrasi. (2014). Unpublished data obtained in Kendari, Sulawesi Tenggara, Indonesia, 30 March 2015.

Durand, F. (1995). Farmer strategies and agricultural development: The choice of cocoa in eastern Indonesia. In *Cocoa cycles: The economics of cocoa supply*. Cambridge, UK: Elsevier.

Erasmi, S., & Twele, A. (2009). Regional land cover mapping in the humid tropics using combined optical and SAR satellite data—A case study from Central Sulawesi, Indonesia. *International Journal of Remote Sensing, 30*, 2465–2478. https://doi.org/10.1080/01431160802552728

Erasmi, S., Twele, A., Ardiansyah, M., Malik, A., & Kappas, M. (2004). Mapping deforestation and land cover conversion at the rainforest margin in Central Sulawesi, Indonesia. *EARSel eProceedings, 3*, 388–397.

Fairhead, J., & Leach, M. (1996). *Misreading the African landscape: Society and ecology in a Forest-Savanna Mosaic*. Cambridge University Press.

Garrity, D. P., Soekardi, M., van Noordwijk, M., de la Cruz, R., Pathak, P. S., Gunasena, H. P. M., ... Majid, N. M. (1996). The Imperata grasslands of tropical Asia: Area, distribution, and typology. *Agroforestry Systems, 36*, 3–29. https://doi.org/10.1007/BF00142865

Gorelick, N., Hancher, M., Dixon, M., Ilyushchenko, S., Thau, D., & Moore, R. (2017). Google Earth Engine: Planetary-scale geospatial analysis for everyone. *Remote Sensing of Environment*. https://doi.org/10.1016/j.rse.2017.06.031

Hansen, M. C., Potapov, P. V., Moore, R., Hancher, M., Turubanova, S. A., Tyukavina, A., ... Townshend, J. R. G. (2013). High-resolution global maps of 21st-century forest cover change. *Science, 342*, 850–853. https://doi.org/10.1126/science.1244693

Henley, D. (2002). Population, economy and environment in island Southeast Asia: An historical view with special reference to northern Sulawesi. *Singapore Journal of Tropical Geography, 23*, 167–206. https://doi.org/10.1111/1467-9493.00124

Jamal, S., & Pomp, M. (1993). Smallholder adoption of tree crops: A case study of cocoa in Sulawesi. *Bulletin of Indonesian Economic Studies, 29*, 69–94. https://doi.org/10.1080/00074919312331336461

Kelley, L. C. (2018). The politics of uneven smallholder cacao expansion: A critical physical geography of agricultural transformation in Southeast Sulawesi, Indonesia. *Geoforum, 97*, 22–34. https://doi.org/10.1016/j.geoforum.2018.10.006

Kelley, L. C., Evans, S. G., & Potts, M. D. (2017). Richer histories for more relevant policies: 42 years of tree cover loss and gain in Southeast Sulawesi, Indonesia. *Global Change Biology, 23*, 830–839. https://doi.org/10.1111/gcb.13434

Lave, R., Wilson, M. W., Barron, E. S., Biermann, C., Carey, M. A., Duvall, C. S., ... Dyke, C. V. (2014). Intervention: Critical physical geography. *The Canadian Geographer/Le Géographe canadien, 58*, 1–10. https://doi.org/10.1111/cag.12061

Lave, R., Biermann, C., & Lane, S. N. (2018). Introducing critical physical geography. In R. Lave, C. Biermann, & S. N. Lane (Eds.), *The Palgrave handbook of critical physical geography* (pp. 3–21). Cham: Springer International Publishing. https://doi.org/10.1007/978-3-319-71461-5_1

Leiter, J., & Harding, S. (2004). Trinidad, Brazil, and Ghana: Three melting moments in the history of cocoa. *Journal of Rural Studies, 20*, 113–130. https://doi.org/10.1016/S0743-0167(03)00034-2

Lukas, M. (2014). Eroding battlefields: Land degradation in Java reconsidered. *Geoforum, 56*, 87–100.

McCarthy, J. F. (2004). Changing to gray: decentralization and the emergence of volatile socio-legal configurations in Central Kalimantan, Indonesia. World Dev. 32, 1199–1223.

Melamba, B. (2014). Sagu (Tawaro) dan kehidupan etnik Tolaki di Sulawesi Tenggara. *Paramita: Historical Studies Journal, 24*. https://doi.org/10.15294/paramita.v24i2.3125

Midgley, S., Rimbawanto, A., Mahfudz, I., Anies Fuazi, I., & Brown, A. (2007). *Options for teak industry development in Southeast Sulawesi*. Australian Center for International Agricultural Research.

Neilson, J. (2007). Global markets, farmers and the state: Sustaining profits in the Indonesian cocoa sector. *Bulletin of Indonesian Economic Studies, 43*, 227–250. https://doi.org/10.1080/00074910701408073

Nieston, E. T., Rice, R. E., Ratay, S. M., Paratore, K., Hardner, J. J., & Fernside, P. M. (2004). *Commodities and conservation: The need for greater habibtat protection in the tropics*. Washington, DC: Conservation International.

Peluso, N. L., & Vandergeest, P. (2011). Political ecologies of war and forests: Counterinsurgencies and the making of National Natures. *Annals of the Association of American Geographers, 101*, 587–608. https://doi.org/10.1080/00045608.2011.560064

Pomp, M., & Burger, K. (1995). Innovation and imitation: Adoption of cocoa by Indonesian smallholders. *World Development, 23*, 423–431. https://doi.org/10.1016/0305-750X(94)00134-K

Potter, L., Lee, J. (1998). Tree Planting Trends in Indonesia: Trends, Impacts and Directions. Occasional Paper No. 18. The Center for International Forestry Research, Bogor, Indonesia.

Reid, A. (2011). *To nation by revolution: Indonesia in the twentieth century*. Singapore: NUS Press.

Robbins, P. (2001). Fixed categories in a portable landscape: The causes and consequences of landcover categorization. *Environment and Planning A, 33*, 161–179.

Ruf, F., & Schroth, G. (2004). Chocolate forests and monocultures: A historical review of cocoa growing and its conflicting role in tropical deforestation and forest conservation. In: Schroth, G., da Fonseca, G.A.B., Harvey, C.A., Gascon, C., Vasconcelos, H.L., Izac, A.N. (Eds.), *Agroforestry and biodiversity conservation in tropical landscapes* (pp. 107–134). Washington, DC: Island Press.

Ruf, F., & Siswoputranto, P. S. (1995). *Cocoa cycles: The economics of cocoa supply*. Elsevier.

Ruf, F., & Yoddang, C. T. (1996). Smallholder Cocoa in Indonesia: why a Cocoa Boom in Sulawesi? In: Clarence-Smith, W.G. (Ed.), *Cocoa Pioneer Fronts Since 1800* (pp. 212–231). Palgrave Macmillan, UK.

Ruf, F., & Yoddang, C. T. (1999). *The impact of the economic crisis on Indonesia's cocoa sector*. ACIAR Indonesia Research Project. Working Paper 99.20.

Sayre, N. F. (2017). *The politics of scale*. Chicago, IL: University of Chicago Press.

Steffan-Dewenter, I., Kessler, M., Barkmann, J., Bos, M. M., Buchori, D., Erasmi, S., ... Tscharntke, T. (2007). Tradeoffs between income, biodiversity, and ecosystem functioning during tropical rainforest conversion and agroforestry intensification. *PNAS, 104*, 4973–4978. https://doi.org/10.1073/pnas.0608409104

Tarimana, A. (1989). Kebudayaan Tolaki (Tolaki Culture), Seri Etnografi Indonesia No. 3. Balai Pustaka, Jakarta.

Turner, M. D. (2003). Methodological reflections on the use of remote sensing and geographic information science in human ecological research. *Human Ecology, 31*, 255–279.

Wallace, A. R. (1869). *The Malay archipelago*. The Floating Press.

Watts, M. J. (1987). *Silent violence: Food, famine, and peasantry in northern Nigeria*. University of Georgia Press.

Whitmore, T. C., & Burnham, C. P. (1975). *Tropical rain forests of the Far East*. Oxford, England: Oxford Science Publications.

Whitten, T., & Henderson, G. S. (2012). *Ecology of Sulawesi*. Tuttle Publishing.

Whitten, T., Mustafa, M., Henderson, G. S. (1987). The Ecology of Sulawesi. Gadjah Mada University Press, Yogyakarta, Indonesia.

Wulder, M. A., Masek, J. G., Cohen, W. B., Loveland, T. R., & Woodcock, C. E. (2012). Opening the archive: How free data has enabled the science and monitoring promise of Landsat. *Remote Sensing of the Environment, 122*, 2–10.

Reframing the Competition for Land between Food and Energy Production in Indonesia

Chong Seok Choi, Iskandar Z. Siregar, and Sujith Ravi

Introduction

As global population grows, so does the demand for energy and food to sustain socioeconomic development and human welfare (Edenhofer et al., 2011). To meet this ever increasing demand, global annual total primary energy supply (TPES), the sum of energy contained in raw fuels, has increased from 6101 million tons of oil equivalent (Mtoe) to 13,699 Mtoe between 1973 and 2014 (International Energy Agency, 2016). Similarly, the projected increase global food demand from 2005 to 2050 varies from 100 to 110% (Tilman, Balzer, Hill, & Befort, 2011). Since food production and energy production both require land and water, the increase and food and energy demand will lead to a competition between the two sectors over land and water (Harvey & Pilgrim, 2011; Tilman et al., 2009, 2011).

In turn, this competition between the two sectors may undermine sustainable developmental goals in the developing world including mitigating climate change, controlling deforestation, and improving the quality of life (Bebbington et al., 2018). Case in point is Indonesia, which currently has one of world's highest deforestation rates that result from an over-reliance on abundant natural resources for food and energy production (Bebbington et al., 2018; Portela & Rademacher, 2001; Sweeney et al., 2004). At the same time, Indonesia is pursuing a national development policy for improving economic growth, calorie consumption and energy accessibility in rural areas. In the context of climate change, one of the key

C. S. Choi (✉) · S. Ravi
Department of Earth & Environmental Science, Temple University, Philadelphia, PA, USA
e-mail: chong.seok.choi@temple.edu

I. Z. Siregar
Department of Silviculture, Faculty of Forestry, IPB University (Bogor Agricultural University), Bogor, Indonesia
e-mail: siregar@apps.ipb.ac.id

© Springer Nature Switzerland AG 2020
S. J. Walsh et al. (eds.), *Land Cover and Land Use Change on Islands*, Social and Ecological Interactions in the Galapagos Islands,
https://doi.org/10.1007/978-3-030-43973-6_11

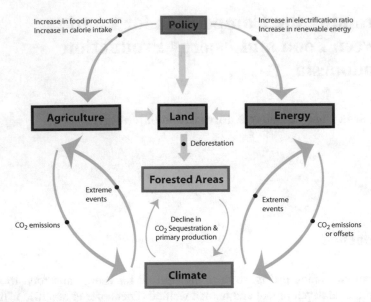

Fig. 1 Feedbacks among accelerated land use for food-energy development, deforestation and climate

challenges for Indonesia to achieving this agenda is to meet the land requirements for food and energy production while fulfilling self-imposed climate and conservation commitments (Kementrian PPN/Bappenas, 2014; Government of the Republic of Indonesia, United Nations in Indonesia, The United Nations Partnership for Development Framework (UNPDF), 2016).

Deforestation and forest degradation lead to reduction in key ecosystem services, including reduction in primary production and carbon sequestration with feedbacks to climate (Fig. 1). Deforestation in Indonesia is often related to the land expansion for both high value crops (e.g. essential oils, biofuel feedstocks, food) and energy development (Abood, Lee, Burivalova, Garcia-Ulloa, & Koh, 2015; Liebman et al., 2019; Tsujino, Yumoto, Kitamura, Djamaluddin, & Darnaedi, 2016; Wicke, Sikkema, Dornburg, & Faaij, 2011; Wijaya, Sugardiman Budiharto, Tosiani, Murdiyarso, & Verchot, 2015). Even with renewable technologies, energy production is often land intensive, and in many cases the additional demand for land will lead to additional deforestation, resulting in increased greenhouse gas emissions (Abood et al., 2015; Fthenakis & Kim, 2009). Moreover, energy resources such as coal, oil, natural gas and geothermal often overlap with forest areas, and Indonesia's widespread forest degradation can be attributed to its heavy reliance on fossil fuels (Abood et al., 2015; International Energy Agency, 2015a). Most of the current electricity use in rural communities is derived from diesel generators, adding to the GHG emissions from land use change and deforestation (Blum, Sryantoro Wakeling, & Schmidt, 2013). On the other hand, lack of energy accessibility is a major bottleneck for other employment generating activities, such as processing agricultural produce and other small-scale industries, leading to further exploitation of natural resources and forest degradation (Zahnd & Kimber, 2009). Even though low carbon emission technologies such as solar energy may provide pathways for sustainable energy production to meet the current and future energy requirements, deploying large-scale renewable energy infrastructure on such a large scale has socioeconomic constraints and may even negatively impact land and water resources in some regions (Fthenakis & Kim, 2009, 2010; Hernandez et al., 2014; Ravi, Lobell, & Field, 2014).

Current Context

Geography

Indonesia is located in Southeastern Asia, between the Indian ocean and the Pacific ocean (Central Intelligence Agency, 2017). It lies on the equator and has a tropical climate (Pegels, 2012). As an archipelago of 17,504 islands, it is the world's largest country comprised entirely of islands. Its total land area is 1.92 million km², which is 15th largest in the world (Dharmawan et al., 2015; Tsujino et al., 2016). Even though the country's 265 million people are spread over approximately 900 permanently inhabited islands (Central Intelligence Agency, 2017), the majority (about 80%) of the population is located in the western Islands of Java and Sumatra (Boer et al., 2016).

Indonesia is highly vulnerable to the effects of climate change because it has an extensive coastline that houses most of the population and economic activity, and much of its economy depends on natural resources sensitive to climate such as agriculture (International Energy Agency, 2015a, 2015b). The estimated quarter of the population who live under or near the poverty line are especially susceptible to the harsher impacts of the climate change (Tharakan, 2015). Furthermore, the land subsidence due to extensive groundwater mining, coupled with the sea level rise due to the climate change, has resulted in frequent high-magnitude floods that have exacerbated current environmental and socioeconomic problems in the nation's capital Jakarta (Abidin et al., 2001; Bakr, 2015; Budiyono, Aerts, Tollenaar, & Ward, 2016; Douglass, 2010; Kennedy, 2019; Rahman, Sumotarto, & Pramudito, 2018).

Between 2000 and 2018, Indonesia population rose from 211 million to 268 million and the GDP from 165 billion USD to 1.04 trillion USD, which translates to an astounding population growth of more than 25% and a GDP growth of more than 5.2% per annum over the last two decades (World Bank, 2019a, 2019b). The service and the industry sectors are the two largest sources of GDP (45.4% and 41%, respectively), and the agricultural sector contributes a considerably smaller share (13.7%) (Fig. 2). For its relatively small contribution to the GDP, however, the agricultural sector employs 32% of the Indonesian population, whereas the industry sector only employs 21% of the population while contributing the largest share to the Indonesian GDP (Central Intelligence Agency, 2019). In addition to the large

Fig. 2 Share of GDP by
sector (in percent) (Central
Intelligence Agency, 2019)

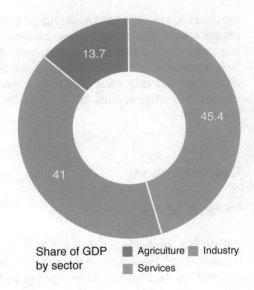

share of employment in agriculture, Indonesia's economy is still strongly dependent
on its natural resources. As of 2014, natural resources still accounted for 25% of
Indonesia's GDP, and the country's top exports in 2017 were palm oil, bituminous
and other coals, natural gas, and petroleum oil (Dutu, 2015; The World Bank
Group, 2019).

Land Use Change as a Major Contributor to GHG Emissions

As a country located in the equatorial Southeast Asia, Indonesia is one of the most
biologically diverse countries and home to the largest area of the tropical rain forest
biome in the world (Tsujino et al., 2016). It has been estimated that nearly all
(99.2%) of Indonesia's land was forested in the past (MacKinnon, 1997). Today,
only 49.9% of the total land area is forested (FAO, 2018), and only half of the forested area is thought to be natural forests, and the remainder of the forests are
thought to be either plantation or secondary (Tsujino et al., 2016) (Fig. 3). The fate
of the remaining major forest stocks largely rests on the domestic and international
demand for cheap oil palm, coal, and forestry products, since 34% (26.8 Mha) of the
combined forest stocks of Kalimantan, Sumatra, Sulawesi, Moluccas, and Papua
exist within the concessions for these goods (Abood et al., 2015). The loss of forests
comes with a steep cost not only for Indonesia's economy (Kurniawan & Managi,
2018; Mumme, Jochum, Brose, Haneda, & Barnes, 2015; Rahman, Sunderland,
Roshetko, Basuki, & Healey, 2016), but also for the country's effort to reduce its
GHG emissions (Fig. 4).

Adding further pressure to the remaining forest is the increasing Indonesian population that will likely require additional resources. Case in point is the linkage

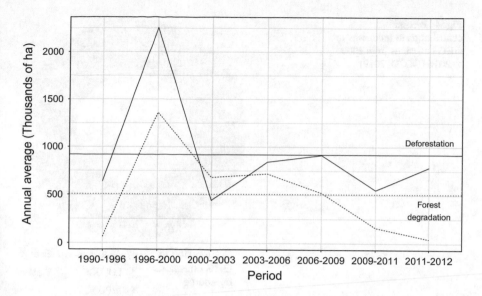

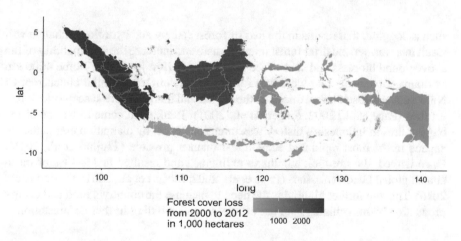

Fig. 3 Top: Average annual deforestation and forest degradation in Indonesia between 1990 and 2012 (Dharmawan et al., 2015). Solid lines represent deforestation and dotted lines represent forest degradation. Horizontal lines are the average over the dataset. Bottom: Forest loss between 2000 and 2012, in thousands of hectares. Sumatra and Kalimantan are major areas of forest loss (Margono, Potapov, Turubanova, Stolle, & Hansen, 2014)

between Indonesia's increasing demand for food in the late 1990s and forest fires, as the government-promoted Mega Rice Project (MRP) in Central Kalimantan resulted in an extensive forest degradation that primed the area for an extensive forest fire (Page & Hooijer, 2016; Tsujino et al., 2016). While naturally resistant to forest fires, wet tropical forests, particularly peat swamp forests that are prevalent in Indonesia, can become vulnerable to fire through degradation. Any human activity,

Fig. 4 Percent
contribution to Indonesia's
GHG emissions from 2000
to 2016 (OECD, 2019)

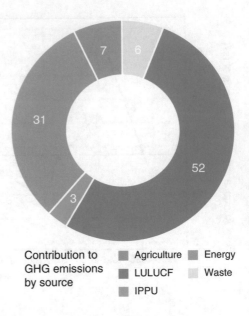

Contribution to   ▉ Agriculture ▉ Energy
GHG emissions   ▉ LULUCF   ▉ Waste
by source   ▉ IPPU

such as logging, that results in the loss of forest canopy and hydrologic change will result in a warmer and drier forest microclimate and increase the flammability of the aboveground biomass and subsurface peat (Page & Hooijer, 2016). Some 4000 km of drainage channels that were dug for the MRP, coupled with the concurrent El Niño Southern Oscillation, dried out the region and made it more susceptible to forest fires (Page et al., 2002; Siegert et al., 2001). Then, what came to be one of the biggest fires in Indonesia's history was intentionally set by plantation companies to expand in the most rapid and economical manner possible (Tsujino et al., 2016). Once ignited, the fire took months to extinguish and resulted in 13–40% of mean annual global GHG emissions (Page et al., 2002; Siegert et al., 2001; Tsujino et al., 2016). This fire further highlights the need to increase the country's food and energy production by maximizing land use efficiency and avoiding further deforestation.

Current Energy Mix, Emissions, and Future Targets

Indonesia's energy demand, which has already risen by nearly 65% between 2000 and 2014, is estimated to rise between 7 and 10% every year (Blum et al., 2013; Tharakan, 2015; International Renewable Energy Agency (IRENA), 2017). In 2017, Indonesia accounted for 35% of Southeast Asia's total energy demand (International Energy Agency (IEA), 2017). The total final consumption (TFC) in 2016 was approximately 164 million tons of oil equivalent (Mtoe), and total primary energy supply (TPES) in the same year was approximately 230 Mtoe (IEA, 2018a). Indonesia, previously a net oil exporter and now one of the biggest exporters of coal,

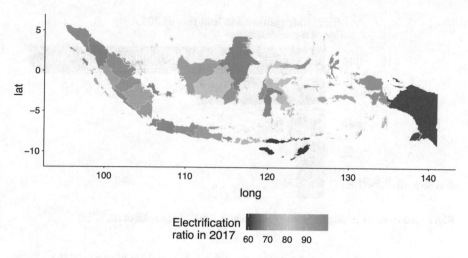

Fig. 5 Electrification rates of Indonesian provinces (2017) (McKenzie, 2018)

has vast reserves in oil, gas, and coal (Agung Wahyuono & Magenika Julian, 2018). Therefore, it comes as no surprise that majority of the total primary energy supply is attributed to coal, crude oil, oil products, and natural gas (66.31%), while renewable energy only accounted for 33.69% of the TPES (IEA, 2018a).

Moreover, electricity demand alone is estimated to triple between 2010 and 2030, and almost quadruple by 2050 (Breyer et al., 2018; Tharakan, 2015). As a result of the expanding demand, Indonesia is prompted to resolve the multidimensional problems of its growing energy sector. While Indonesia's energy sector has been effective in increasing the electrification ratio from 80% in 2015 to nearly 95% in 2018 (International Energy Agency, 2015a, 2018b) (Fig. 5), it has not shown very much progress in resolving other problems, such as primary dependence on fossil fuels, increasing dependence on imported oil, and the need to dissociate economic growth with greenhouse gas emissions in the face of climate change (Blum et al., 2013; International Energy Agency, 2015a). As a major producer and net exporter for fossil fuels, Indonesia is largely dependent on fossil fuels for electricity generation (Fig. 6), and the two largest sources of electricity are coal and natural gas, which accounted for respectively 57.2% and 24.8% of total generated electricity in 2017 (Kamuradin et al., 2018). Adding to these problems is the archipelagic geography of Indonesia that renders most of the country's non-electrified regions too remote and costly to electrify by extension of established electrical grids (Blum et al., 2013).

At the same time, the Indonesian government has acknowledged its role in mitigating the impacts of climate change when it agreed to reduce the national GHG emissions by 29% by 2030 under the Paris agreement (Hasudungan & Sabaruddin, 2018). In an effort to diversify its energy sources, improve environmental sustainability, and maximize the usage of domestic energy resources, the Indonesian government has set a medium-term and long-term targets for the share of renewable

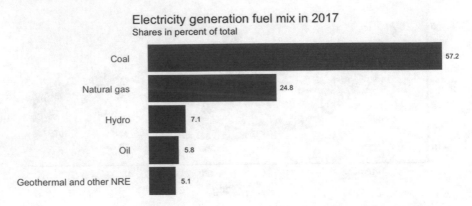

Fig. 6 Indonesia's fuel mix for power generation in 2017 (Kamuradin et al., 2018)

energy as a percent share of the total primary energy supply, which are 23% by 2025 and 31% by 2050, respectively, from the base of 6% in 2014 (Maulidia, Dargusch, Ashworth, & Ardiansyah, 2019; Tharakan, 2015). However, a slower transition from the fossil fuel is anticipated (Tharakan, 2015).

Indonesia and Land Use Problems

Globally, GHG from deforestation is considered to be the second largest after those from the energy sector. For Indonesia, however, land use change and deforestation (including peat fires) is the largest contributor to the national GHG emissions (Liebman et al., 2019; OECD, 2019; Wijaya et al., 2017). Indonesia is the world's sixth largest emitter of GHG and the largest contributor of emissions related to forests, and while emissions from the energy sector is significant, emissions due to LULUCF and agriculture comprise more than 50% of Indonesia's total GHG emissions (Hallik, Niinemets, & Wright, 2009; Liebman et al., 2019; OECD, 2019; Wijaya et al., 2017) (Fig. 3). Because of Indonesia's vulnerability to climate change and their considerable contribution to the global GHG emissions, it is important to examine the land use changes that are relevant to the GHG emissions.

It has been reasonably estimated that original forest cover occupied 99.2% of the land area (MacKinnon, 1997). As of 2018, agricultural land and forest occupied 31.5% and 49.9% of Indonesia's land area (FAO, 2018). The earliest record by the Indonesian Forest Service on the forest cover in Indonesia shows that, in 1950, the forest covered 84% of Indonesia's land area, and that only 16% of the total land area was occupied by rice farming (FWI/GFW, 2002). The pressure on Indonesia's forest happens in a number of different industries at different scales, ranging from large-scale logging operations to forest clearance by smallholder farming operations (FWI/GFW, 2002; Tsujino et al., 2016). The concerns over deforestation began in the mid-1960s to early 1970s as large-scale commercial logging concessions were established for the first time, followed by a rapid growth in timber and pulp production (FWI/GFW, 2002). The growth in timber production was driven by the international demand for wood or pulp, and the increased logging concessions were determined to be a major contributor to the reduction of the forest cover to 119.7 Mha (or 63% of Indonesia's total land area) (Sunderlin & Resosudarmo, 1996; Tsujino et al., 2016). Coinciding with the logging concessions was the government-sponsored transmigration program (Tsujino et al., 2016). This program was the Indonesian government's effort to move landless people living in densely populated Java and Madura islands to less populated areas in the country, which included Sumatra, Kalimantan, Sulawesi, and Papua (Tsujino et al., 2016). Between 1950 and 1979, an annual average of 6570 transmigrating families were recorded, and between 1980 and 1984, the annual average increased to 73,200 families (Sunderlin & Resosudarmo, 1996). Because of the fact that the transmigration occurred simultaneously with the increased forest logging concessions in the areas of low population density, the transmigration program served as a supply channel for the labor force in these areas, and many international non-governmental organizations in the 1980s viewed the program as a contributing factor to the deforestation in Indonesia (Sunderlin & Resosudarmo, 1996). The deforestation rates reached 2 Mha/year after 1996, and Sulawesi, Sumatra, and Kalimantan all lost more than 20% of its forests between mid-1980s and 1997 (FWI/GFW, 2002).

In the wake of the Asian Financial Crisis that began in 1997, the government of Indonesia and IMF reached an agreement for an economic reform, and the policy changes that followed resulted in a transition of power to a new national government, which then began distributing the authorities of the Ministry of Forestry to districts (Hansen et al., 2009). During this chaotic period, the confusion in forest management practices brought on by lack of proper transition of administrative power during decentralization of the government lead to a vacuum in oversight, which then resulted in increased forest-related conflicts such as timber theft and forest clearance (Tsujino et al., 2016). Furthermore, IMF-recommended elimination of all restrictions on plywood export and reduced roundwood and sawn timber export tariffs accelerated deforestation and forest degradation in Indonesia (Tsujino et al., 2016). International demand for wood, notably that of China's, continued, and the decreased production from the legal logging operations were subsidized by illegal logging, which had already been significant in Indonesia's forestry sector (Pagiola, 2000; Tsujino et al., 2016). For example, Greenpeace stated that Indonesian timber export to China in 2002 was 0.336 million m^3 according to Indonesian data, while Chinese data indicated that it imported 1.22 million m^3 from Indonesia (Tsujino et al., 2016). The difference of 0.86 million m^3 was illegally exported timber, which far out-scaled legal logging exports despite the Indonesian government's logging bans and quotas (Lang & Chan, 2006). Ministry of Forestry stated that the supply shortage of approximately 40 million m^3 was met with illegal logging in 2006 (Obidzinski & Chaudhury, 2009).

In addition to logging, the palm oil industry is often highlighted as the major cause for deforestation. Indonesia has had the world's lowest palm oil production cost in the world even before the Asian Financial Crisis, and the financial crisis

Table 1 Areas of industrial concessions within five Indonesian islands with the biggest forest stocks

Region	Land area (ha)	Area of industrial sectors (1000 ha)					
		Oil palm	Logging	Fiber	Mining	Mixed concessions	All industries
Kalimantan	53,602	8367	9192	4243	2538	4737	29,077
Sumatra	47,640	3099	1368	4468	1584	768	11,287
Papua	41,506	417	10,443	1412	N/A	276	12,547
Sulawesi	18,738	249	1664	442	N/A	24	2379
Maluku	7.885	0	1326	44	N/A	0	1370
Total	169,371	12,132	23,992	10,609	4122	5805	56,660

Abood et al. (2015)

boosted Indonesia's cost advantage even further (Tsujino et al., 2016). In fact, this demand led to heightened rate of conversion from forest to oil palm plantations between 2009 and 2011, and oil palm concessions became the second largest type of concession (the first was logging) (Abood et al., 2015; Tsujino et al., 2016). However, the relative contribution of the oil palm industry to the deforestation and land degradation in Indonesia has been contested, as the land use of the oil palm concessions in Indonesia is neither the largest in area nor the most damaging to the land in terms of remaining carbon stock (Abood et al., 2015). Logging concessions are the largest in area (Table 1), and coal mining concessions, while just over a fourth of logging concessions in size, result in the smallest remaining carbon stock per unit area of land (Abood et al., 2015).

Recent trend shows increased conversion of logged forests instead of natural forests to oil palm plantations (Casson, 1999), and latest regulations on illegal logging, prohibition of certain wood exports, promotion of certified timber production and export, and re-forestation efforts have resulted in reduced deforestation rate in the 2000s (Tsujino et al., 2016).

Current Energy and Electricity and Land Use for Energy

In order to estimate land occupation for electricity generation in Indonesia, a few assumptions can be made about the energy density per unit land area. Table 2 lists the assumptions of unit land area transformed for unit electricity. The current land use dedicated to electricity generation is mostly dominated by coal, hydroelectricity, and biomass gasification. However, more land is expected to be dedicated to renewable electricity generation as renewable energy gains larger shares of the grid mix (Table 2).

Table 2 Current and future electricity generation by source and respective and footprint of each

Energy type	Land footprint assumption (m² GWh⁻¹)	Usage (GWh/yr)				Land footprint (ha/yr)			
		2017 Indonesian use	2030 Projection, BAU	2030 Projection, Remap	2050 Projection	2017 energy land footprint	2030 Projection, BAU	2030 Projection, Remap	2050 Projection
Coal, surface average	425	145,600	N/A	N/A	0	61,880	N/A	N/A	0
Natural gas	320	63,100	N/A	N/A	0	20,192	N/A	N/A	0
PV, SW	340	0	13,000	66,200	833,000	0	4420	22,508	283,220
Wind, Germany	2100	0	5700	9100	0	0	11,970	19,110	0
Hydro, reservoir, CO	4100	18,000	106,900	113,200	25,000	73,800	438,290	464,120	102,500
Biomass, willow gasification, NY	12,600	4600	32,500	49,800	33,000	57,960	409,500	627,480	415,800
Total	N/A	254,500	158,100	238,300	891,000	213,832	864,180	1,133,218	801,520

2030 business-as-usual projections were made by IRENA (International Renewable Energy Agency (IRENA), 2017), and the 2050 projection from Breyer et al. (2018). The assumptions for land footprint used to calculate the total land footprints were from Fthenakis and Kim (2009)

Possible Mitigation Through Smart Deployment
of Renewable Energy

While coal is an attractive resource to a rapidly growing economy (Qi, Stern, Wu, Lu, & Green, 2016), coal usage has been linked with health issues stemming from coal combustion's impact on the environment (Koplitz, Jacob, Sulprizio, Myllyvirta, & Reid, 2017). Furthermore, diminishing coal reserves, environmental degradation of coal extraction and combustion, and greenhouse emission reduction goals prompt Indonesia to increase the generation capacity of renewable energy sources (Kennedy, 2018). However, not all renewable energies are equally suited for Indonesia. While hydropower and geothermal currently generate the largest portion of electricity (7% and 5%, respectively) in Indonesia (Kamuradin et al., 2018), their suitability for rural electrification projects leaves much to be desired. For hydropower, the technical potential is high, but its energy resources is usually too far from demand centers (International Energy Agency, 2015a). As for geothermal, the development is may be counterintuitive to the purpose of reducing forest degradation and deforestation as 42% of Indonesia's geothermal energy is located in forest conservation areas (International Energy Agency, 2015a). In addition to spatial constraints, both technologies typically require significant up-front investment and are difficult to scale.

On the other hand, solar photovoltaic technology is considered an excellent candidate for rural electrification in Indonesia because of Indonesia's high average daily insolation ($4.5–5.1$ kWh/m^2) (International Energy Agency, 2015a; Tharakan, 2015). Additionally, the non-electrified portion of the population are often scattered over remote islands and communities with limited to no grid connections, and solar photovoltaic electricity is less constrained by geographical location and easier to scale to smaller rural populations than coal (Tharakan, 2015). Given Indonesia's archipelagic geography and its vested interest in abating the effects of climate change, distributed solar PV systems are an excellent candidate for electrification of the rural population of Indonesia. Furthermore, the cost of generation has decreased significantly in the last decade to complement the Indonesian government's decision to increase the share of renewables (IRENA, 2019). Further, recent observations from impacts of natural disasters indicate that solar PV electricity infrastructure may be more resilient to extreme events, as it is easier to reestablish solar electricity micro-grids in areas affected by natural disasters (D'Cunha, 2017; Mooney, 2017). However, the feasibility and the benefits of solar PV-based rural village grids are contingent on the profitability of such grids along with local participation and employment generation.

Some experts believe that PV could constitute as much as 88% of Indonesia's total power supply by 2050 (Breyer et al., 2018). In reality, however, the growth of PV has been slower than expected, and many experts cite the government's unclear policy and unfavorable renewable energy subsidy structure as reasons for the sluggish growth in the renewables (International Renewable Energy Agency (IRENA), 2017; Kamuradin et al., 2018; Maulidia et al., 2019; Kennedy, 2018). The roadblocks to wider implementation of this technology are as follows: First, deployment of large-scale solar PV infrastructure may have negative impacts on land and water resources: for example, farmlands and grazing lands are land covers that are occupied most extensively by PV infrastructures in California, and these areas of food production are displaced by high-density energy production (Hernandez, Hoffacker, Murphy-Mariscal, Wu, & Allen, 2015). Moreover, occupation by PV may alter the microclimate, carbon cycling, and other variables that control the land's ability to perform various ecosystem services and its ability to produce food, including heightened erosion of the soil under the PV panels (Armstrong, Ostle, & Whitaker, 2016; Cook & McCuen, 2013; Hernandez et al., 2014). These changes in the PV-occupied land presents a greater concern in the context of our current discussion on deforestation and competing land uses: Conventional deployment of community-scale PV facilities may take up valuable agricultural lands or require additional deforestation, which may be counter-productive to the development of rural economy or adversely impact the local environment and the effort to reduce carbon emissions. Despite these obstacles, there are several strategies that may effectively avoid additional conversions of forests and mitigate adverse impacts of solar PV land use or disturbance of agriculturally or ecologically important land. First, PV arrays could be built in already developed environments, such as rooftops or sound barriers to avoid any direct occupation by PV. Second, utility scale solar PV systems can also be deployed without affecting food productivity on degraded lands, such as brownfields, salt-affected lands, landfills, mine sites, and other types of contaminated lands that are ill-suited for agriculture or forestry (Hernandez et al., 2014). Another emerging option is co-location of PV array with agriculture to increase the land-use efficiency (Dupraz et al., 2011; Macknick, Beatty, & Hill, 2013; Ravi et al., 2014, 2016). In the co-location scheme, crops (ranging from cash crops and biofuels to native vegetation) are placed under or around the PV modules (Macknick et al., 2013). By co-locating crop or vegetation with the PV, land reclamation efforts can be integrated into construction and maintenance of PV systems via on-site landscaping with native plants and soil amendments without incurring loss in food productivity and avoid land occupation in critical forested areas (Hernandez et al., 2015).

To put the avoided land occupation by PV into perspective, a more recent study in the energy transition scenario projects that as much as 750 TWh of Indonesia's electricity may come from solar PV by 2050 (Breyer et al., 2018). At mean power density of 400 kW ha^{-1} (Ravi et al., 2016) and average annual electricity output of 1376 kWh/W$_p$ (World Bank Group, 2017), conventional deployment of PV of such scale would require approximately 1.4 million hectares of new land (2% of Indonesia's total agricultural land or 1% of Indonesia's remaining forest cover). Even though this land footprint is small compared to the area of forest degradation from other causes, it may still contribute significantly to the overall impact of deforestation for a few reasons: (1) magnitudes of some environmental responses to deforestation is non-linear to the areal extent of deforestation, and (2) distinguishing the types of environmental response that will be linear from those that will be non-linear remains difficult (Lavigne & Gunnell, 2006). (3) Furthermore, the indirect and regional effects due to the habitat fragmentation caused by conventional solar facilities are also difficult to quantify and mitigate, as repatriation and translocation programs have low success rates (<20%) (Hernandez et al., 2014). Therefore, extra caution is needed for estimating the environmental cost of additional electrification projects in remote areas.

Co-location of PV with vegetation are expected to bring several socioeconomic benefits, such as dual income stream for farmers via wholesale of electricity, additional employment opportunities at the solar facility for the management of the crops, electrification of rural areas, avoided health costs from fire combustion, and promotion of secondary enterprises that require electricity (e.g. locally processing agricultural products) (Ravi et al., 2014). Furthermore, the replacement of firewood with electricity may lead to reduced deforestation or forest degradation, and it may also eliminate the time needed for procuring wood, which could be spent on education or secondary economic activities (World Bank, 2008). The co-location may also result in environmental benefits, such as the decreased demand of firewood that may ultimately reduce deforestation and forest degradation for energy. Additionally, the co-location design may also increase the water-use efficiency by recycling the water used for cleaning the PV modules and dust suppression for irrigation of the crops, suppress erosion from PV facilities, and increase the PV efficiency via cooling of the PV modules (Macknick et al., 2013; Ravi et al., 2016). The PV modules have also been shown to create a favorable microclimatic environment for underlying plants in a temperate climate, which would further increase the land use efficiency by increasing the crop yield (Hassanpour Adeh, Selker, & Higgins, 2018). However, the techno-economic feasibility and environmental co-benefits of the off-grid co-located systems at different scales and agroclimatic conditions are poorly understood, due to the lack of reliable local data and pilot field experiments of co-location.

The challenge to optimizing these socioeconomic and environmental co-benefits is choosing the right approach to co-location. A crop ideal for co-location with PV should be short as not to shade the PV modules, shade-tolerant plant so that it can survive the canopy created by the PV modules, and require little to no mechanization in its cultivation to minimize the disturbance of the PV array with heavy machinery. Furthermore, the crop should be one that is already cultivated locally and should have established processing facilities and marketing channels (Ravi et al., 2016).

In developing countries like Indonesia, large solar infrastructures are rather often impractical as they are expansive and land intensive. In these cases, "*crop centric*" or "*life-style centric*" approaches are required to integrate renewable energy services into rural communities. These approaches will involve implementation of low-density PV over existing crops or processing facilities (Fig. 7). Even though the intermittency of solar electricity is a concern, it can be designed for agricultural activities where intermittency is not a concern (e.g. fan-assisted drying of produces, aeration of aquaculture ponds) and for retrofitting existing diesel generators. Optimistic scenarios for future solar PV expansion in Indonesia will require significant direct land transformation. Assuming that all these new capacity additions

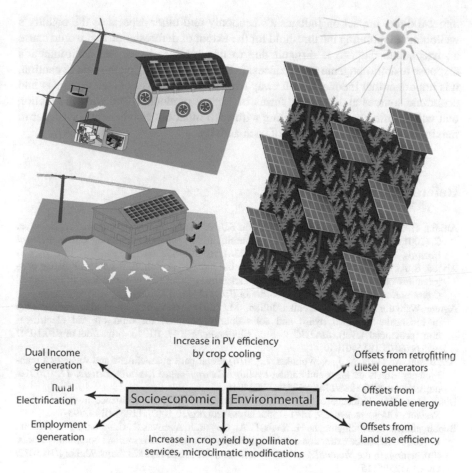

Fig. 7 Proposed potential solar-agriculture co-location scenarios for Indonesia. Clockwise from top: Solar PV over a produce drying/processing shed, low density solar PV over an ideal high value crop, and solar PV over an integrated poultry-aquaculture system

could be integrated with existing crops or other agricultural activities, no additional land transformation will be required.

Conclusion

Despite the recent trend of reduced deforestation rate, Indonesia's forests have been largely at the mercy of domestic and international demands for various agricultural goods, timber, and fuel. As Indonesia's population increases, such demands will only increase, and the country risks losing more of its forests, which serves a critical role in maintaining the country's biodiversity and the ecosystem services that have direct a direct impact on Indonesia's economy and other aspects of the country's wellbeing. Pinpointing the threshold for the extent of deforestation that would cause an irreparable damage is difficult due to the nonlinearity of the environment's response to anthropogenic disturbances. In the spirit of erring on the side of caution, it is important that Indonesia shift away from its vast reserves of coal for power and economic sources and consider forms of energy production and food production that will not merely avoid competing with each other but also benefit each other to maximize the land use efficiency of each activity.

References

Abidin, H. Z., Djaja, R., Darmawan, D., Hadi, S., Akbar, A., Rajiyowiryono, H., … Subarya, C. (2001). Land subsidence of Jakarta (Indonesia) and its geodetic monitoring system. *Natural Hazards, 23*, 365–387. https://doi.org/10.1023/A:1011144602064

Abood, S. A., Lee, J. S. H., Burivalova, Z., Garcia-Ulloa, J., & Koh, L. P. (2015). Relative contributions of the logging, fiber, oil palm, and mining industries to forest loss in Indonesia. *Conservation Letters, 8*, 58–67. https://doi.org/10.1111/conl.12103

Agung Wahyuono, R., & Magenika Julian, M. (2018). Revisiting renewable energy map in Indonesia: Seasonal hydro and solar energy potential for rural off-grid electrification (provincial level). *MATEC Web of Conferences, 164*, 01040. https://doi.org/10.1051/matecconf/201816401040

Armstrong, A., Ostle, N. J., & Whitaker, J. (2016). Solar park microclimate and vegetation management effects on grassland carbon cycling. *Environmental Research Letters, 11*, 074016. https://doi.org/10.1088/1748-9326/11/7/074016

Bakr, M. (2015). Influence of groundwater management on land subsidence in deltas. *Water Resources Management, 29*, 1541–1555. https://doi.org/10.1007/s11269-014-0893-7

Bebbington, A. J., Bebbington, D. H., Sauls, L. A., Rogan, J., Agrawal, S., Gamboa, C., … Verdum, R. (2018). Resource extraction and infrastructure threaten forest cover and community rights. *Proceedings of the National Academy of Sciences, 115*, 13164–13173. https://doi.org/10.1073/pnas.1812505115

Blum, N. U., Sryantoro Wakeling, R., & Schmidt, T. S. (2013). Rural electrification through village grids—Assessing the cost competitiveness of isolated renewable energy technologies in Indonesia. *Renewable and Sustainable Energy Reviews, 22*, 482–496. https://doi.org/10.1016/j.rser.2013.01.049

Boer, R., Siagian, U. W. R., Dewi, R. G., Ginting, G. E., Hendrawan, I., & Yuwono, B. B. (2016). Pathways to deep decarbonization in the agriculture, forest and other land use sector in Indonesia. Retrieved from http://deepdecarbonization.org/wp-content/uploads/2017/02/DDPP_AFOLU_ID.pdf.

Breyer, C., Bogdanov, D., Aghahosseini, A., Gulagi, A., Child, M., Oyewo, A. S., … Vainikka, P. (2018). Solar photovoltaics demand for the global energy transition in the power sector. *Progress in Photovoltaics: Research and Applications, 26*, 505–523. https://doi.org/10.1002/pip.2950

Budiyono, Y., Aerts, J. C. J. H., Tollenaar, D., & Ward, P. J. (2016). River flood risk in Jakarta under scenarios of future change. *Natural Hazards and Earth System Sciences, 16*, 757–774. https://doi.org/10.5194/nhess-16-757-2016

Casson, A. (1999). The Hesitant boom: Indonesia's oil palm sub-sector in an era of economic crisis and political change, Center for International Forestry Research (CIFOR), Bogor, Indonesia. https://doi.org/10.17528/cifor/000625.

Central Intelligence Agency. (2019). The World Factbook. Retrieved from October 9, 2019, from https://www.cia.gov/library/publications/resources/the-world-factbook/index.html.

Central Intelligence Agency. (2017). The World Factbook. Retrieved February 4, 2017, from https://www.cia.gov/library/publications/the-world-factbook/geos/id.html.

Cook, L. M., & McCuen, R. H. (2013). Hydrologic response of solar farms. *Journal of Hydrologic Engineering, 18*, 536–541. https://doi.org/10.1061/(ASCE)HE.1943-5584.0000530

D'Cunha, S.D. (2017) How off-grid renewable energy came to the rescue in India's flood zones. https://www.forbes.com/sites/suparnadutt/2017/09/15/how-off-grid-renewable-energy-came-to-the-rescue-in-indias-floodzones/#523632a0bc49. Accessed 26 August 2019

Dharmawan, A., Budiman, A., Wijaya, A., Margono, B. A., Budiharto, D. Martinus, … T. Rusolono. (2015). National Forest reference emissions level for REDD+ In the Context of Decision 1/CP.16 Paragraph 70 UNFCCC, the Directorate General of Climate Change (DG-PPI). Jakarta, Indonesia. Retrieved from https://redd.unfccc.int.

Douglass, M. (2010). Globalization, mega-projects and the environment. *Environment and Urbanization ASIA, 1*, 45–65. https://doi.org/10.1177/097542530900100105

Dupraz, C., Marrou, H., Talbot, G., Dufour, L., Nogier, A., & Ferard, Y. (2011). Combining solar photovoltaic panels and food crops for optimising land use: Towards new agrivoltaic schemes. *Renewable Energy, 36*, 2725–2732. https://doi.org/10.1016/j.renene.2011.03.005

Dutu, R. (2015). Making the most of natural resources. 97–134. https://doi.org/10.1787/eco_surveys-idn-2015-6-en.

Edenhofer, O., Pichs-Madruga, R., Sokona, Y., Seyboth, K., Eickemeier, P., Matschoss, P., … Von Stechow, C. (2011). IPCC, 2011: Summary for policymakers. In IPCC Special Report on *Renewable energy sources and climate change mitigation.* https://doi.org/10.5860/CHOICE.49-6309.

FAO. (2018). Land use. FAOSTAT. Retrieved from August 14, 2019, from http://www.fao.org/faostat/en/#data/EL.

Fthenakis, V., & Kim, H. C. (2009). Land use and electricity generation. A life-cycle analysis. *Renewable and Sustainable Energy Reviews, 13*, 1465–1474. https://doi.org/10.1016/j.rser.2008.09.017

Fthenakis, V., & Kim, H. C. (2010). Life-cycle uses of water in U.S. electricity generation. *Renewable and Sustainable Energy Reviews, 14*, 2039–2048. https://doi.org/10.1016/j.rser.2010.03.008

FWI/GFW. (2002). The state of the Forest: Indonesia, Global Forest Watch/World Resources Institute, Washington, DC.

Government of the Republic of Indonesia, United Nations in Indonesia, The United Nations Partnership for Development Framework (UNPDF). (2016). 2016–2020.

Hallik, L., Niinemets, Ã., & Wright, I. J. (2009). Are species shade and drought tolerance reflected in leaf-level structural and functional differentiation in Northern Hemisphere temperate woody flora? *The New Phytologist, 184*, 257–274. https://doi.org/10.1111/j.1469-8137.2009.02918.x

Hansen, M. C., Stehman, S. V., Potapov, P. V., Arunarwati, B., Stolle, F., & Pittman, K. (2009). Quantifying changes in the rates of forest clearing in Indonesia from 1990 to 2005 using remotely sensed data sets. *Environmental Research Letters, 4*, 034001. https://doi.org/10.1088/1748-9326/4/3/034001

Harvey, M., & Pilgrim, S. (2011). The new competition for land: Food, energy, and climate change. *Food Policy, 36*, S40–S51. https://doi.org/10.1016/j.foodpol.2010.11.009

Hassanpour Adeh, E., Selker, J. S., & Higgins, C. W. (2018). Remarkable agrivoltaic influence on soil moisture, micrometeorology and water-use efficiency. *PLoS One, 13*, e0203256. https://doi.org/10.1371/journal.pone.0203256

Hasudungan, H. W. V., & Sabaruddin, S. S. (2018). Financing renewable energy in Indonesia: A CGE analysis of feed-in tariff schemes. *Bulletin of Indonesian Economic Studies, 54*, 233–264. https://doi.org/10.1080/00074918.2018.1450961

Hernandez, R. R., Easter, S. B., Murphy-Mariscal, M. L., Maestre, F. T., Tavassoli, M., Allen, E. B., … Allen, M. F. (2014). Environmental impacts of utility-scale solar energy. *Renewable and Sustainable Energy Reviews, 29*, 766–779. https://doi.org/10.1016/j.rser.2013.08.041

Hernandez, R. R., Hoffacker, M. K., Murphy-Mariscal, M. L., Wu, G. C., & Allen, M. F. (2015). Solar energy development impacts on land cover change and protected areas. *Proceedings of the National Academy of Sciences of the United States of America, 112*, 13579–13584. https://doi.org/10.1073/pnas.1517656112

International Energy Agency (IEA). (2015a). Energy policies beyond IEA Countries—Indonesia, 2015. https://doi.org/10.1787/9789264065277-en.

International Energy Agency (IEA). (2015b). Southeast Asia Energy Outlook. World Energy Outlook Spec. Report 131. https://doi.org/10.1787/weo-2013-en.

International Energy Agency (IEA). (2016). Key world energy statistics. Paris. Retrieved from https://doi.org/10.1787/key_energ_stat-2016-en.

International Energy Agency (IEA). (2017). Southeast Asia Energy Outlook, OECD, 2017. Retrieved from www.iea.org/t&c/.

International Energy Agency (IEA). (2018a). World Energy Balances 2018.

International Energy Agency (IEA). (2018b). World Energy Outlook 2018, OECD. https://doi.org/10.1787/weo-2018-en.

International Renewable Energy Agency (IRENA). (2017). Renewable Energy Prospects: Indonesia. In *REmap 2030* (p. 106), Abu Dhabi. https://doi.org/10.1145/347642.347800.

IRENA. (2019). Renewable power generation costs in 2018.

Kementrian PPN/Bappenas. (2014). National Nutrition Strategy Paper of Indonesia.

Kennedy, M. (2019). Indonesia plans to move its capital out of Jakarta, a city that's sinking, NPR. Retrieved October 14, 2019, from https://www.npr.org/2019/04/29/718234878/indonesia-plans-to-move-its-capital-out-of-jakarta-a-city-thats-sinking.

Kennedy, S. (2018). Indonesia's energy transition and its contradictions: Emerging geographies of energy, finance, and land use. *Energy Research and Social Science, 41*, 230–237. https://doi.org/10.1016/J.ERSS.2018.04.023

Koplitz, S. N., Jacob, D. J., Sulprizio, M. P., Myllyvirta, L., & Reid, C. (2017). Burden of disease from rising coal-fired power plant emissions in Southeast Asia. *Environmental Science & Technology, 51*, 1467–1476. https://doi.org/10.1021/acs.est.6b03731

Kurniawan, R., & Managi, S. (2018). Economic growth and sustainable development in Indonesia: An assessment. *Bulletin of Indonesian Economic Studies, 4918*, 1–31. https://doi.org/10.1080/00074918.2018.1450962

Lang, G., & Chan, C. H. W. (2006). China's impact on forests in Southeast Asia. *Journal of Contemporary Asia, 36*, 167–194. https://doi.org/10.1080/00472330680000111

Lavigne, F., & Gunnell, Y. (2006). Land cover change and abrupt environmental impacts on Javan volcanoes, Indonesia: A long-term perspective on recent events. *Regional Environmental Change, 6*, 86–100. https://doi.org/10.1007/s10113-005-0009-2

Liebman, A., Reynolds, A., Robertson, D., Nolan, S., Argyriou, M., & Sargent, B. (2019). Green Finance in Indonesia: Barriers and solutions. In *Handbook of Green Finance: Energy security and sustainable development* (pp. 1–30). Singapore, Singapore: Springer. https://doi.org/10.1007/978-981-10-8710-3_5-1

MacKinnon, J. (1997). Protected areas systems review of the Indo-Malayan Realm. https://doi.org/10.5962/bhl.title.44928.

Macknick, J., Beatty, B., & Hill, G. (2013). Overview of opportunities for co-location of solar energy technologies and vegetation, Golden, Colorado. Retrieved from http://www.nrel.gov/docs/fy14osti/60240.pdf.

Margono, B. A., Potapov, P. V., Turubanova, S., Stolle, F., & Hansen, M. C. (2014). Primary forest cover loss in Indonesia over 2000–2012. *Nature Climate Change, 4*, 730–735. https://doi.org/10.1038/nclimate2277

Maulidia, M., Dargusch, P., Ashworth, P., & Ardiansyah, F. (2019). Rethinking renewable energy targets and electricity sector reform in Indonesia: A private sector perspective. *Renewable and Sustainable Energy Reviews, 101*, 231–247. https://doi.org/10.1016/j.rser.2018.11.005

McKenzie, B. (2018). Indonesian Government Publishes 2017 Cost of Generation (BPP) Figures, Jakarta. Retrieved from https://www.lexology.com/library/detail.aspx?g=a19a0702-4e86-4d98-ae45-9cf32e877725.

Mooney, C. (2017) Severe power failures in Puerto Rico and across the Caribbean spur new push for renewable energy. https://www.washingtonpost.com/news/energy-environment/wp/2017/09/28/storm-driven-power-failures-in-thecaribbean-spur-new-interest-in-renewable-energy/. Accessed 26 August 2019

Mumme, S., Jochum, M., Brose, U., Haneda, N. F., & Barnes, A. D. (2015). Functional diversity and stability of litter-invertebrate communities following land-use change in Sumatra, Indonesia. *Biological Conservation, 191*, 750–758. https://doi.org/10.1016/j.biocon.2015.08.033

Obidzinski, K., & Chaudhury, M. (2009). Transition to timber plantation based forestry in Indonesia: Towards a feasible new policy. *International Forestry Review, 11*, 79–87. https://doi.org/10.1505/ifor.11.1.79

OECD. (2019). OECD Green Growth Policy review of Indonesia 2019, OECD. https://doi.org/10.1787/1eee39bc-en.

Page, S. E., & Hooijer, A. (2016). In the line of fire: The peatlands of Southeast Asia. *Philosophical Transactions of the Royal Society B, 371*, 20150176. https://doi.org/10.1098/rstb.2015.0176

Page, S. E., Siegert, F., Rieley, J. O., Boehm, H.-D. V., Jaya, A., & Limin, S. (2002). The amount of carbon released from peat and forest fires in Indonesia during 1997. *Nature, 420*, 61–65. https://doi.org/10.1038/nature01131

Pagiola, S. (2000). Land use change in Indonesia. Washington, DC: World Bank Environment Department.

Pegels, K. (2012) Solar energy in Indonesia. Internship Report, University of Twente, Enschede, The Netherlands

Portela, R., & Rademacher, I. (2001). A dynamic model of patterns of deforestation and their effect on the ability of the Brazilian Amazonia to provide ecosystem services. *Ecological Modelling, 143*, 115–146. https://doi.org/10.1016/S0304-3800(01)00359-3

Kamuradin, Y., Natakusumah, G., Then, L. (2018) Power In Indonesia: Investment and Taxation Guide. 6th ed. PwC Publications.

Qi, Y., Stern, N., Wu, T., Lu, J., & Green, F. (2016). China's post-coal growth. *Nature Geoscience, 9*, 564–566. https://doi.org/10.1038/ngeo2777

Rahman, S., Sumotarto, U., & Pramudito, H. (2018). Influence the condition land subsidence and groundwater impact of Jakarta coastal area. *IOP Conference Series: Earth and Environmental Science, 106*, 012006. https://doi.org/10.1088/1755-1315/106/1/012006

Rahman, S. A., Sunderland, T., Roshetko, J. M., Basuki, I., & Healey, J. R. (2016). Tree culture of smallholder farmers practicing agroforestry in Gunung Salak Valley, West Java, Indonesia. *Small-Scale Forestry, 15*, 433–442. https://doi.org/10.1007/s11842-016-9331-4

Ravi, S., Lobell, D. B., & Field, C. B. (2014). Tradeoffs and synergies between biofuel production and large solar infrastructure in deserts. *Environmental Science & Technology, 48*, 3021–3030. https://doi.org/10.1021/es404950n

Ravi, S., Macknick, J., Lobell, D., Field, C., Ganesan, K., Jain, R., ... Stoltenberg, B. (2016). Colocation opportunities for large solar infrastructures and agriculture in drylands. *Applied Energy, 165*, 383–392. https://doi.org/10.1016/j.apenergy.2015.12.078

Siegert, F., Boehm, H.-D. V., Rieley, J. O., Page, S. E., Jauhiainen, J., Vasander, H., & Jaya, A. (2001, August 22–23). Peat fires in Central Kalimantan, Indonesia: Fire impacts and carbon release. In *International Symposium Tropical Peatland* (pp. 1–13).

Sunderlin, W. D., & Resosudarmo, I. A. P. (1996). Rates and causes of deforestation in Indonesia: Towards a resolution of the ambiguities. Center for International Forestry Research (CIFOR). https://doi.org/10.17528/cifor/000056.

Sweeney, B. W., Bott, T. L., Jackson, J. K., Kaplan, L. A., Newbold, J. D., Standley, L. J., ... Horwitz, R. J. (2004). Riparian deforestation, stream narrowing, and loss of stream ecosystem services. *Proceedings of the National Academy of Sciences, 101*, 14132–14137. https://doi.org/10.1073/pnas.0405895101

Tharakan, P. (2015). Summary of Indonesia's Energy Sector Assessment (ADB Papers on Indonesia No. 9), Adb. 40. Retrieved from https://www.adb.org/sites/default/files/publication/178039/ino-paper-09-2015.pdf.

The World Bank Group. (2019). Indonesia Trade Summary 2017 Data, World Integrated. Trade Solution. Retrieved from October 14, 2019, from https://wits.worldbank.org/CountryProfile/en/Country/IDN/Year/LTST/Summary.

Tilman, D., Balzer, C., Hill, J., & Befort, B. L. (2011). Global food demand and the sustainable intensification of agriculture. *Proceedings of the National Academy of Sciences, 108*, 20260–20264. https://doi.org/10.1073/pnas.1116437108

Tilman, D., Socolow, R., Foley, J. A., Hill, J., Larson, E., Lynd, L., … Williams, R. (2009). Beneficial biofuels—The food, energy, and environment trilemma. *Science, 325*(5938), 270–271. https://doi.org/10.1126/science.1177970

Tsujino, R., Yumoto, T., Kitamura, S., Djamaluddin, I., & Darnaedi, D. (2016). History of forest loss and degradation in Indonesia. *Land Use Policy, 57*, 335–347. https://doi.org/10.1016/j.landusepol.2016.05.034

Wicke, B., Sikkema, R., Dornburg, V., & Faaij, A. (2011). Exploring land use changes and the role of palm oil production in Indonesia and Malaysia. *Land Use Policy, 28*, 193–206. https://doi.org/10.1016/j.landusepol.2010.06.001

Wijaya, A., Chrysolite, H., Ge, M., Wibowo, C. K., Pradana, A., Utami, A. F., & Austin, K. (2017). How can Indonesia achieve its climate change mitigation goal? An analysis of potential emissions reductions from energy and land-use policies. Retrieved from https://www.wri.org/publication/how-can-indonesia-achieve-its-climate-goal.

Wijaya, A., Sugardiman Budiharto, R. A., Tosiani, A., Murdiyarso, D., & Verchot, L. V. (2015). Assessment of large scale land cover change classifications and drivers of deforestation in Indonesia. *ISPRS—International Archives of the Photogrammetry Remote Sensing and Spatial Information Sciences, XL-7/W3*, 557–562. https://doi.org/10.5194/isprsarchives-XL-7-W3-557-2015

World Bank. (2008). The welfare impact of rural electrification: A reassessment of the costs and benefits. https://doi.org/10.1596/978-0-8213-7367-5.

World Bank. (2019a). Population, total, World Development Indicators. Retrieved from August 8, 2019, from data.worldbank.org/indicator/SP.POP.TOTL?locations=ID.

World Bank. (2019b). GDP growth (annual %), World Development Indicators. Retrieved from August 8, 2019, from data.worldbank.org/indicator/NY.GDP.MKTP.KD.ZG?locations=ID.

World Bank Group. (2017). Solar resource and photovoltaic potential of Indonesia, 86. Retrieved from http://documents.worldbank.org/curated/en/729411496240730378/Solar-resource-and-photovoltaic-potential-of-Indonesia.

Zahnd, A., & Kimber, H. M. (2009). Benefits from a renewable energy village electrification system. *Renewable Energy, 34*, 362–368. https://doi.org/10.1016/j.renene.2008.05.011

The Carbon Balance of Tropical Islands: Lessons from Soil Respiration

Sarah G. McQueen, Diego A. Riveros-Iregui, and Jia Hu

Introduction

The study of tropical islands has long shaped our understanding of the earth system as a whole by facilitating small-scale observations of a range of mechanisms and processes central to multiple disciplines, including biogeography, geoscience, population dynamics, and human-environment science. The isolation of islands promotes endemism and highlights the role of changes in biophysical resources in mediating the structure and function of natural systems (Adler, 1992; Briggs, 1966), allowing scientists to study a wide range of geological and biological phenomena over relatively small spatial scales. However, despite their smaller size, tropical islands have proven difficulty to characterize, particularly in the face of disturbance and rapidly changing land use and land cover (LULC). Tropical islands are highly heterogeneous, complex, and often exhibit confounding variables due to juxtaposed processes, mechanisms, and environmental threats (Lal, Harasawa, & Takahashi, 2002; Meehl, 1996).

One aspect in which tropical islands remain difficult to characterize is in the role they play in the global carbon cycle and the magnitude and direction of carbon exchanged between the land and the atmosphere. Because land cover of islands is highly variable and rapidly changing, carbon-related processes remain difficult to quantify or even measure at the scale of entire islands. Furthermore, many tropical

S. G. McQueen
Department of Biology, University of North Carolina at Chapel Hill, Chapel Hill, NC, USA

D. A. Riveros-Iregui (✉)
Department of Geography, University of North Carolina at Chapel Hill,
Chapel Hill, NC, USA
e-mail: diegori@unc.edu

J. Hu
School of Natural Resources and the Environment, University of Arizona, Tucson, AZ, USA

© Springer Nature Switzerland AG 2020
S. J. Walsh et al. (eds.), *Land Cover and Land Use Change on Islands*, Social
and Ecological Interactions in the Galapagos Islands,
https://doi.org/10.1007/978-3-030-43973-6_12

islands are composed of basaltic and ultramafic rocks, which despite having a small surface area, have a disproportionately greater capacity to sequester atmospheric CO_2 (McGrail et al., 2006). Thus, the extent to which tropical islands sequester or emit atmospheric carbon remains highly uncertain. Large vegetation that is characteristic of the highest, wettest, and often the most productive areas in tropical islands is often removed and replaced by crops that provide food and fodder. While farming in tropical islands remains a small-scale activity, farming leads to extremely fragmented landscapes with multiple and varied land uses, posing consequences on soil carbon sequestration potential.

Tropical islands are also known for being affected by the proliferation of invasive species (Kueffer et al., 2010). Invasive plants alter ecosystems through the modification of physical, chemical and biological properties, including soil structure, soil water holding capacity, and soil chemical composition. Pacific Islands, for example, have experienced the highest rates of invasive plant species proliferation per unit area in the world (Van Kleunen et al., 2015). Through changes in soil nutrient content, invasive plant species can in turn impact soil chemical composition, carbon content, and carbon exchange (Mack, Schuur, Bret-Harte, Shaver, & Chapin, 2004). To complicate matters further, the irregular progression of invasive plant encroachment—which is often driven by socio-economic factors—makes it difficult to characterize both the spatial extent of land cover changes and the ecophysiological traits that the newly encroaching plants impose over the entire ecosystem.

In this chapter, we provide an example of how field observations of one of the major components of the carbon cycle—soil respiration—can be useful in identifying major patterns in carbon fluxes across a heterogeneous island. We conclude by arguing that not only is the study of the carbon processes of tropical islands necessary to characterize the carbon status of these important ecosystems, but also a viable opportunity to improve our understanding of carbon balance and exchange in highly heterogeneous landscapes in other parts of the planet.

Background

Invasive plants are known to facilitate changes in biological processes and properties (Vitousek, Walker, Whiteaker, Mueller-Dombois, & Matson, 1987), soil chemistry (Baruch & Goldstein, 1999; Rodríguez-Caballero et al., 2017), and physical changes like soil erosion (Greenwood & Kuhn, 2014) and soil compaction (Kyle, Beard, & Kulmatiski, 2007). Invasive plant species have the capacity to drastically alter ecosystems and affect soil-plant interactions, thereby affecting the rate of carbon exchange between land and the atmosphere. High rates of invasion are believed to be the result of available ecological niches that the native plant species are unable to fill (Van Kleunen et al., 2015). For example, invasive plant species are associated with changes in the plant and soil relationship, and can show higher plant-soil nutrient concentration in nutrient poor environments (Sardans et al., 2017). This means that invasive plants may be able to more efficiently conserve and use resources, especially in environments where the soil generally lacks nutrients (Sardans et al., 2017).

Among the Pacific Islands, the Galápagos Islands are of exceptional interest because of their rich history in the study of evolution (Darwin, 1859). Although the Galápagos Islands are generally believed to be a pristine environment, these islands actually have an extraordinarily high rate of invasive plant species. In particular, it has been reported that 70% of the highlands of San Cristobal Island are covered in invasive species (Villa & Segarra, 2010; Percy et al., 2016). The overpopulation of introduced animals, including goats, in combination with introduced plants, are the main mechanisms responsible for the spread of invasive plant species and the eradication of other native vegetative cover. As the feral goat population increased, there were major changes in the vegetation across the entire archipelago (Coblentz, 1978).

The Galápagos Islands alone have a total of 750 non-native plant species, including both managed and non-managed invasive species, which were introduced by settlers for agricultural purposes (Coffey, Froyd, & Willis, 2011). One important way in which non-native plants have altered vegetation structure on the Galápagos Islands is the growth of larger trees that have created shade over smaller plants (Jager, 2015), changing the microclimate and nutrient cycling of native plants. Yet, there is little information regarding the role of non-native trees on regulating soil nutrient content or mediating the local carbon balance. Agricultural activities have been shown to cause an increase in soil nutrient fluxes and soil organic matter, especially soil organic carbon, causing the quality of the soil to greatly decrease (McLauchlan, Hobbie, & Post, 2006). Biomass alterations and tillage have some of the greatest potential to alter soil carbon content rapidly and facilitate erosion (Lal, 2004), diminishing the overall quality of the soil.

Volcanic soils—such as those in the Galápagos Islands—are typically low in soil carbon content and soil nutrients (Yoshitake et al., 2013), and decomposing plants represent the primary source for nutrients. If non-native plant species are proliferating faster than native plants, they will contribute more to plant decomposition than native plants, enhancing nutrient turnover in the soil. In turn, enhanced soil nutrient turnover can increase the release of carbon from soils (Mack et al., 2004). Soil nutrients also impact the structure of the soil as well as the organic matter that is produced by the decomposition of plant litter. High rates of organic matter accumulation are known to drive an increase in soil respiration when controlled for temperature, soil moisture, and substrate conditions (Raich & Tufekciogul, 2000).

Research in the Hawaiian Archipelago has shown that photosynthetic nitrogen use by invasive species is 15% more efficient than that of native species, resulting in greater net CO_2 assimilation rates (Baruch & Goldstein, 1999). These results suggest that invasive species use resources more efficiently than native species, which may potentially explain their higher growth rates and rapid spread in island settings (Baruch & Goldstein, 1999) and their advantage to outgrow and ultimately outcompete native plant species across the Pacific islands. These observations highlight important differences in the use of nutrients by invasive and native plant species, but more direct observations, particularly with regards to carbon processes, are needed.

Soil CO_2 efflux, also known as soil respiration, is a major contributor to atmospheric CO_2 and is influenced by soil nutrients, precipitation, temperature, and vegetation type (Dube et al., 2013). Different vegetation types influence the rate of soil respiration because each plant influences the soil microclimate and the structure of the soil differently (Raich & Tufekciogul, 2000). This study seeks to examine the effects of invasive and introduce plant species on one of the main components of the carbon cycle, soil respiration, and in particular across different microclimates in island environments. We hypothesized that areas with non-native plant species will have higher respiration rates than areas with native plant species, and that these differences would be consistent across different microclimatic zones.

Methods

Study Area

This study took place on San Cristobal, the fifth largest and easternmost island of the Galápagos Archipelago. In total, there are twenty islands that make up the Galápagos Archipelago and they are located approximately 1000 km off the coast of Ecuador. We selected three sites that spanned both leeward and windward sides of the island. These sites were called Mirador (leeward), Cerro Alto (leeward), and El Junco (windward), located at altitudes of 320 m, 520 m, and 690 m, respectively. At each of these three sites we established 16 ~1 m^2 plots, eight of which contained native plant species and eight of which contained non-native plant species, for a total of 48 plots (Table 1). There are two seasons in the Galápagos Islands. The wet season extends from December to May (ranging from 600 mm of precipitation in

Table 1 Species name and prevalence at Mirador (320 m), Cerro Alto (520 m), and El Junco (690 m) sites

	Species name	Mirador	Cerro Alto	El Junco
Native	*Bursera graveolens*	Y		
	Hippomane mancinella	Y		
	Zanthoxylum fagara	Y	Y	
	Scalesia gordilloi	Y		
	Toslnefortia pubescens	Y		
	Leocarpus darwini	Y		
	Heliotropium indicum		Y	
	Chiococca alba		Y	
	Miconia robinsoniana			Y
Non-native	*Psidium guajava*	Y		Y
	Citrus sinensis	Y	Y	
	Rubus niveus		Y	
	Pennisetum purpureum		Y	

the lowlands to over 1200 mm in the highlands), whereas the dry season extends from June to November (ranging from less than 100 mm precipitation in the lowlands to about 400 mm in the highlands) (Trueman & D'Ozouville, 2010). Our study was limited to the dry season; however, during our field campaign, we observed rainfall at each site, most notably at El Junco.

The native and non-native vegetation among the three sites varied. At the leeward sites, Mirador and Cerro Alto, some of the dominant native species included *Zanthoxylum fagara* and non-native plants included *Citrus sinensis*. At the windward island site, El Junco, the dominant native plant was *Miconia robinsoniana* and non-native plant was *Psidium guajava* (Table 1).

Meteorological Data

At each elevation, a previously installed weather variable station recorded weather variables, including rainfall, air temperature, relative humidity, solar radiation, volumetric soil water content, wind speed, and wind directions at 15-min intervals. These stations were installed between June and July of 2015. Here we report on precipitation and air temperature in combination with direct measurements of soil respiration. In addition, we measured average volumetric soil water content for the top 12 cm of soil using the HydroSense II Soil Moisture Measurement System model HS2 (Campbell Scientific Inc., Logan, Utah) three times per plot. Measurements were typically taken between 0800 and 1400 LT.

Measurements of Soil Respiration

We measured soil respiration at each microclimate using an infrared gas analyzer (EGM4, PP Systems, Amesbury, MA) connected to the soil respiration chamber SRC-1. Soil respiration was measured in triplicate at each plot and measurements were averaged every day. Care was taken to gently remove all non-vegetative material covering each plot, leaving all roots intact to minimize the disturbance to the soil. Each measurement was collected over the course of ~120 s and chamber measurements were collected every 3–7 days at each plot. Measurements were collected from 27 June 2017 until 24 July 2017. Over the course of a month, each plot in each microclimate was visited five times. Table 2 shows the composition of native and non-native plant species that were found at each of the plots sampled.

Table 2 Species name and plot number for native and invasive plants

	Mirador		Cerro Alto		El Junco	
	Plot	Plant	Plot	Plant	Plot	Plant
Native	A2	*Bursera graveolens*	A1	*Heliotropium indicum*	A1	*Miconia robinsoniana*
	A3	*Hippomane mancinella*	A2	*Zanthoxylum fagara*	A2	*Miconia robinsoniana*
	BI	*Zanthoxylum fagara*	A3	*Heliotropium indicum*	A3	*Miconia robinsoniana*
	B2	*Hippomane mancinella*	A4	*Zanthoxylum fagara*	A4	*Miconia robinsoniana*
	B3	*Scalesia gordilloi*	A5	*Chiococca alba*	AI	*Miconia robinsoniana*
	B4	*Tournefortia pubescens*	BI	*Chiococca alba*	B2	*Miconia robinsoniana*
	C3	*Scalesia gordilloi*	B2	*Chiococca alba*	B3	*Miconia robinsoniana*
	C4	*Leocarpus darwini*	B3	*Heliotropium indicum*	B4	*Miconia robinsoniana*
Non-native	D2	*Psidium guajava*	D1	*Rubus niveus*	D1	*Psidium guajava*
	D4	*Citrus sinensis*	D2	*Pennisetum purpureum*	D2	*Psidium guajava*
	D5	*Citrus sinensis*	D3	*Pennisetum purpureum*	D3	*Psidium guajava*
	E3	*Psidium guajava*	D4	*Pennisetum purpureum*	D4	*Psidium guajava*
	E4	*Citrus sinensis*	D5	*Citrus sinensis*	D5	*Psidium guajava*
	E5	*Psidium guajava*	E1	*Pennisetum purpureum*	E1	*Psidium guajava*
	F4	*Citrus sinensis*	E2	*Pennisetum purpureum*	E2	*Psidium guajava*
	F5	*Citrus sinensis*	E3	*Pennisetum purpureum*	E3	*Psidium guajava*

Results

Meteorological Data

Over the study period, precipitation was considerably higher for the high-elevation site, El Junco, with a cumulative total of 149.3 mm, compared to mid elevation site, Cerro Alto with 47.0 mm, and lower elevation site, Mirador with 59 mm. Mean air temperature for Mirador was 18.90 °C, whereas mean air temperature was 18.20 °C for Cerro Alto. Air temperature for El Junco was not available due to instrument malfunction during the study period.

Volumetric soil water content was highest at El Junco (>35% VWC), followed by Mirador (20–25% VWC), and then Cerro Alto (10–17% VWC). We did not find a

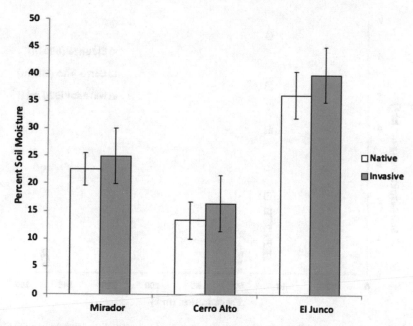

Fig. 1 Mean soil moisture (%) for native and invasive plants at Mirador, Cerro Alto, and El Junco. Error bars represent two standard deviations

difference in volumetric water content between plots with non-native vegetation and plots with native vegetation at all sites (Fig. 1).

Soil Respiration Rates

A major finding of our study was that soil respiration exhibited a negative relationship with precipitation (Fig. 2). Observations from the Cerro Alto (driest site) showed a higher and broader range of soil respiration rates than Mirador and El Junco (Fig. 2). Maximum soil respiration rates for Cerro Alto was 6.5 g m^{-2} hr^{-1} (average of three measurements on a single day). In particular, four plots exhibited the highest soil respiration values: B2, B3, E2, and E3 (Table 2). B2 and B3 were plots with native vegetation represented by *Chiococca alba* and *Heliotropium indicum*, whereas E2 and E3 contained a non-native grass *Pennisetum purpureum* (Table 2). In contrast, Mirador and El Junco showed much smaller ranges of soil respiration rates and higher cumulative precipitation over the study period.

A second finding of our study was that sites with invasive vegetation showed consistently higher soil respiration rates than sites with native vegetation, although these differences were statistically significant only at the low elevation site (i.e., Mirador, $p < 0.05$; Fig. 3). When we examined a day-by-day comparison of soil respiration rates at Mirador, we found that of the 5 measuring days, there were

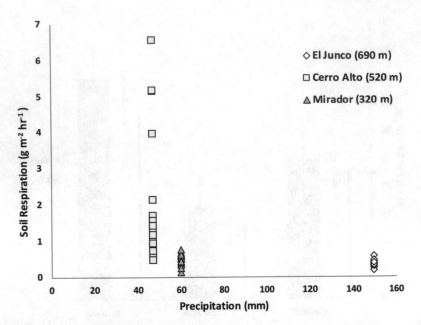

Fig. 2 Average soil respiration rates for each site compared to cumulative precipitation for Cerro Alto, Mirador, and El Junco (study period 24 June 2017 to 24 July 2017)

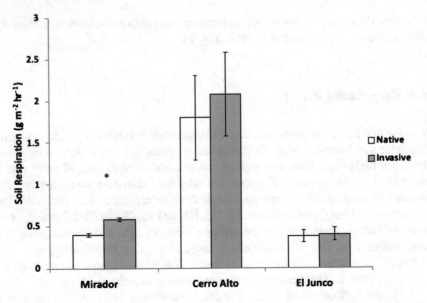

Fig. 3 Soil respiration rates for native and invasive plants at Mirador, Cerro Alto, and El Junco. Only one statistically significant difference (denoted with ∗) was found (Mirador site; $p < 0.05$). Error bars represent two standard deviations

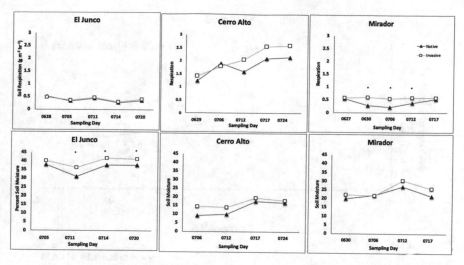

Fig. 4 Average soil respiration and soil moisture measurements for native and invasive plants by day, by site. (∗ t-test, P (I > N) < .0001). Observations in plots with invasive plant species are represented by white boxes, whereas observations in plots with native plant species are represented by black triangles

differences between invasive and native plant plots on 3 days: 30 June 2017 (p < 0.0001, I > N), 6 July 2017 (p < 0.0001, I > N), and 12 July 2017 (p = 0.0001, I > N), with invasive plants consistently having higher soil respiration.

We also found that plots with invasive plant species consistently had greater soil moisture, although this difference was only statistically significant at the high-elevation site El Junco (p < 0.0001, I > N; Fig. 4). Note that soil moisture was not measured during the first week of fieldwork. We found negative trends between soil respiration and soil moisture (Fig. 5), although none of these relationships were statistically significant. At Cerro Alto, the negative trend can be explained in part by the large soil respiration values observed at plots B2, B3, E2, and E3, as explained above. While consistent in time, these large soil respiration values could be the result of local variability caused by recent fires in the area. When examining at the species level, we found that in plots with *Citrus sinesis*, a non-native plant species, there was a negative relationship between soil respiration and soil moisture (Fig. 6) (p < 0.05). This negative relationship between soil moisture and soil respiration in plots with *Citrus sinesis* might explain the negative trend between soil moisture and respiration at the Cerro Alto site.

To further investigate the effects of soil moisture on soil respiration, we separated the driest and wettest days (Fig. 7) of the study period, using the cumulative precipitation during the 3 days prior to each soil respiration measurement. This analysis unveiled a statistically significant relationship between soil respiration and soil moisture (p < 0.0001) for the Cerro Alto site during days in which rainfall had not fallen during the 3 days prior.

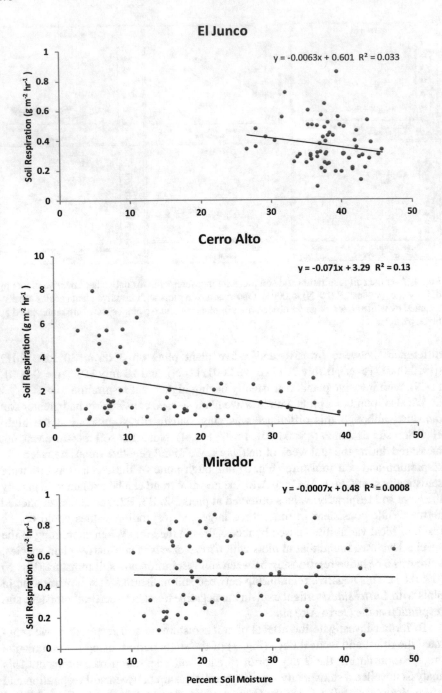

Fig. 5 Comparison of instantaneous soil moisture and soil respiration across all three sites, El Junco, Cerro Alto, Mirador

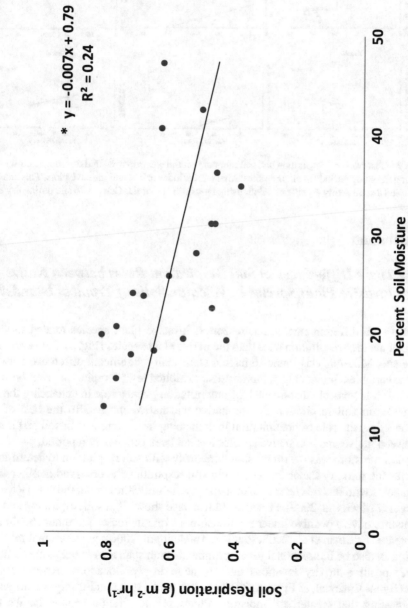

Fig. 6 Comparison of soil moisture and soil respiration for *Citrus sinesis* only at the low-elevation site, Mirador ($R^2 = 0.24$; $p < 0.05$)

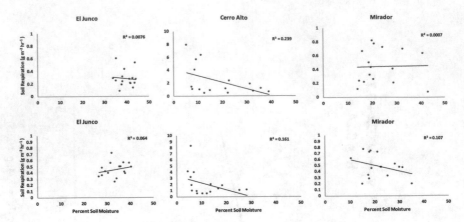

Fig. 7 Observed soil respiration and soil moisture on the wet (top row) and dry (bottom row) days during the study period for all three sites. Analysis includes native and invasive plots. This analysis unveiled a statistically significant relationship (p < 0.0001) for the Cerro Alto site during dry days (bottom row)

Discussion

Are There Differences in Soil Respiration Rates between Native and Invasive Plant Species in Volcanic Soils of Tropical Islands?

Across the different microclimatic zones, invasive plant species tended to show higher average respiration rates than the native plant species (Fig. 3). However, only one site, Mirador, was found to have a statistically significant difference between these variables. Interestingly, this site also exhibited low precipitation over the study period. It is possible that the lack of precipitation plays a role in enhancing the differences in soil respiration by non-native and native species. To the best of our knowledge, the role of precipitation in enhancing the differences in soil respiration between native and non-native species has not been previously reported.

Soil moisture was measured simultaneously with soil respiration to determine if it was the primary factor for increasing soil respiration, as observed in other environments with drastic changes in moisture availability throughout the year (Riveros-Iregui & McGlynn, 2009). Previous studies have shown that soil respiration and soil moisture have a positive linear relationship: as one increases the other should also increase (Freeman et al., 2008; Raich & Tufekciogul, 2000); however, that relationship can also be bidirectional for environments than span a wide range in soil moisture: positive in dry locations and negative in wet locations (Riveros-Iregui, McGlynn, Emanuel, & Epstein, 2012). In our study, that relationship was negative, suggesting that greater soil moisture affects soil respiration negatively. We thus examined whether specific plant species were responsible for this negative relationship and found that the managed, non-native, *Citrus sinesis* exhibited a negative relationship for soil moisture and soil respiration at Mirador (Fig. 6). Previous studies have suggested that *Citrus* plants have a unique relationship between air temperature and soil respiration. One study in particular reported that when temperature is held constant, respiration rates can acclimate to the controlled temperature (Bryla, Bouma, Hartmond, & Eissenstat, 2001). The same study found that the response of this plant to changes in soil moisture was temperature dependent (Bryla et al., 2001). However, to the best of our knowledge a negative relationship between soil respiration and soil moisture in *Citrus* plants had not been previously reported.

Plots E2, E3, B2, and B3 at Cerro Alto were also individually analyzed due to their unusually high respiration rates and low soil moisture rates. E2 and E3 are the non-native plant *Pennisetum purpureum*, commonly known as Elephant Grass. Grasses are known to have higher respiration rates due to their rapid root turnover rates and higher fine root density (Raich & Tufekciogul, 2000; Tufekciogul, Raich, Isenhart, & Schultz, 2001). However, parts of this site, including plots B2 and B3, were also affected by a recent fire which could explain the large rates of soil respiration that were measured.

When compared by site, invasive plant species appeared to have slightly higher soil respiration rates than the native plants on most days, particularly for the low-elevation site (Mirador). These differences could be the result of low moisture availability at this site and may indicate better adaptation capacity of invasive plants to low moisture, higher temperature conditions. These differences could also be affected by higher soil N content at Mirador relative to other sites (Hu, Qubain, & Riveros-Iregui, in press), as land management differences also exist across all sites. Our results suggest that while there are clear differences in soil respiration rates between non-native plants and native plants, these differences are highly sensitive to moisture availability and appear more complex and nuanced than previously understood, at least for island environments with drastic gradients in biophysical variables. It is likely that the observed responses of soil respiration to changes in soil moisture or plant cover, are the result of unique combinations of environmental factors with a highly fragmented land use/land cover. This conclusion points at challenges when trying to generalize the effects of native vs. invasive plant species on the carbon cycle of island environments and highlights the opportunity that islands offer in the study of mechanisms and processes driven by rapid changes in vegetative cover.

How Do Soil Respiration Fluxes Differ Across Different Microclimates?

The microclimates of our sampling sites were compared using weather variables, including precipitation, air temperature, and soil moisture. We found that the main variability was imposed by cumulative precipitation during the study period and soil moisture. Our results further suggest that there was greater variability in soil respiration rates with decreasing precipitation (Fig. 2). Cerro Alto showed the lowest cumulative rainfall of all sites over the study period (47 mm) and also the greatest range in soil respiration rates, spanning almost an order of magnitude. Given that soil moisture was lowest at Cerro Alto, it is likely that vegetative cover and microbes are more sensitive and thereby more spatially heterogeneous in their response to new precipitation inputs at this site. This heterogeneity and responsive behavior has been observed in other moisture-limited environments (Cable & Huxman, 2004; Riveros-Iregui, McGlynn, Epstein, & Welsch, 2008). Taken together, our findings suggest that soil respiration in tropical islands may be highly responsive to soil moisture, and may even exhibit the bidirectional behavior reported in previous studies (Riveros-Iregui et al., 2012; Skopp, Jawson, & Doran, 1990) when compared across different microclimates. However, further examination is required to fully explore the emergence of such bidirectional behavior at the landscape scale in island settings.

Concluding Remarks and Lessons from Soil Respiration

Our results suggest that precipitation and soil moisture play an important role in the magnitude of soil respiration fluxes across space and time in tropical islands. In particular, our findings suggest that carbon fluxes such as soil respiration are very sensitive to changes in moisture availability in dry locations or during dry periods. More importantly, our results suggest that soil respiration fluxes are strongly mediated by the type of vegetative cover. Given the extent of land fragmentation in tropical islands—and the rate at which invasive plant encroachment occurs—this particular finding suggests that characterizing the magnitude of carbon fluxes in island settings is difficult and it depends on the tight coupling between landscape position, microclimate, and land cover.

In our study, we hypothesized that areas with non-native plant species would have higher respiration rates than areas with native plant species, and that these differences would be consistent across different microclimatic zones. We found this to be the case *only* at the low-elevation site, Mirador. Other forms of land uses that were not the focus on this study, including presence of livestock or agricultural management practices, may also affect the relationship between soil respiration and water availability. In particular, given the relatively young age of soils in the Galapagos Islands, modifications that agriculture and land use have on soil quality and soil nutrient status may be amplified and their effects further exacerbate soil respiration dynamics.

Finally, we suggest that the study of the carbon cycle in tropical environments could benefit from assessment of mechanisms and processes taking place on islands. Highly fragmented landscapes, including landscapes with multiple and variable land uses and ages, that remain difficult to characterize today, could be modeled after island settings and thus facilitate the characterization of their carbon sink/source status. In concert with research on continental environments, tropical islands can provide valuable insight into how changing climates and human-driven land use change may affect the carbon balance of other terrestrial environments around the world.

References

Adler, G. H. (1992). Endemism in birds of tropical Pacific islands. *Evolutionary Ecology, 6,* 296–306.

Baruch, Z., & Goldstein, G. (1999). Leaf construction cost, nutrient concentration, and net CO_2 assimilation of native and invasive species in Hawaii. *Oecologia, 121*(2), 183–192.

Briggs, J. C. (1966). Oceanic Islands, endemism, and marine paleotemperatures. *Systematic Biology, 15*(2), 153–163.

Bryla, D. R., Bouma, T. J., Hartmond, U., & Eissenstat, D. M. (2001). Influence of temperature and soil drying on respiration of individual roots in citrus: Integrating greenhouse observations into a predictive model for the field. *Plant, Cell & Environment, 24*(8), 781–790.

Cable, J. M., & Huxman, T. E. (2004). Precipitation pulse size effects on Sonoran Desert soil microbial crusts. *Oecologia, 141*(2), 317–324.

Coblentz, B. E. (1978). The effects of feral goats (Capra hircus) on island ecosystems. *Biological Conservation, 13*(4), 279–286. https://doi.org/10.1016/0006-3207(78)90038-1

Coffey, E., Froyd, C., & Willis, K. (2011). When is invasive not an invasive? Macrofossil evidence of doubtful native plant species in the Galapagos Islands. *Ecology.* https://doi.org/10.1890/10-1290.1.

Darwin, C. R. (1859). *On the origin of species.* London: John Murray Publisher.

Dube, F., Thevathasan, N., Stolpe, N., Zagal, E., Gordon, A., Espinosa, M., & Saez, K. (2013). Selected carbon fluxes in Pinus ponderosa-based silvopastoral systems, exotic plantations and natural pastures on volcanic soils in the Chilean Patagonia. *Agroforestry Systems, 87*(3), 525–542. https://doi.org/10.1007/s10457-012-9574-9

Freeman, J. C. & Orchard, V. (2008). Relationships between soil respiration and soil moisture, *Soil Biology and Biochemistry, 40*(5), 1013–1018, https://doi.org/10.1016/j.soilbio.2007.12.012

Greenwood, P., & Kuhn, N. J. (2014). Does the invasive plant, Impatiens glandulifera, promote soil erosion along the riparian zone? An investigation on a small watercourse in northwest Switzerland. *Journal of soils and sediments, 14*(3), 637–650.

Hu, J., Qubain, C., & Riveros-Iregui, D. A. (in press). How do non-native plants influence soil nutrients along a hydroclimate gradient on San Cristobal Island? In S. J. Walsh, D. Riveros-Iregui, & J. Acre-Nazario (Eds.), *Land cover and land use change on islands: Social & ecological threats to sustainability.*

Jager, H. (2015). Biology and impacts of Pacific island invasive species. *Pacific Science, 69*(2), 133–153. https://doi.org/10.2984/69.2.1

Kueffer, C., Daehler, C. C., Torres-Santana, C. W., Lavergne, C., Meyer, J. Y., Otto, R., & Silva, L. (2010). A global comparison of plant invasions on oceanic islands. *Perspectives in Plant Ecology, Evolution and Systematics, 12*(2), 145–161.

Kyle, G. P., Beard, K. H., & Kulmatiski, A. (2007). Reduced soil compaction enhances establishment of non-native plant species. *Plant Ecology, 193*(2), 223–232.

Lal, M., Harasawa, H., & Takahashi, K. (2002). Future climate change and its impacts over small island states. *Climate Research, 19,* 179–192.

Lal, R. (2004). Carbon sequestration in soils of central Asia. *Land Degradation & Development, 15*(6), 563–572.

Mack, M. C., Schuur, E. A., Bret-Harte, M. S., Shaver, G. R., & Chapin, F. S. (2004). Ecosystem carbon storage in arctic tundra reduced by long-term nutrient fertilization. *Nature, 431*(7007), 440–443.

McGrail, B. P., Schaef, H. T., Ho, A. M., Chien, Y. J., Dooley, J. J., & Davidson, C. L. (2006). Potential for carbon dioxide sequestration in flood basalts. *Journal of Geophysical Research: Solid Earth, 111*(B12). https://doi.org/10.1029/2005JB004169

McLauchlan, K. K., Hobbie, S. E., & Post, W. M. (2006). Conversion from agriculture to grassland builds soil organic matter on decadal timescales. *Ecological Applications, 16*(1), 143–153.

Meehl, G. A. (1996). Vulnerability of freshwater resources to climate change in the tropical Pacific region. *Water Air Soil Pollution, 92,* 203–213.

Percy, M. S., Schmitt, S. R., Riveros-Iregui, D. A., & Mirus, B. B. (2016). The Galápagos archipelago: A natural laboratory to examine sharp hydroclimatic, geologic and anthropogenic gradients. *WIREs Water, 3*(4), 587–600. https://doi.org/10.1002/wat2.1145

Raich, J. W., & Tufekciogul, A. (2000). Vegetation and soil respiration: Correlations and controls. *Biogeochemistry, 48*, 71–90.

Riveros-Iregui, D. A., & McGlynn, B. L. (2009). Landscape structure control on soil CO_2 efflux variability in complex terrain: Scaling from point observations to watershed scale fluxes. *Journal of Geophysical Research, 114*, G02010. https://doi.org/10.1029/2008JG000885

Riveros-Iregui, D. A., McGlynn, B. L., Emanuel, R. E., & Epstein, H. E. (2012). Complex terrain leads to bidirectional responses of soil respiration to inter-annual water availability. *Global Change Biology, 18*(2), 749–756. https://doi.org/10.1111/j/1365-2486.2011.02556.x

Riveros-Iregui, D. A., McGlynn, B. L., Epstein, H. E., & Welsch, D. L. (2008). Interpretation and evaluation of combined measurement techniques for soil CO_2 efflux: Discrete surface chambers and continuous soil CO_2 concentration probes. *Journal of Geophysical Research, 113*, G04027. https://doi.org/10.1029/2008JG000811

Rodríguez-Caballero, G., Caravaca, F., Alguacil, M. M., Fernández-López, M., Fernández-González, A. J., & Roldán, A. (2017). Striking alterations in the soil bacterial community structure and functioning of the biological N cycle induced by Pennisetum setaceum invasion in a semiarid environment. *Soil Biology and Biochemistry, 109*, 176–187.

Sardans, J., Bartrons, M., Margalef, O., Gargallo-Garriga, A., Janssens, I. A., Ciais, P., … Peñuelas, J. (2017). Plant invasion is associated with higher plant–soil nutrient concentrations in nutrient-poor environments. *Global Change Biology, 23*(3), 1282–1291.

Skopp, J., Jawson, M. D., & Doran, J. W. (1990). Steady-state aerobic microbial activity as a function of soil-water content. *Soil Science Society of America Journal, 54*(6), 1619–1625.

Trueman, M., & D'Ozouville, N. (2010). Characterizing the Galapagos terrestrial climate in the face of global climate change. *Galapagos Research, 67*, 26–37.

Tufekciogul, A., Raich, J. W., Isenhart, T. M., & Schultz, R. C. (2001). Soil respiration with riparian buffers and adjacent crop fields. *Plant and Soil, 229*, 117–124.

Van Kleunen, M., Dawson, W., Essl, F., Pergl, J., Winter, M., Weber, E., … Antonova, L. A. (2015). Global exchange and accumulation of non-native plants. *Nature, 525*(7567), 100–103.

Villa, C., & Segarra, P. (2010). Changes in land use and vegetation cover in the rural areas of Santa Cruz and San Cristóbal. *Galapagos Report 2009–2010*, 85–91.

Vitousek, P. M., Walker, L. R., Whiteaker, L. D., Mueller-Dombois, D., & Matson, P. A. (1987). Biological invasion by Myrica faya alters ecosystem development in Hawaii. *Science, 238*(4828), 802–804.

Yoshitake, S., Fujiyoshi, M., Watanabe, K., Masuzawa, T., Nakatsubo, T., & Koizumi, H. (2013). Successional changes in the soil microbial community along a vegetation development sequence in a subalpine volcanic desert on Mount Fuji, Japan. *Plant and Soil, 364*(1–2), 261–272.

Impacts and Management of Invasive Species in the UK Overseas Territories

Nicola Weber and Sam Weber

Introduction

For centuries, humans have transported plants, animals and other organisms beyond their natural ranges, either deliberately or unintentionally, as trade and colonisation have expanded around the globe. With ever increasing globalisation and the growth in international travel over recent decades, the potential pathways for the spread of biological agents have continued to multiply (Early, Bradley, Dukes, et al., 2016). Not all introduced species cause problems in their new locations and indeed some generate considerable benefits for societies and economies, including in agriculture, horticulture and forestry. However, those that become established and proliferate in unintended ways can cause substantial, and often irreversible, ecological damage to the environments that receive them (Blackburn et al., 2011). These 'invasive, non-native species' (INNS) are recognised as one of the principle drivers of biodiversity loss worldwide (Díaz et al., 2019) and pose major threats to food security, human livelihoods and ecosystem service provision (Paini et al., 2016; Reaser, Meyerson, Cronk, et al., 2007). Many of the most pervasive INNS share a common set of attributes, such as high fecundity and growth rates and being well-adapted to surviving in disturbed or human-modified environments (Rejmanek & Richardson, 1996; Van Kleunen et al., 2010; Jelbert et al., 2019); nevertheless, it is often difficult to predict which species will establish themselves in a given environment (Early et al., 2016; Mack et al., 2000).

N. Weber (✉) · S. Weber
Centre for Ecology and Conservation, College of Life and Evironmental Sciences, University of Exeter, Penryn Campus, Cornwall, UK
e-mail: N.L.Weber2@exeter.ac.uk

© Springer Nature Switzerland AG 2020
S. J. Walsh et al. (eds.), *Land Cover and Land Use Change on Islands*, Social and Ecological Interactions in the Galapagos Islands, https://doi.org/10.1007/978-3-030-43973-6_13

277

The impacts of INNS have been particularly acute on islands where they threaten native species through a combination of competition, predation, habitat modification, disease transmission, disruption of ecosystem functions and changing trophic dynamics (Doherty, Glen, Nimmo, Ritchie, & Dickman, 2016; Simberloff, Martin, Genovesi, et al., 2013). A number of characteristics of islands leave them vulnerable to this kind of biological invasion; many have relatively simplistic ecosystems with limited functional diversity, leaving niches open for colonisation (Moser, Lenzner, Weigelt, et al., 2018). Insular species and communities have also often evolved in isolation over long periods of time leading to "ecological naivety" (or "island tameness"; Carthey & Banks, 2014; Cooper, Pyron, & Garland, 2014) and placing them at competitive disadvantage to vigorous, exotic competitors released from the constraints of pests, predators and diseases that regulate their populations in their native ranges (e.g. Funk & Throop, 2010). Another consequence of isolation is that islands often support high levels of biological endemism and evolutionarily distinctiveness, leading to high extinction risk (Berglund, Järemo, & Bengtsson, 2009). Despite making up just 5.3% of the Earth's land area, islands are estimated to support around 20% of the world's species (Tershy, Shen, Newton, Homes, & Croll, 2015). Many island endemics are already of conservation concern due to small range and population sizes that render them vulnerable to environmental stochasticity, and this is exacerbated by the threat from INNS (Doherty et al., 2016; Spatz et al., 2017). Indeed, 61% of all known species extinctions and 37% of species regarded as critically endangered are restricted to islands (Tershy et al., 2015).

Ironically, the same attributes of small size and isolation mean that it is also on islands that we stand the best chance of reversing these trends and restoring native ecosystems (Veitch, Clout, & Towns, 2011). In practice, however, small island governments and communities are often ill-equipped to tackle the challenges posed by INNS due to small operating budgets and the wide remits of local environmental agencies and NGOs, despite the potentially far-reaching social, economic and environmental benefits of doing so (e.g. Key & Moore, 2019; Vaas, Driessen, Giezen, van Laerhoven, & Wassen, 2017). The building and reinforcement of INNS management capacity in local organisations is vital to help address the impacts of INNS on islands, as is the development of more strategic approaches and toolkits that can be mainstreamed and shared across different agencies, sectors and island regions (e.g. Soubeyran et al., 2015). Grassroots support and community-led actions are also crucial in such situations in order to address fluctuations in capacity, which are a persistent problem for many small island communities (Balchin, Duncan, Key, & Stevens, 2019). In this chapter we explore these themes using case studies drawn from the UK Overseas Territories (UKOTs); a constellation of 14 predominantly islands scattered across four of the world's oceans, stretching from the ice in Antarctica to shallow tropical seas, and encompassing temperate, tropical and polar biomes. The UKOTs together account for 94% of the UK's unique biodiversity and as such make a significant contribution to global biodiversity (Churchyard et al., 2014; Key & Moore, 2019). Some are near pristine wildernesses, while others support vibrant communities whose livelihoods and cultural identities are inextricably linked to the natural heritage of the islands and their surrounding oceans, typifying many of the challenges faced by small island states.

Invasive vertebrates, including rodents (rats [*Rattus spp.*] and house mice [*Mus musculus*]), feral cats (*Felis catus*), the Asian mongoose (*Herpestes javanicus*), goats (*Capra hircus*) and rabbits (*Oryctolagus cuniculus*) have been identified as having a negative effect on 191 islands in the UKOTs (Dawson et al., 2015). The impact of invasive plants is likely to be equally severe although they have yet to be formally assessed across these Territories. Eradicating INNS from islands has become a mainstream conservation practice, with at least 1192 mammalian eradication attempts and an 88% success rate on 792 islands (DIISE, 2018). The ability to effectively remove INNS from islands is improving as techniques continue to be refined (Veitch et al., 2011), opening up more potential opportunities for eradication campaigns. Such management interventions are, however, highly costly in terms of financial investment, personnel and time and must be balanced against the potential benefits and likelihood of success. There is an increasing body of literature assessing the conservation outcomes of mammal eradications on islands and a variety of prioritisation tools have been developed that consider variables such as conservation value (e.g. number of native species, percentage of the species' global/ regional breeding population), the feasibility of eradicating invasive species based on the island's area and human population size, and natural reinvasion risk (Dawson et al., 2015). More difficult to quantify but also important considerations include social feasibility of eradication, anthropogenic reinvasion risk, terrain complexity and seasonal factors.

To explore these points using relevant case studies, this chapter is organized as follows (1) We begin on Ascension Island in the South Atlantic, where the eradication of feral cats has led to a resurgence in seabird nesting but where rampant biological invasions continues to threaten the Island's native flora and unique natural character. (2) We then move to the Caribbean to explore the challenges of ecological management on the island of Montserrat with its active volcano facilitating the establishment and spread of invasive species. (3) To consider in more detail what determines a successful eradication we consider two very different territories—the sub-Antarctic island of South Georgia in the south Atlantic and Pitcairn in the tropical south Pacific, united by their attempts to deal with their invasive rat populations. (4) Finally, we finish with a trip back to the South Atlantic where we discuss how St Helena is leading the way in INNS prevention through their community-backed biosecurity initiative.

Ascension Island

Ascension Island (7°57'S, 14°22'W, 97 km²) is located in the heart of the South Atlantic, 1660 km from Africa and 2250 km from South America. Formed approximately 1 million years ago from the eruptions of an underwater volcano along the mid-Atlantic Ridge, Ascension Island is both remote and geologically young; characteristics that ensured that at the time of human discovery in 1501 the Island's terrestrial ecosystems were in a relatively early stage of development, dominated by an assemblage of mobile, "pioneer" species. Today, Ascension Island has been more heavily impacted by the introduction of INNS than almost any other island on Earth with >90% of vascular plant species and likely similar numbers of invertebrates estimated to have been introduced, along with all nine extant species of land mammals and birds. All terrestrial habitats have been subject to encroachment by introduced species to a considerable extent, and virtually nothing still exists that could be described as truly "native habitat", with the possible exception of some areas of extremely barren coastal desert and some relict fragments of upland vegetation on exposed, misty slopes (Ascension Island Government, 2015). Indeed 9 of the 13 terrestrial species action plans lodged in the Island's National Biodiversity Action Plan cite INNS as being of high or medium risk to species recovery and survival (Ascension Island Government, 2015). Two major episodes of species introductions, discussed below, have left an indelible imprint on Ascension's natural environment with which conservation managers are still wrestling: the successive introductions of rats and cats in the Eighteenth and Nineteenth Centuries and the Victorian era experiment in ecosystem engineering of the Island's montane habitats to create an environment more hospitable for human life.

Case Study 1: Of Rats and Cats

Rats rank among the most successful human-mediated colonists of islands beyond their native ranges, and their impacts have been particularly well-documented (Harper & Bunbury, 2015; Jones et al., 2008; Towns, Atkinson, & Daugherty, 2006). It is not known precisely when rats reached Ascension Island, although they appear to have been well-established by the 1720s (Ritsema, 2010). Various theories have been proposed for the route of their introduction, including the spread of a founder population from the shipwreck of the notorious pirate and Privateer William Dampier who ran aground off the Island in 1701 (Ashmole & Ashmole, 2000). Few biological records are available from this pre-settlement era and the impacts of rats on Ascension's original seabird community must be largely inferred from contemporary sources. It seems likely that rats would have been a persistent presence among Ascension's mainland seabird colonies and may have decimated populations of smaller, burrow-nesting species (Jones et al., 2008) such as band-rumped storm petrels which are now confined to nesting on Boatswainbird Island and inaccessible cliffs. It is also possible that they contributed to the extinction of Ascension's two known endemic land birds (Bourne et al., 2003) and even the extirpation of now-absent members of the seabird community, such as Audubon's shearwater (*Puffinus iherminieri*) which is still frequently sighted around the Island (e.g. Bourne & Loveridge, 1978) and whose sub-fossil remains have been discovered amongst the Island's abandoned "ghost colonies" (Olson, 1977).

Rats do not appear to have had a catastrophic impact on Ascension's larger breeding seabirds, whose impressive mainland colonies were described by visiting naturalists right through into the mid-nineteenth century (Stonehouse, 1960). However, the subsequent introduction of domestic cats *Felis silvestris catus* in a bid to control the burgeoning rodent population following human occupation of the Island in 1815 led to the rapid decline of all but one of the remaining seabird species on the mainland (Ashmole et al. 1994; Fig. 1). By the 1950s the relatively small relict populations that remained survived only on inaccessible cliff ledges and off-shore stacks, including the largest, Boatswainbird Island, with the exception of the vast sooty tern colony that persisted by virtue of their numbers and seasonal migrations away from the island (Stonehouse, 1960).

In 2001, a seabird restoration project was initiated with the aim of eradicating feral cats from Ascension Island and reinstating seabird breeding colonies to the mainland (Ratcliffe et al., 2010). In ecological terms the programme was a resounding success. The last known feral cat was removed from the mainland in March 2004 and the island was declared feral cat free in 2006. Seabird recolonisation of accessible mainland sites began almost immediately in 2002 and numbers have increased steadily since (Ascension Island Government Conservation Department [AIGCD], *unpublished data*). The first species to recolonise were the masked booby (Fig. 1), brown booby, brown noddy and tropicbirds followed by the endemic Ascension frigatebird ten years later. However, the eradication programme also exemplified challenges encountered by similar programmes elsewhere. During the project, 38% of domestic cats were killed accidently, which alienated sections of the local community and caused public consternation that has persisted to the present day among some of those affected (Ratcliffe et al., 2010). Collateral and concerns over animal welfare are an inherent risk of eradications on inhabited islands, hence methods to reduce this and to better engage communities should be a key consideration in such campaigns.

Fig. 1 Ascension. (**a**) A feral cat larder with the remains of Ascension Frigatebirds (credit: Philip and Myrtle Ashmole). (**b**) Masked boobies nesting on the mainland Letterbox Peninsula following the feral cat removal project (credit: Nicola Weber 2014)

On Ascension Island, there is evidence to suggest that populations of rats are rebounding following the feral cat eradication (Hughes, 2014). While quantitative data are lacking, historical accounts suggest that rats were confined to upland areas in the 1950s–1990s and rarely observed in low-lying areas or around seabird colonies (Ashmole, 1963; Duffey, 1964): areas where they now achieve highest abundances (AIGCD, *unpublished data*). Examples of such "mesopredator release" and other unintended trophic cascades are not uncommon on islands following the eradication of introduced apex predators (Bergstrom et al., 2009; Rayner, Hauber, Imber, Stamp, & Clout, 2007) and must be carefully considered during the planning stage when weighing up overall conservation benefits. While there is no doubt that the feral cat eradication has had a net benefit to Ascension's native biodiversity, rodent impacts need to be carefully monitored going forward to ensure that this is not reversed. Rat eradication on Ascension has been ruled out in the short term as the island's tropical climate, rugged terrain, permanent human population and abundance of land crabs that will compete for bait all reduce the likelihood of success (*Black Rat Action Plan* Ascension Island Government, 2015; Holmes et al., 2015) and the scientific case of impacts is not sufficiently well-developed to justify the significant investment of resources that it would involve.

Case Study 2: What Do You Do with a Problem like Green Mountain?

Rising 867 m above sea level, Green Mountain is Ascension Island's highest peak and—as its name implies—is a world apart from the arid expanses of volcanic cinder, lava and ash that dominate the coastal lowlands below. Declared a National Park in 2005 under local legislation, virtually every aspect of Green Mountain's ecology is affected by regular immersion in the banks of cloud that frequently envelope its summit slopes (Ascension Island Government, 2015). These mists, formed as moisture-laden air carried on the south-easterly trade winds is forced to rise sharply over the mountainous eastern portion of the Island, condense on exposed surfaces (occult precipitation, or "fog drip") and ensure a continuously damp environment that supports an abundance of plant life (Ashmole & Ashmole, 2000). To nineteenth century mariners landing on Ascension's barren shores, Green Mountain's cool misty slopes represented a lifeline and were the stage for a grand Victorian experiment in terraforming, the legacy of which continues to challenge conservation managers to this day (Duffey, 1964; Cronk, 1980; Ashmole & Ashmole, 2000; Gray, Pelembe & Stroud, 2005) (Figs. 1 & 2).

Although Green Mountain's ecosystem shares many of the features of a true cloud forest, the original vegetation of this zone was entirely devoid of trees. Mass introductions of exotic species have since fundamentally altered the ecology of this area and many details of the original plant and animal communities are consequently lacking. However, based on historical collections and observations of the

Fig. 2 Green Mountain. (**a**) An early drawing of Green Mountain by William Allen, a Navy lieutenant passing through Ascension Island on a vessel in the early 1800s (credit: Ascension Island Heritage Society). (**b**) The summit of Green Mountain in the centre of the Island with recreational trails to the "Dew Pond" and "Elliot's Pass" (credit: AIG Conservation Department 2015). (**c**) A moisture-laden Ficus microcarpa tree at the summit of Green Mountain; an introduced exotic that supports epiphytic populations of native and endemic flora (credit: Sam Weber 2014). (**d**) A small thicket of introduced bamboo at the summit of Green Mountain (around the man-made "Dew Pond") with their nodes supporting unique bryophyte "balls" comprising of native bryophytes, and the endemic vascular plant Stenogrammatis ascensionensis (credit: Jeff Duckett 2014)

few remaining fragments of native vegetation, we can surmise that the original biota of this area was relatively species poor, with a fauna comprised entirely of invertebrates and a flora dominated by ferns, bryophytes and lichens, along with club mosses (*Lycopodium spp.*) and a few grasses (Ashmole & Ashmole, 1997; Cronk, 1980; Duffey, 1964; Lambdon et al., 2009). Nineteenth century accounts invariably describe the summit of Green Mountain as being overgrown with a carpet of ferns, probably comprised of species such as *Histiopteris incisa*, *Christella dentata* and *Ptisana purpurascens* which still form a vigorous community on a few exposed, south-facing slopes (Duffey, 1964; Lambdon et al., 2009). The extinct *Dryopteris ascensionis* is also thought to have formed part of this community (Lambdon et al., 2009). Lycopods, bryophytes and smaller, lithophytic ferns and grasses probably became important on steeper banks and more exposed outcrops where they still occur today in a few places (see Lambdon et al., 2009 for a more detailed discussion of community types). Although limited in terms of richness, the mist zone flora included a number of endemic species, including at least five ferns, one grass and 10 bryophytes, along with an unknown number of lichens (Ascension Island Government, 2015). The grass *Sporobolus caespitosus* and the ferns *Anogramma ascensionis, Asplenium ascensionis, Stenogrammitis ascensionensis* and *Ptisana purpurascens* still occur either exclusively or partially in this area, as do 10 endemic mosses, one liverwort and the recently described frilly hornwort *Anthoceros cristatus* (Villarreal, Duckett, & Pressel, 2017). Unfortunately, no early accounts of the fauna of Green Mountain exist and the subsequent invasion of this area by introduced species largely vitiates any attempt to reconstruct the original animal community, which would have been entirely comprised of invertebrates. At least one endemic flightless moth has described from a small area on Green Mountain ("Windy Ridge") and others may have been lost or await discovery (Davis & Mendel, 2013).

Today, the original habitats of Green Mountain have largely disappeared, replaced by a man-made mosaic of grasses, shrubs and trees. Efforts to secure a more reliable supply of food, timber and freshwater began soon after the establishment of a permanent military garrison on the Island in 1815, and Green Mountain with its favourable climate was at the heart of these efforts. The most significant event in this transformation was brought about at the instigation of the eminent botanist Joseph Hooker, who, following a visit to the Island in 1843 and with the help of his childhood friend Dr John Lindley, advocated an ambitious programme of planting aimed at creating pasture, reducing erosion and greatly increasing mist interception, soil development and groundwater recharge (Ashmole & Ashmole, 2000). What ensued was a remarkable feat of "ecological engineering" that has fundamentally and irrevocably altered the natural environment of Green Mountain. In the years following Hooker's visit more than 220 exotic plant species were imported from around the world, many of which subsequently naturalised and still grow side-by-side in a wild "botanical garden" of sorts. A small forest of bamboo (*Bambusa spp.*) and groves of exotic evergreen trees now occupy the summit ridge and some of the more sheltered slopes, while swathes of shell ginger (*Alpinia zerumbet*) and expanses of non-native grassland and shrub land cover many of the more exposed faces. Only on a few of the most mist-exposed ridges and banks do the last vestiges of the original, fern-dominated communities remain, and these are increasingly overgrown. Along with imports of exotic vegetation and topsoil, large numbers of exotic invertebrates were inevitably introduced and these now dominate the montane fauna (AIGCD, *unpublished data*).

Green Mountain nowadays represents a paradigmatic example of a "novel ecosystem"—a self-sustaining, living system that has been heavily modified by human agency and lacks analogues in the natural world (Hobbs, Arico, Aronson, et al., 2006). Much has been written about the complex philosophical and management questions that such transformed landscapes pose for conservationists and land managers (Evers et al., 2018; Hobbs, Higgs, & Harris, 2009; Miller & Bestlemeyer, 2016) and the novel ecosystem concept continues to polarise opinion, including on Ascension Island. To some Green Mountain's manmade cloud forest is an extraordinary example of ecological terraforming, with a new, functioning ecosystem having been created almost entirely from scratch (Wilkinson, 2004); to others it is simply another example of INNS degrading a previously pristine island environment (Gray, 2004). Such debates are now largely academic as restoring Ascension Island's mist zone to anything approaching a "natural" state will be practically, if not technically, impossible (Cronk, 1980). Nevertheless, the question of "what to do about Green Mountain" continues to challenge conservation managers. Most of the ecosystem services that the cloud forest once provided are now redundant: since the 1960s the Island's water needs have been met entirely through desalination and fresh food and building materials are either imported or supplemented by hydroponically grown produce. However, Green Mountain National Park continues to be an important recreation space for islanders and the small numbers of visitors. Despite the changes wrought by INNS, it remains the centre of botanical diversity in the Territory, supporting 15 out of the 18 known endemic plant species. Many of these species are now severely endangered and some, such as the Ascension Island parsley fern (*Anogramma ascensionis*)—"rediscovered" clinging to a remote cliff face in 2009 (Baker et al., 2014)—teeter on the brink of extinction. The Ascension Island Government is committed under multilateral environmental agreements to conserve the island's endemic flora and the habitats they depend upon to the greatest extent possible, but it presents a daunting task. In the short term, the immediate priority has been to secure viable *ex situ* living collections and seedbanks to mitigate against extinction in wild and act as a source of material for reintroduction into the wild. However, there is a recognised need for a long-term vision for the mist zone (Ascension Island Government, 2015).

The large number of invasive plant species now present in the mist zone represents the greatest threat to the conservation interest and recreational value of this area. Unfortunately, the list of introductions includes some of the most notorious pest species in the tropics, comprising many extremely vigorous and competitive threats. *Alpinia zerumbet, Buddleja madagascarensis* and several other thicket-forming, invasive shrubs are a particularly pervasive problem. Monocultures of these species have already smothered large tracts of apparently prime native habitat and greatly diminish the amenity value of Green Mountain National Park, choking footpaths and obscuring otherwise panoramic views of the Island. There is no evidence to suggest that their advance has slowed; if anything, the spread of certain ascendant, shrubby species such as *Vitex trifolia* and *Clerodendrum chinense* appears to be accelerating (AIGCD, *unpublished data*). The increasing scarcity of open, sparsely-vegetated habitat is of particular concern for the conservation of the early successional endemic flora. *Sporobolus caespitosus, Stenogrammitis ascensionensis, Anogramma ascensionis* and *Asplenium ascensionis*, are all at least partially dependent on a dwindling number of exposed, rocky banks. Some of these species now survive in only a handful of locations where they are vulnerable to encroachment by shrubs, maidenhair ferns (*Adiantum spp.*) and many small weedy species, including Koster's curse

(*Clidemia hirta*), goat weed (*Ageratum conyzoides*), buttonweed (*Spermacoce verticillata*) and cape grass (*Sporobolus africanus*), to name a few. Faced with such rampant biological invasions, efforts to restore the remaining fragments of native habitat and establish new populations of endemic species have so far either met with limited success or have required intensive ongoing maintenance, made more challenging by the steep, unstable terrain and dense vegetation that characterises this area (AIGCD, *unpublished data*). The limited functional diversity amongst the early successional native flora is also a major obstacle to any sustainable restoration scheme as the original fern and bryophyte-dominated communities typically lack the canopy-forming and ground cover species needed to exclude vigorous, invasive weeds. Given these constraints, it seems increasingly unlikely that "traditional" ecological restoration techniques will be successful in all but a few accessible and well managed sites. Green Mountain is not unique in this predicament. Human-modified ecosystems have become a dominant feature of the Anthropocene and the field of restoration ecology has had to race to adapt. Current thinking among many restoration ecologists is that rather than recreate what was once there, we need to work within the ecological realities of novel ecosystems to achieve a "hybrid state" that conserves the desirable and beneficial elements of both the old and the new (Hobbs et al., 2006, 2009). These two strategies are advocated in *Mist Region Habitat Action Plan for Green Mountain* (Ascension Island Government, 2015).

One option could be to explore biological control options for permanently reducing the competitiveness of particularly aggressive invasive plant species. The increasing dominance of Ascension's montane flora by a few vigorous, invasive weeds may be partly attributed to a lack of the natural enemies that normally regulate their populations: clear evidence that Green Mountain is not the stable, functioning ecosystem that has sometimes been portrayed (Wilkinson, 2004). Introducing specialised pests and diseases from these species native ranges could therefore be a sustainable and cost-effective way of achieving lasting control and tipping the balance in favour of native species. This type of targeted biological control has already been successfully applied to some invasive plants on Ascension Island and "off-the-shelf" solutions are available for some of the more problematic weeds currently affecting the mist region (Maczey, Tanner, & Shaw, 2012).

A second strategy that has gained traction in recent years is to actively enlist some of the more benign, non-native species as surrogates for missing elements of the native vegetation in order to supplement the development of stable, mature communities (i.e. ecological substitution: Griffiths & Harris, 2010; Parker et al., 1999; Schlaepfer et al., 2011). This integration has already happened naturally to some extent. For example, the stands of exotic *Ficus*, *Podocarpus* and *Elaeodendron* trees that inhabit Green Mountain's summit ridge now contain important populations of several native and endemic species which occur as epiphytes (e.g. *Stenogrammitis ascensionensis* and a number of bryophyte species; Fig. 2) or colonise patches of permanently damp ground or dead wood in the shaded understory beneath them (e.g. *Ptisana purpurascens*).

Because the canopy effectively excludes the vast majority of invasive shrubs, weeds and grasses, the resultant communities are relatively stable and require little or no management. For many, these old growth trees with their dripping aerial roots and moss-laden branches have become synonymous with Hooker's cloud forest; but they are in fact a relatively rare and precarious habitat. Some key species such as *Ficus microcarpa* (Fig. 2) are sterile, lacking the specialised pollinators that they need reproduce, while others struggle to gain a foothold in the dense scrub that surrounds them. As the original specimens die, they are often replaced by low-grade shrubland—a reminder of the lack of well-defined successional cycles in Green Mountain's "Frankenstein" ecosystem. Maintaining and expanding the current cloud forest would therefore seem to be a viable, long-term conservation strategy for a number of native and endemic species, if not for the communities and habitats that they once formed. In 2014, AIGCD established a small trial restoration plot in an area of scrub known as the "Weather Gardens" aimed at doing exactly this. While it will take decades until the saplings planted are large enough to harbour populations of native and endemic species, it nevertheless represents a shift in thinking about the goals of restoration and the means by which they can be achieved. Perhaps rather than trying to reverse the legacy of Hooker's experiment, the future of Green Mountain will be to guide its evolutionary trajectory in ways that restore ecological balance and enable the old and the new to co-exist.

Montserrat: INNS Management in the Wake of a Natural Disaster

Montserrat (16°45′N, 62°12′W, 198 km²) lies at the northern end of the Lesser Antilles in the eastern Caribbean Sea. The terrain is mountainous, with a rugged coastline and a densely forested mountainous interior spread across three distinct ranges: The Silver Hills (450 m) in the north, the Centre Hills (740 m) and the active volcanic region of the Soufrière Hills (900 m) in the south. Virtually every aspect of the island, from its natural environment to its human population, infrastructure and economic development, have been heavily impacted by natural disasters including hurricanes and volcanic activity. The eruption of the Soufriere Hills volcano between 1995 and 1999, was the most destructive in the Caribbean volcanic arc since 1902 (Kokelaar, 2002). In 1997 alone almost two-thirds of the island was devastated by pyroclastic flows, leading to loss of life, the complete destruction of the capital city, and extensive damage to natural habitats. The volcano has been less active in recent years and is monitored continuously by the Montserrat Volcano Observatory. However, the island remains divided into a number of zones: a Safe Zone (33% of the land mass) in the north where all of the human population currently resides along with five terrestrial and two coastal Hazard Level System Zones in the south (the "exclusion zone") where access is tightly regulated. The aftermath of the volcanic activity and continued limitations on access have had wide-ranging effects on the human population and also on the management of the natural environment.

In common with many small islands, Montserrat's biodiversity faces multiple pressures from development, habitat loss and fragmentation, INNS, and, increasingly, from climate change (Millennium Ecosystem Assessment, 2005); however, their impact has been significantly exacerbated by the volcanic situation. Montserrat lost approximately 45% of its forest during the 1995–1997 eruptions, directly threatening the species that depend upon them. There have also been indirect impacts in the aftermath which continue to threaten native biodiversity, in particular by habitat degradation and the spread of invasive species in the south of the Island. While functioning as a de facto protected area from human activity by ongoing access restrictions, the presence of largely unmanaged feral livestock in the southern exclusion zone is one of the major concerns for the conservation of the native habitat in that area. These mammals, including pigs, goats, sheep, cattle, horses, donkeys and chickens were abandoned during the mass evacuation and relocation that followed the eruption. They have since established self-sustaining, feral populations that threaten indigenous flora and fauna by grazing, browsing and trampling native vegetation, preventing regeneration and reducing resistance to encroachment by invasive weeds. Similar impacts have been observed in the Silver Hills in the north of the island where forest clearance and overgrazing have resulted in degraded scrub vegetation with little or no regeneration (Oppel et al., 2015). With no management actions in place, very little hunting pressure in the South Soufrière Hills exclusion zone (some illegal hunting is reported to take place) and ever-increasing feral mammal populations, anecdotal evidence suggests that the forest understory is heavily depleted and virtually no regeneration occurs. Feral pigs also affect forest habitats by destroying large clumps of *Heliconia caribaea* (the preferred plant in which the Montserrat oriole constructs its nests), and may be significant predators of mountain chickens (*Leptodactylus fallax*; a shared endemic frog with Dominica), the Montserrat galliwasp (*Diploglossus montisserrati*), and sea turtle eggs on beaches (Oppel et al., 2015).

While there are dedicated efforts underway to control feral mammal numbers in the Centre Hills forest in the habited part of the island (already made difficult by the rugged terrain), the unmanaged populations in the South are likely to be significantly impeding progress by acting as a reservoir for these species. Of all the islands and islets in the UKOTs (>2000), Montserrat has been identified as the one that would benefit most in ecological terms if all invasive vertebrates could be eradicated (Dawson et al., 2015). However, it ranked considerably lower (23rd) when considering only those INNS for which technically and logistically feasible eradication techniques existed (in 2013) due to the situation with access (Dawson et al., 2015). It is widely accepted that conservation measures to protect the forests on Montserrat should focus on the removal of invasive mammals (Allcorn et al., 2012; UKOTCF, 2018). However, sufficient robust data do not currently exist to evaluate and prioritise management areas and actions. Despite anecdotal reports from island residents, it is difficult to estimate accurately the size and distribution of the various feral livestock populations in the south and therefore to propose and evaluate appropriate control measures. Additionally, eradication programs can be controversial and sometimes face substantial public opposition, often relating to concerns about animal welfare, and a lack of evidence for the threat caused by INNS (Blackburn et al., 2010). On Montserrat, there are difficulties with the cultural significance of feral livestock, with for example, the feral donkeys holding a religious significance (Oppel, Beaven, Bolton, Vickery, & Bodey, 2011). As it is highly unlikely that the areas in the exclusion zone will become habitable within the foreseeable future then biodiversity conservation that works with improving livelihoods, for example through the eco-tourism potential is an approach that could be used to try to secure funding to carry out such a campaign. There are further complications with the preservation of biodiversity in the south too in that the lands in these areas are privately owned and so such socio-economic factors need to be considered.

A Tale of Two Islands: The Contrasting Fortunes of Rodent Eradications on South Georgia and Henderson Island

Regarded as one of the most destructive INNS, the removal of introduced rats from islands is an increasingly important tool for the conservation of island biodiversity. The success rate and scale of such eradications has improved as tools and methods have become more refined, and rodent eradications are now one of the most cost-effective methods of island preservation (Jones, Holmes, Dutchart, et al., 2016). However, there have also been many failed eradication attempts that have provided opportunities to reflect on the causes of failure and improve practice where appropriate (Giffiths et al., 2019; Holmes et al., 2015; Keitt et al., 2015). To date, there have been 1120 eradication operations targeting rodents on 831 islands of which 63% (704 operations on 569 islands) have been declared successful (a further 186 eradications were successful but then followed by reinvasion) (DIISE, 2018). Several quantitative and qualitative reviews have sought to identify factors that influence success rates of such eradication campaigns (Giffiths et al., 2019; Holmes et al., 2015). Here, we examine some of these factors drawing on two recent rat eradication attempts in the UKOTs: the sub-Antarctic Island of South Georgia (54°15′S, 36°45′W, 3500 km²) which was recently the site of the world's largest successful rat eradication scheme and Henderson Island (24°20′S, 128°19′W, 43 km²), part of the Pitcairn group in the tropical South Pacific, which coincidentally experienced a frustrating failed eradication attempt (Amos, Nichols, Churchyard, & Brooke, 2016) (Fig. 3).

Although ecologically very different, South Georgia and Henderson Island share a number of features in common. Both are remote, oceanic islands that lack permanent human populations, and both are noted for their ornithological importance which has made them high priorities for rodent eradication (Dawson et al., 2015; Hilton & Cuthbert, 2010). Henderson Island is a UNESCO World Heritage site and has been described as "one of the petrel capitals of the world" holding regionally

Fig. 3 Henderson. (**a**) The flat coral island of Henderson in the South Pacific (credit: Steffen Oppel 2015). (**b**) An invasive rat on Henderson Island given a unique identifying ear tag to document movement patterns to inform management practices (credit: Steffen Oppel 2015)

important populations of four gadfly petrel species—including the endemic and globally-endangered Henderson petrel (*Pterodroma atrata*)—along with four endemic land birds (Brooke et al., 2010). Predation by introduced Polynesian rats (*Rattus exulans*), or kiore, has been linked to long term declines in several of these species and represents a major threat to their long-term survival (Brooke et al., 2010; Brooke, 2010). The sub-Antarctic island of South Georgia supports one of the world's most abundant and diverse seabird communities, comprising of approximately 30 million pairs of 23 species, along with five waterbird and a single endemic passerine, the South Georgia pipit (*Anthus antarcticus*). Many of these species were threatened by brown rats (*Rattus norvegicus*) introduced in the late 1700s and had been either extirpated or severely depleted in areas where they were present (Poncet, 2006; Pye & Bonner, 1980). Attempts to eradicate rats were initiated on both islands in 2011 following fundraising campaigns organised by the RSPB for Henderson (~£1.5 million) and by the South Georgia Heritage Trust (~£7.5 million). Both operations adopted international best practice involving aerial broadcast of poison baits containing the active ingredient brodifacoum (Amos et al., 2016; Martin & Richardson, 2019). However, while South Georgia was declared officially rat free in 2018, the Henderson Island eradication failed to remove all rats and within 2 years the population had recovered to near pre-eradication levels (Amos et al., 2016; Bond et al., 2019).

The most obvious difference between Henderson Island and South Georgia is climate, and this distinction was undoubtedly significant in the contrasting outcomes of rodent eradication. A meta-analysis of 216 eradication campaigns found that mean annual temperature was the single most important factor determining success, with tropical and sub-tropical islands such as Henderson having a significantly higher failure rates than those lying at higher latitudes (Holmes et al., 2015; Keitt et al., 2015). This climatic effect has been attributed to limited seasonality, year-round breeding of rats and more consistent availability of preferred natural food sources in the tropics, making it difficult to time eradication campaigns to coincide with periods when bait uptake is likely to be highest (Holmes et al., 2015). In contrast, the successful eradication South Georgia was able to deploy baits in the austral autumn when rats are not breeding and alternative food sources are less plentiful (Amos et al., 2016). Several other predictors of failure, such as the presence of coconut palms, bait competition with native land crabs and high intra-annual variation in precipitation also affect Henderson Island and may have contributed to the outcome (e.g. Cuthbert et al., 2012; Giffiths et al., 2019).For example, it has been suggested that an exceptional abundance of fruit linked to elevated rainfall preceding the eradication campaign may have reduced bait palatability (Giffiths et al., 2019). A more detailed understanding of ecology of invasive rats, in particular the factors that drive natural fluctuations in abundance, resource availability and diet, will be vital to maximise the success of future eradication campaigns on tropical and sub-tropical islands like Henderson (Bond et al., 2019); Fig. 3.

Besides climate, area consistently emerges as a strong predictor of rat eradication success, with large islands typically having higher failure rates than small islands (Holmes et al., 2015). With an area >100 times higher than Henderson Island and eight times larger as any other island cleared to date (the previous record-holder being Macquarie Island in Australia), South Georgia apparently runs contrary to this trend; however, climate and geography were in its favour. Much of interior of South Georgia is permanently covered in ice with numerous coastal glaciers that act as natural barriers to rodent movement, effectively dividing the territory into a series of habitat 'islands' that could be treated separately without risk of reinvasion (Cook, Poncet, Cooper, Herbert, & Christie, 2010; Pye & Bonner, 1980). The eradication could therefore by carried out in phases over multiple seasons, each covering 120–580 km^2 (Amos et al., 2016). Indeed, the rapid retreat of these glaciers due to anthropogenic climate change was a strong incentive to expedite the operation as increasing habitat connectivity would likely have rendered it unfeasible in the near future (Cook et al., 2010). Due to the different climatic constraints affecting eradications on polar/temperate and tropical islands, different area thresholds are likely to apply (Giffiths et al., 2019). At 43 km^2 Henderson Island was the largest tropical or sub-tropical island to have been subjected to a rat eradication operation, and this alone may explain its failure by increasing the likelihood that a few individuals were not exposed to bait. Indeed, genetic approaches show that there was no reinvasion of Henderson from external sources and have revealed how close the operation came to succeeding, suggesting that the population was reduced to as few as 60–80 individuals (Amos et al., 2016).

The contrasting experiences of South Georgia and Henderson Island highlight many of the challenges associated with eradicating rodents from islands and the constraints and opportunities imposed by climate and geography. However, they also underscore many of the potential benefits of such schemes. Recolonization of eradicated areas on South Georgia by the endemic pipit began almost immediately following the removal of invasive rats and there was evidence of prospecting by burrow-nesting seabirds (Amos et al., 2016). While the Henderson Island campaign ultimately failed, at least one endemic land bird, the Henderson reed warbler (*Acrocephalus taiti*), experienced a substantial population increase following the eradication operation, which may have been facilitated by the temporary reduction of the Pacific rat population (Bond, Brooke, Cuthbert, et al., 2019).

Prevention is Better than Cure: Development Biosecurity on St Helena

Preventing the arrival and establishment of INNS through rigorous biosecurity is known to be more effective and economical than the removal of these species once they have established (Key & Moore, 2019; Oppel et al., 2019). In theory, small islands provide model systems for the implementation of rigorous biosecurity programmes as the number of potential entry points for novel INNS (e.g. ports or airports) are generally few and can therefore be exhaustively monitored by relevant authorities. However, many island countries and territories lack biosecurity regulations or practical biosecurity measures and successfully establishing such mechanisms will often require significant financial investment and societal change. As with the prioritisation of eradication campaigns, a similar strategic framework can be applied to guide these investments towards islands where native biodiversity is at highest risk from potential invasions and to identify which islands would benefit the most from establishing or improving biosecurity. This approach was demonstrated by Oppel et al. (2019) who carried out such an evaluation for 318 islands in the Caribbean UKOTs and Bermuda. Newly eradicated islands, such as the South Georgia example above are obvious priorities for enhanced biosecurity measures to prevent reintroductions. However, on inhabited islands that rely on the movement of goods and people, biosecurity schemes can be more challenging to implement and enforce.

St Helena Island, a sub-tropical UK Overseas Territory in the South Atlantic (15°56′S, 05°43′W, 122 km^2), provides an excellent case study of an inhabited island where a functioning biosecurity regime has been successfully installed through a gradual process of policy and behavioural change. Arguably most famous as the home of exiled French emperor Napoleon Bonaparte, St Helena is also a hotspot of botanical and invertebrate diversity, supporting around one-third of the endemic species found in the UKOTs (~500 species). Historically, the remote location of the island serviced by a single supply ship (up to 25 times a year) acted as a natural barrier to INNS arrival and border control focussed largely on the agricultural sector. However, the construction and opening of an international airport on the island and commencement of commercial flights in 2017 in a bid to boost tourism and financial self-sufficiency has exposed the island to a potentially wider range of threats and invasion pathways. St Helena has a limited number of existing cosmopolitan pests and is very vulnerable to new introductions that may be harmful to the economy, community health, environment and the recent investments in tourism development (Pryce, 2015). In anticipation of the changes brought by greater accessibility, St Helena authorities introduced a national biosecurity framework and associated policy under the banner of *Biosecurity St Helena*. Balchin et al. (2019) detail the extensive processes that led to development and implementation of this policy that applies international standards implemented by two full-time biosecurity officers who work closely at the border with Customs and detection dogs. However, faced with the limited human and financial resources typical of many small island nations, the delivery of St Helena's biosecurity strategy also relies heavily on public awareness and community support to promote voluntary compliance. Community acceptance of more stringent biosecurity measures was not immediate and initial attempts were met with considerable resistance. However, an active communication and public information strategy that engages all community sectors in the need for and benefits of biosecurity has gradually built a broad base of grassroots support which is integral to its success (Balchin et al., 2019; Key & Moore, 2019). Biosecurity St Helena has now established a strong brand on the island and benefits from a high level of compliance which may serve as a model for other small island nations.

Conclusions

Unequivocally recognised as a serious threat to island ecosystem integrity, the fields of invasion biology and restoration ecology are ever expanding to combat this. With previously unrecognised impacts now being documented, research into managing and combatting the effects is yielding improvements in traditional methods as discussed above, and also the advent of several new approaches, including the use of species-specific genetic and pheromonal methods (Campbell, Saah, Brown, et al., 2019; Serr, Heard, & Godwin, 2019). There is also the train of thought amongst some that novel ecosystems are so inevitable that they constitute "the new normal" (Marris, 2010). We will not address that debate here, but point in the direction of relevant literature e.g. Pearce, 2013; Murcia et al., 2014; Simberloff, 2015. All ecosystems are constantly changing and thus modern restoration ecology generally works to re-establish historical trajectories of an ecosystem prior to human influences rather than recreating the past (Simberloff, 2015). As discussed in this chapter, eradication is the gold-standard for restoring islands subjected to the effects of INNS, and biosecurity for their conservation in the long-term. However, for a number of different reasons this is not always practically possible and so is not a universal solution to preserve global island biodiversity. Regardless of approach, with the often limited resources available on small islands, the development and implementation of a common framework and catalysing of regional cooperation is likely to result in the greatest and most sustained levels of progress and action.

Acknowledgements The writing in this chapter has been drawn from the authors' experiences and documents produced during their time working with colleagues in the UK Overseas Territories, in particular those at Ascension Island Government Conservation Department. We have learnt much of what know we from the many talented ecologists and conservationists who we have had the privilege of working with both in the Territories themselves and also internationally and are fortunate enough to still be working as part of the UKOT extended network.

References

Allcorn, R. I., Hilton, G. M., Fenton, C., Atkinson, P. W., Bowden, C. G. R., Gray, G. A. L., … Oppel, S. (2012). Demography and breeding ecology of the Critically Endangered Montserrat oriole. *The Condor, 114*, 227–235.

Amos, W., Nichols, H. J., Churchyard, T., & Brooke, M. L. (2016). Rat eradication comes within a whisker! A case study from the South Pacific. *Royal Society Open Science, 3*, 160110.

Ascension Island Government. (2015). *The Ascension Island Biodiversity Action Plan*. Ascension Island Government Conservation Department, Georgetown, Ascension Island.

Ashmole, N. P. (1963). The biology of the Wideawake or sooty tern on Ascension Island. *Ibis, 103*, 297–364.

Ashmole, N. P., & Ashmole, M. J. (1997). The land fauna of Ascension Island: New data from caves and lava flows, and a reconstruction of the prehistoric ecosystem. *Journal of Biogeography, 24*, 549–589.

Ashmole, N. P., & Ashmole, M. J. (2000). *St Helena and Ascension Island: A natural history*. Oswestry: Anthony Nelson.

Ashmole, N.P., Ashmole, M.J.. & Simmons, K.E.L. (1994). Seabird conservation on Ascension Island. In Seabirds on Islands: Threats, Case Studies and Action Plans (eds Nettleship, D.N., Burger, J. & Gochfeld, M.), pp. 94–121. BirdLife International, Cambridge, UK.

Baker, K., Lambdon, P., Jones, E., Pellicer, J., Stroud, S., Renshaw, O., … Sarasan, V. (2014). Rescue, ecology and conservation of a rediscovered island endemic fern (*Anogramma ascensionis*): *Ex situ* methodologies and a road map for species reintroduction and habitat restoration. *Botanical Journal of the Linnean Society, 174*, 461–477.

Balchin, J. R., Duncan, D. G., Key, G. E., & Stevens, N. (2019). Biosecurity on St Helena island—A socially inclusive model for protecting small island nations from invasive species. In C. R. Veitch, M. N. Clout, A. R. Martin, J. C. Russell, & C. J. West (Eds.), *Island invasives: Scaling up to meet the challenge* (Occasional Paper SSC no. 62) (pp. 468–472). Gland, Switzerland: IUCN.

Berglund, H., Järemo, J., & Bengtsson, G. (2009). Endemism predicts intrinsic vulnerability to nonindigenous species on islands. *The American Naturalist, 174*, 94–101.

Bergstrom, D. M., Lucieer, A., Kiefer, K., Wasley, J., Belbin, L., Pedersen, T. K., & Chown, S. L. (2009). Indirect effects of invasive species removal devastate World Heritage Island. *Journal of Applied Ecology, 46*, 73–81.

Blackburn, T. M., Pettorelli, N., Katzner, T., Gompper, M. E., Mock, K., Garner, T. W. J., … Gordon, I. J. (2010). Dying for conservation: Eradicating invasive alien species in the face of opposition. *Animal Conservation, 13*, 227–228.

Blackburn, T. M., Pyšek, P., Bacher, S., Carlton, J. T., Duncan, R. P., Jarošík, V., … Richardson, D. M. (2011). A proposed unified framework for biological invasions. *Trends in Ecology and Evolution, 26*, 333–339.

Bond, A. L., Brooke, M. L., Cuthbert, R., et al. (2019). Population status of four endemic land bird species after an unsuccessful rodent eradication on Henderson Island. *Bird Conservation International, 29*, 124–135.

Bond, A. L., Cuthbert, R. J., McClelland, G. T. W., Churchyard, T., Duffield, N., Harvey, S., ... Oppel, S. (2019). Recovery of introduced Pacific rats following a failed eradication attempt on subtropical Henderson Island, South Pacific Ocean. In C. R. Veitch, M. N. Clout, A. R. Martin, J. C. Russell, & C. J. West (Eds.), *Island invasives: Scaling up to meet the challenge* (Occasional Paper SSC no. 62) (pp. 167–174). Gland, Switzerland: IUCN.

Bourne, W. R. P., & Loveridge, A. (1978). Small shearwaters from Ascension and St Helena, South Atlantic Ocean. *Ibis, 120*, 65–66.

Bourne, W., Ashmole, N., & Simmons, K. (2003). A new subfossil night heron and a new genus for the extinct rail from Ascension Island, central tropical Atlantic Ocean. *Ardea, 91*(1), 45–51.

Brooke, M. D., O'Connell, T. C., Wingate, D., Madeiros, J., Hilton, G. M., & Ratcliffe, N. (2010). Potential for rat predation to cause decline of the globally threatened Henderson petrel *Pterodroma atrata*: Evidence from the field, stable isotopes and population modelling. *Endangered Species Research, 11*, 47–59.

Campbell, K. J., Saah, J. R., Brown, P. R., et al. (2019). A potential new tool for the toolbox: Assessing gene drives for eradicating invasive rodent populations. In C. R. Veitch, M. N. Clout, A. R. Martin, J. C. Russell, & C. J. West (Eds.), *Island invasives: Scaling up to meet the challenge* (Occasional Paper SSC no. 62) (pp. 6–15). Gland, Switzerland: IUCN.

Carthey, A. J. R., & Banks, P. B. (2014). Naïveté in novel ecological interactions: Lessons from theory and experimental evidence. *Biological Reviews, 89*, 932–949.

Churchyard, T., Eaton, M., Hall, J., Millett, J., Farr, A., Cuthbert, R., & Stringer, C. (2014). *The UK's Wildlife Overseas: A stocktake of nature in our Overseas Territories*. Sandy, UK: Royal Society for the Protection of Birds.

Cook, A. J., Poncet, S., Cooper, A. P. R., Herbert, D. J., & Christie, D. (2010). Glacier retreat on South Georgia and implications for the spread of rats. *Antarctic Science, 22*, 255–263.

Cooper, W. E., Pyron, R. A., & Garland, T. (2014). Island tameness: Living on islands reduces flight initiation distance. *Proceedings of the Royal Society B: Biological Sciences, 281*, 20133019.

Cronk, Q. C. B. (1980). Extinction and survival in the endemic vascular flora of Ascension Island. *Biological Conservation, 17*, 207–219.

Cuthbert, R. J., Brooke, M. de L., Torr, N. (2012). Overcoming hermit-crab interference during rodent-baiting operations: a case study from Henderson Island, South Pacific. *Wildlife Research 39* (1):70.

Davis, D. R., & Mendel, H. (2013). The genus *Erechthias merick* of Ascension Island, including discovery of a new brachypterous species (Lepidoptera, Tineidae). *ZooKeys, 341*, 1–20.

Dawson, J., Oppel, S., Cuthbert, R., Holmes, N., Bird, J. P., Butchart, S., ... Tershy, B. (2015). Prioritizing islands for the eradication of invasive vertebrates in the UK Overseas Territories. *Conservation Biology, 29*, 143–153.

Díaz, S., Settele, J., Brondízio, E. S., Ngo, H. T., Agard, J., Arneth, A., et al. (2019). Pervasive human-driven decline of life on earth points to the need for transformative change. *Science, 366*, 6471.

DIISE. (2018). The Database of Island Invasive Species Eradications, developed by Island Conservation, Coastal Conservation Action Laboratory UCSC, IUCN SSC Invasive Species Specialist Group, University of Auckland and Landcare Research New Zealand. Retrieved from July 22, 2019, from http://diise.islandconservation.org.

Doherty, T. S., Glen, A. S., Nimmo, D. G., Ritchie, E. G., & Dickman, C. R. (2016). Invasive predators and global biodiversity loss. *Proceedings of the National Academy of Sciences of the United States of America, 113*, 11261–11265.

Duffey, E. (1964). The terrestrial ecology of Ascension Island. *Journal of Applied Ecology, 1*, 219–251.

Early, R., Bradley, B. A., Dukes, J. S., et al. (2016). Global threats from invasive alien species in the twenty-first century and national response capacities. *Nature Communications, 7*, e12485.

Evers, C. R., Wardropeer, C. B., Branoff, B., Granek, E. F., Hirsch, S. L., Link, T. E., ... Wilson, C. (2018). The ecosystem services and biodiversity of novel ecosystems: A literature review. *Global Ecology and Conservation, 13*, e00362.

Funk, J. L., & Throop, H. L. (2010). Enemy release and plant invasion: Patterns of defensive traits and leaf damage in Hawaii. *Oecologia, 162*, 815–823.

Giffiths, R., Brown, D., Tershy, B., Pitt, W. C., Cuthbert, R. J., Wegmann, A., ... Howald, G. (2019). Successes and failures of rat eradications on tropical islands: A comparative review of eight recent projects. In *Island invasives: Scaling up to meet the challenge* (pp. 120–130). Gland, Switzerland: IUCN.

Gray, A. (2004). The parable of Green Mountain: Massaging the message. *Journal of Biogeography, 31*, 1549–1550.

Gray, A., Pelembe, T., Stroud, S. (2005). The conservation of the endemic vascular flora of Ascension Island and threats from alien species. Oryx 39 (04):449.

Griffiths, C. J., & Harris, S. (2010). Prevention of secondary extinctions through taxon substitution. *Conservation Biology, 24*, 645–646.

Harper, G. A., & Bunbury, N. (2015). Invasive rats on tropical islands: Their population biology and impacts on native species. *Global Ecology and Conservation, 3*, 607–627.

Hilton, G. M., & Cuthbert, R. (2010). The catastrophic impact of invasive mammalian predators on birds of the UK Overseas Territories: A review and synthesis. *Ibis, 152*, 443–458.

Hobbs, R. J., Arico, S., Aronson, J., et al. (2006). Novel ecosystems: Theoretical and management aspects of the new ecological world order. *Global Ecology and Biogeography, 15*, 1–7.

Hobbs, R. J., Higgs, E., & Harris, J. A. (2009). Novel ecosystems: Implications for conservation and restoration. *Trends in Ecology & Evolution, 24*, 599–605.

Holmes, N. D., Griffiths, R., Pott, M., Alifano, A., Will, D., Wegmann, A. S., & Russell, J. C. (2015). Factors associated with rodent eradication failure. *Biological Conservation, 185*, 8–16.

Hughes, J. (2014). *Breeding and population ecology of sooty terns on Ascension Island.* PhD Thesis, University of Birmingham, UK.

Jelbert, K., Buss, D., McDonald, J., Townley, S., Franco, M., Stott, I., Jones, O., Salguero-Gómez, R., Buckley, Y., Knight, T., Silk, M., Sargent, F., Rolph, S., Wilson, P., Hodgson, D. (2019). Demographic amplification is a predictor of invasiveness among plants. Nature Communications 10 (1).

Jones, H. P., Holmes, N. D., Butchart, S. H. M., et al. (2016). Invasive mammal eradication on islands results in substantial conservation gains. *Proceedings of the National Academy of Sciences of the United States of America, 113*, 4033–4038.

Jones, H. P., Tershy, B. R., Zavaleta, E. S., Croll, D. A., Keitt, B. S., Finkelstein, M. E., & Howald, G. R. (2008). Severity of the effects of invasive rats on seabirds: A global review. *Conservation Biology, 22*, 16–26.

Keitt, B., Griffiths, R., Boudjelas, S., Broome, K., Cranwell, S., Millett, J., ... Samaniego-Herrera, A. (2015). Best practice for rat eradication on tropical islands. *Biological Conservation, 185*, 17–26.

Key, G. E., & Moore, N. E. (2019). Tackling invasive non-native species in the UK Overseas Territories. In C. R. Veitch, M. N. Clout, A. R. Martin, J. C. Russell, & C. J. West (Eds.), *Island invasives: Scaling up to meet the challenge* (Occasional Paper SSC no. 62) (pp. 468–472). Gland, Switzerland: IUCN.

Kleunen M. V., Weber, E., Fischer, M. (2010). A meta-analysis of trait differences between invasive and non-invasive plant species. Ecology Letters 13 (2):235–245.

Kokelaar, B. P. (2002). Setting, chronology and consequences of the eruption of Soufrière Hills Volcano, Montserrat (1995–1999). In T. H. Druitt & B. P. Kokelaar (Eds.), *The eruption of Soufriere Hills Volcano, Montserrat, from 1995 to 1999* (Vol. 21, pp. 1–43). London: Geological Society.

Lambdon, P., Stroud, S., Clubbe, C., Gray, A., Hamilton, M., Niissalo, M., ... Renshaw, O. (2009). A plan for the conservation of endemic and native flora on Ascension Island.

Mack, R. N., Simberloff, D., Lonsdale, W. M., Evans, H., Clout, M., & Bazzaz, F. A. (2000). Biotic invasions: Causes, epidemiology, global consequences, and control. *Ecological Applications, 10*, 689–710.

Maczey, N., Tanner, R., & Shaw, R. (2012). *Understanding and addressing the impact of invasive non-native species in the UK Overseas Territories in the South Atlantic: A review of the potential for biocontrol. Preliminary results: Ascension Island.* Unpublished report. CABI (ref: TR10086).

Marris, E. (2010). The new normal. *Conservation Magazine, 11*, 7–13.

Martin, A. R., & Richardson, M. G. (2019). Rodent eradication scaled up: Clearing rats and mice from South Georgia. *Oryx, 53*, 27–35.

Millennium Ecosystem Assessment. (2005). *Ecosystems and well-being: Biodiversity synthesis.* Washington, DC: World Resources Institute.

Miller, J. R., & Bestlemeyer, B. T. (2016). What's wrong with novel ecosystems, really? *Restoration Ecology, 24*, 577–582.

Moser, D., Lenzner, B., Weigelt, P., et al. (2018). Remoteness promotes biological invasions on islands worldwide. *Proceedings of the National Academy of Sciences, 115*, 9270–9275.

Murcia, C., Aronson, J., Kattan, G. H., Moreno-Mateos, D., Dixon, K., & Simberloff, D. (2014). A critique of the 'novel ecosystem' concept. *Trends in Ecology and Evolution, 29*, 548–553.

Olson, S. L. (1977). Additional notes on subfossil bird remains from Ascension Island. *Ibis, 119*, 37–43.

Oppel, S., Beaven, B., Bolton, M., Vickery, J. A., & Bodey, T. W. (2011). Eradication of invasive mammals on islands inhabited by humans and domestic animals. *Conservation Biology, 25*, 232–240.

Oppel, S., Gray, G., Daley, J., Mendes, S., Fenton, C., Galbraith, G., … Millett, J. (2015). Important bird areas: Montserrat. *British Birds, 108*, 80–96.

Oppel, S., Havery, S. J., John, L., Bambini, L., Varnham, K., Dawson, J., & Radford, E. (2019). Maximising conservation impact by prioritising islands for biosecurity. In C. R. Veitch, M. N. Clout, A. R. Martin, J. C. Russell, & C. J. West (Eds.), *Island invasives: Scaling up to meet the challenge* (pp. 659–665). Gland, Switzerland: IUCN.

Paini, D. R., Sheppard, A. W., Cook, D. C., Barro, P. J. D., Worner, S. P., & Thomas, M. B. (2016). Global threat to agriculture from invasive species. *Proceedings of the National Academy of Sciences of the United States of America, 113*, 7575–7579.

Parker, I. M., Simberloff, W. M., Lonsdale, W. M., Goodell, K., Wonham, M., Kareiva, P. M., … Goldwasser, L. (1999). Impact: Toward a framework for understanding the ecological effects of invaders. *Biological Invasions, 1*, 3–19.

Pearce, F. (2013). On a remote island, lessons in how ecosystems function. *Environment 360*. Retrieved from http://e360.yale.edu/feature/on_a_remote_island_lessons__in_how_ecosystems_function/2683/.

Poncet, S. (2006). South Georgia and the South Sandwich Islands. In S. M. Sanders (Ed.), *Important bird areas in the United Kingdom Overseas Territories* (pp. 21–226). Sandy, UK: Royal Society for the Protection of Birds.

Pryce, D. (2015). St Helena Pest and biocontrol species list. Retrieved from http://www.sainthelena.gov.sh/integrated-pest-management/.

Pye, T., & Bonner, W. N. (1980). Feral Brown rats, *Rattus norvegicus*, in South Georgia (South Atlantic Ocean). *Journal of Zoology, 192*, 237–255.

Ratcliffe, N., Bell, M., Pelembe, T., Boyle, D., Benjamin, R., White, R., … Sanders, S. (2010). The eradication of feral cats from Ascension Island and its subsequent recolonization by seabirds. *Oryx, 44*, 20–29.

Rayner, M. J., Hauber, M. E., Imber, M. J., Stamp, R. K., & Clout, M. N. (2007). Spatial heterogeneity of mesopredator release within an oceanic island system. *Proceedings of the National Academy of Sciences of the United States of America, 104*, 20862–20865.

Reaser, J. K., Meyerson, L. A., Cronk, Q., et al. (2007). Ecological and socioeconomic impacts of invasive alien species in island ecosystems. *Environmental Conservation, 34*, 98–111.

Rejmanek, M. & Richardson, D. M. (1996). What attributes make some plant species more invasive? *Ecology, 77*, 1655–1661.

Ritsema, A. (2010). *A Dutch Castaway on Ascension Island in 1725* (2nd ed.). Deventer, The Netherlands: A. Ritsema.

Schlaepfer, M. A., Sax, D. F., Olden, J. A. (2011). The Potential Conservation Value of Non-Native Species. *Conservation Biology 25*(3):428–437.

Serr, M., Heard, N., & Godwin, J. (2019). Towards a genetic approach to invasive rodent eradications: Assessing reproductive competitiveness between wild and laboratory mice. In C. R. Veitch, M. N. Clout, A. R. Martin, J. C. Russell, & C. J. West (Eds.), *Island invasives: Scaling up to meet the challenge* (Occasional Paper SSC no. 62) (pp. 64–71). Gland, Switzerland: IUCN.

Simberloff, D. (2015). Non-native invasive species and novel ecosystems. *F1000Prime Reports, 7*, 47.

Simberloff, D., Martin, J.-L., Genovesi, P., et al. (2013). Impacts of biological invasions: What's what and the way forward. *Trends in Ecology and Evolution, 28*, 58–66.

Soubeyran, Y., Meyer, J-M., Lebouvier, M., De Thoisy, B., Lavergne, C., Urtizberea, F., Kirchner, F. (2015). Dealing with invasive alien species in the French overseas territories: results and benefits of a 7-year Initiative. *Biological Invasions 17*(2):545–554.

Spatz, D. R., Zilliacus, K. M., Holmes, N. D., Butchart, S. H. M., Genovesi, P., Ceballos, G., ... Croll, D. A. (2017). Globally threatened vertebrates on islands with invasive species. *Science Advances, 3*, e1603080.

Stonehouse, B. (1960). *Wideawake Island, The story of the B.O.U. centenary expedition to Ascension Island*. London: Hutchinson.

Tershy, B. R., Shen, K.-W., Newton, K. M., Homes, N. D., & Croll, D. A. (2015). The importance of islands for the protection of biological and linguistic diversity. *BioScience, 65*, 592–597.

Towns, D. R., Atkinson, I. A. E., & Daugherty, C. H. (2006). Have the harmful effects of introduced rats on islands been exaggerated? *Biological Invasions, 8*, 863–891.

UK Overseas Territories Conservation Forum (UKOTCF). (2018). Workshop Report: Darwin-Plus supported project: *Maximising the long-term survival prospects of Montserrat's endemic species and ecosystem services*. Fourth workshop for all stakeholders to explore options for the future of the south of Montserrat.

Vaas, J., Driessen, P. P. J., Giezen, M., van Laerhoven, F., & Wassen, M. J. (2017). Who's in charge here anyway? Polycentric governance configurations and the development of policy on invasive species in the semisoverign Caribbean. *Ecology and Society, 22*, 1.

Veitch, C. R., Clout, M. N., & Towns, D. R. (Eds.). (2011). *Island invasives: Eradication and management*. Gland and Auckland: IUCN and CBB.

Villarreal, J. C., Duckett, J. G., & Pressel, S. (2017). Morphology, ultrastructure and phylogenetic affinities of the single-island endemic *Anthoceros cristatus* Steph. (Ascension Island). *Journal of Bryology, 39*, 226–234.

Wilkinson, D. M. (2004). The parable of Green Mountain: Ascension Island, ecosystem construction and ecological fitting. *Journal of Biogeography, 31*, 1–4.

Correction to: Threats of Climate Change in Small Oceanic Islands: The Case of Climate and Agriculture in the Galapagos Islands, Ecuador

Carlos F. Mena, Homero A. Paltán, Fatima L. Benitez, Carolina Sampedro, and Marilú Valverde

Correction to:
Chapter 5 in: S. J. Walsh et al. (eds.), *Land Cover and Land Use Change on Islands*, **Social and Ecological Interactions in the Galapagos Islands,**
https://doi.org/10.1007/978-3-030-43973-6_5

The original version of this chapter was revised due to author name was incorrectly mentioned. This has now been updated as "Homero A. Paltán" in the chapter opening page and front matter of the book.

The updated online version of this chapter can be found at
https://doi.org/10.1007/978-3-030-43973-6_5

© Springer Nature Switzerland AG 2020
S. J. Walsh et al. (eds.), *Land Cover and Land Use Change on Islands*, Social and Ecological Interactions in the Galapagos Islands,
https://doi.org/10.1007/978-3-030-43973-6_14

Index

Printed in the United States
by Baker & Taylor Publisher Services